Macroscopic and Microscopic Thermodynamics: From Fundamentals to Present Applications 2.0

Macroscopic and Microscopic Thermodynamics: From Fundamentals to Present Applications 2.0

Guest Editors

José S. Urieta

Ana M. Mainar

MDPI

Basel • Beijing • Wuhan • Barcelona • Belgrade • Novi Sad • Cluj • Manchester

Guest Editors

José S. Urieta
Group of Applied Thermodynamics and Surfaces (GATHERS)
Universidad de Zaragoza
Zaragoza
Spain

Ana M. Mainar
Group of Applied Thermodynamics and Surfaces (GATHERS)
Universidad de Zaragoza
Zaragoza
Spain

Editorial Office
MDPI AG
Grosspeteranlage 5
4052 Basel, Switzerland

This is a reprint of the Special Issue, published open access by the journal *International Journal of Molecular Sciences* (ISSN 1422-0067), freely accessible at: https://www.mdpi.com/journal/ijms/special_issues/thermodynamics_applications.

For citation purposes, cite each article independently as indicated on the article page online and as indicated below:

Lastname, A.A.; Lastname, B.B. Article Title. *Journal Name* **Year**, *Volume Number*, Page Range.

ISBN 978-3-7258-6924-4 (Hbk)
ISBN 978-3-7258-6925-1 (PDF)
https://doi.org/10.3390/books978-3-7258-6925-1

Cover image courtesy of Ana M. Mainar

Contents

About the Editors

José S. Urieta

José S. Urieta is Professor Emeritus of Physical Chemistry at the University of Zaragoza, where he earned his PhD in 1970 and received the Extraordinary Doctorate Award. A former DAAD fellow, he carried out postdoctoral research at the University of Heidelberg. His scientific career spans more than five decades, with significant contributions to applied thermodynamics, high-pressure fluid systems, electrochemistry, green chemistry, and the ecological field. Founder of the GATHERS research group, he is currently Associate Researcher at the Aragon Institute of Engineering Research (I3A) and Honorary Collaborator at his home university. Author of nearly 200 scientific articles—more than 170 indexed in Web of Science—he is co-inventor of an internationally licensed patent on biopesticide production. He has supervised 15 doctoral theses and contributed to several international academic networks linking Europe and Latin America. He has presented at numerous scientific conferences and led a wide range of research projects. A member of FLUCOMP and various groups within the Royal Spanish Societies of Physics and Chemistry, he also serves as Guest Editor for *IJMS* (MDPI) and is a Full Member of the Royal Academy of Sciences of Zaragoza.

Ana M. Mainar

Ana M. Mainar is an Associate Professor of Physical Chemistry at the Universidad de Zaragoza, where she earned her Ph.D. in 2000 and received the Extraordinary Doctorate Award. Her research focuses on supercritical fluid technologies, applied thermodynamics, and the valorization of natural products for sustainable applications. She is affiliated with the Instituto de Investigación en Ingeniería de Aragón (I3A) and leads the GATHERS research group. She has undertaken research stays at the Universitat Rovira i Virgili, the Instituto Superior Técnico (Lisbon), and the Universidad de Santiago de Compostela, contributing to high-pressure thermodynamics and green extraction processes. She has authored over 100 indexed publications. In 2009, she was awarded the Research Prize in Chemistry by the Real Academia de Ciencias de Zaragoza. Dr. Mainar is an active member of scientific societies including FLUCOMP, and the specialized groups GET (Termodinámica) and GECAT (Análisis Térmico y Calorimetría) of the Reales Sociedades Españolas de Física y Química.

Preface

The present Reprint, "Macroscopic and Microscopic Thermodynamics: From Fundamentals to Present Applications, 2nd Edition", reflects the continuing relevance of thermodynamics as a unifying discipline that bridges fundamental theory and experimental and computational applications. Our motivation as Guest Editors was to provide a forum for researchers addressing thermodynamic questions at different levels of complexity, from classical macroscopic measurements to molecular modeling and simulation.

This Reprint brings together contributions that illustrate the richness of the field: studies of liquid mixtures and phase equilibria, investigations of excess properties and interfacial phenomena, analyses of solubility in innovative solvents, and examinations of biomolecular systems and non-equilibrium processes. This diversity of perspectives underscores the capacity of thermodynamics to adapt and remain central in contemporary science.

This Reprint is intended for a broad scientific audience, including physicists, chemists, engineers, and life scientists, as well as graduate students seeking to deepen their understanding of thermodynamic principles and their practical implications. By compiling these contributions, we aim to encourage further cross-disciplinary dialogue and to highlight the enduring power of thermodynamics to connect scales, disciplines, and applications.

José S. Urieta and Ana M. Mainar
Guest Editors

International Journal of
Molecular Sciences

Editorial

Macroscopic and Microscopic Thermodynamics: From Fundamentals to Present Applications, 2nd Edition

Ana M. Mainar [1,2,*] and José S. Urieta [1,*]

1 Group of Applied Thermodynamics and Surfaces (GATHERS), Aragon Institute for Engineering Research (I3A), Universidad de Zaragoza, 50018 Zaragoza, Spain
2 Departamento de Química Física, Facultad de Ciencias, Universidad de Zaragoza, 50009 Zaragoza, Spain
* Correspondence: ammainar@unizar.es (A.M.M.); urieta@unizar.es (J.S.U.)

1. Some General Points

This Special Issue, "*Macroscopic and Microscopic Thermodynamics: From Fundamentals to Present Applications. 2.0*" was proposed with the aim of highlighting some of the most important and contemporary aspects that the very broad field of thermodynamics offers, both as a doctrine and as a tool for a very wide range of scientific disciplines, including technical and industrial applications.

Since its initial development as a mature science throughout the 19th century, on the basis that thermodynamics primarily concerns topics relating to the interconversion of heat and other forms of energy, both its conceptual field and its use as a tool have expanded greatly. Indeed, throughout its two centuries of history, thermodynamics has acquired a leading reputation due to the style of its own developments and also because of its capacity as a very useful instrument with great potential to facilitate the development of many scientific fields that are relatively closely related to physics and beyond [1,2]. The applications of thermodynamics in cosmology [3], information and computer science [4], and even social sciences such as economic systems [5,6], and traffic flow [7], are examples of the latter.

As is widely acknowledged, the most classical thermodynamic studies, i.e., those corresponding to *macroscopic thermodynamics*, are capable of making accurate and valuable predictions about processes involving an enormous variety of systems without taking into account their intimate structures. Nevertheless, moving beyond the last century, a new qualitative advance has emerged. This advance was made possible thanks to an adequate molecular vision of systems and particularly thanks to the developments in *statistical mechanics*. The aim of this formalism is to explain the physical properties of matter in bulk based on the dynamical behavior of its microscopic constituents. For this purpose, statistics and mechanics are used; subsequently, from that microscopic vision, the thermodynamic properties of a significant variety of systems in equilibrium, or even out of equilibrium, can be deduced. This molecular approach, corresponding to *microscopic thermodynamics*, is especially essential in the developments of chemical thermodynamics and chemical engineering. In general, statistics mechanics is divided into two forms: (i) the form dealing with systems in thermodynamic equilibrium, that is, historical statistical thermodynamics, and (ii) the form dedicated to the study of transport phenomena (the transport of matter, heat, etc.) and chemical reactions, that is, non-equilibrium statistical mechanics. Within these contexts, and in general scientific terminology, the phrase *chemical thermodynamics* refers to studies interrelating energy with chemical reactions or with a physical change of state within the frame of the laws of thermodynamics. In turn, Prausnitz [8] defines *molecular thermodynamics* as a synthesis of classical and statistical thermodynamics, molecular physics, and physical chemistry capable of providing quantitative information on the equilibrium properties of a variety of fluid mixtures. This process is essential in the development of chemical engineering. Conversely, for systems that are changing or may change over time and that are subject continuously or discontinuously to a flow of matter or energy to or from other systems, the theoretical treatments applied to equilibrium need to be implemented, including the laws of evolution regarding the changes of the state variables with time.

It is also remarkable that in recent decades, thanks to the increase in computational calculation capacity, the prediction and correlation of thermodynamic quantities through empirical, semi-empirical, and molecular models have made it possible to find solutions to many practical problems for simple systems or complex mixtures. Thus, the term *computational thermodynamics* has been applied to the use of computers to solve problems, ranging from typically scientific problems, such as those of thermodynamics in materials science [9], to those already mentioned in sociological areas.

With the purpose of systematizing and simplifying the above nomenclature and concepts, thermodynamics conceived as a discipline is usually divided into several fundamental branches, according to the different conceptual models considered, their theoretical or experimental basis, and the different types of systems under study. When addressing these bases, three branches are frequently considered: (*i*) *classical thermodynamics*, which corresponds to what we refer to as *macroscopic thermodynamics*, (*ii*) *statistical thermodynamics*, and (*iii*) *chemical thermodynamics* (including equilibrium thermodynamics and non-equilibrium thermodynamics). The last two branches, (*ii*) and (*iii*), correspond to what we refer to in the present Special Issue as *microscopic thermodynamics*.

When approaching thermodynamic methods as a tool for interpreting the behavior of numerous systems, we find a considerable number of fields in which thermodynamics has and will continue to find very important applications. In fact, thermodynamics has been penetrating the borders of many classical and current scientific fields. It covers broad and important aspects within the developments of engineering [10] (including, for example, chemical engineering [11], engineering biosystems [12], and food processing [13]), electrochemistry [14], energy use and conversion [15], materials science [16], biology [17], vital processes and life science [18], nutrition [19,20], agriculture [21], geology [22], and also in modern aspects of architecture [23], cosmology (the thermodynamics of solar systems [24], black holes [25], and the thermodynamics of space-time [26,27]), the environment [28], sustainability [29], or the economy, ecology and society (ecosystem ecology [30], thermoeconomics and industrial ecology [31], and the thermodynamics of quality [32]).

Given the enormous variety of topics in which thermodynamics can play a leading role, the need to limit the number of topics to be discussed in a Special Issue (SI) such as the one to which this Editorial is dedicated, is obvious. With this purpose of self-limitation and the aim of achieving a certain level of coherence, a series of topics were chosen. These topics represent, bring together, and promote current lines of research, building bridges between the basic aspects of thermodynamic science and new developments in technological processes and materials.

Under the umbrella defined by the proposed scopes, different types of systems, processes, and approaches of interest can be seen in the following SI. A set of nine research articles and two reviews are included, on which some brief comments are made and can be read below.

2. Overview of the Published Articles

The article by Gimeno et al. (Contribution 1) focuses on the thermodynamic behavior of a binary mixture (2-propanol + 1,8-cineole). The mixture is of particular interest both from a theoretical point of view and in light of its potential applications. These factors are corroborated by a detailed bibliographic review performed by the authors. Particularly noteworthy are the current and future applications of cineole in the field of pharmacy and medicine, in addition to the potential use of their mixtures with alcohols as additives for diesel or gasoline [33]. The authors describe the experimental in study of vapor–liquid equilibrium (VLE) performed. The study is complemented by measurements of densities and enthalpies of mixing for the aforementioned system, in addition to adequate theoretical examination of the resulting data. The experimental results correspond, as a whole, to the temperature interval between 278.15 and 323.15 K. By using the VLE data, activity coefficients and excess Gibbs energies were calculated, and the thermodynamic consistency test between excess molar Gibbs energies and excess molar enthalpies was substantiated. VLE data were correlated via Robinson-Mathias and Peng-Robinson-Stryjek-Vera equations of state, both including the Peneloux volume translation correction. These data were also correlated using the SAFT model. This latter approach offers a molecular vision quite suitable for systems with highly non-spherical or associated molecules. Of the three theoretical approaches considered, SAFT is the one that best approximates the volumetric behavior for the present system. Regarding the VLE, the three models provide theoretical values reasonably close to the experimental ones. Such results allow the performance of these approximations to be positively evaluated for the type of mixtures such as the one considered here. From a different perspective, when comparing the actual system with that resulting in the replacement of the cyclic ether with the corresponding lineal ether, a relatively higher value of the OH–O interaction energy between molecules of cineol and propanol can be deduced. This fact was interpreted by the authors based on the 3D structure of cineol and propanol molecules and based on an increase in the electron density of the oxygen atom in the cyclic molecule. Lastly, from the various conclusions derived from this work, it can be highlighted that the study provides suitable data and updates on developments for the simulation and design of separation processes to obtain chemical substances of interest, such as 1,8-cineole. In addition, from a microscopic perspective, the work provides new bases for the elucidation of the interaction forces between molecules carrying ether or alcohol functional groups.

The authors of Contribution 2, Semenov et al., report precise data on one of the non-variant equilibrium four phases of the binary system (CO_2 + H_2O). In relation to the simple p-T phase diagram of carbon dioxide, the introduction of the second component, water, significantly complicates the aforementioned diagram, even if extreme conditions are not considered. The above is particularly due to the appearance of a clathrate hydrate, a carbon dioxide hydrate, that is, a crystalline inclusion compound. This type of compound, clathrate hydrates, is an ice-like material and consists of a solid network of hydrogen-bonded water molecules creating cavities capable of enclosing "guest" molecules, such as the considered carbon dioxide or other non-polar compounds, such as some noble gases, methane, or small hydrocarbon chains [34]. The existence of a liquid phase rich in supercooled water is also considered in this work. Due to the existence of the number of different phases, a series of quadruplet points can appear, for some of which the corresponding values of pressure and temperature have been determined previously and can be found in the literature [35]. Two of these points, Q1 and Q2, fall within pressure and temperature ranges not far from ambient conditions. In their contribution to the present SI, the authors highlight the notable interest in the physicochemical properties of clathrate hydrates and their numerous current and potential applications. Areas of application include seawater desalination, the separation of gas mixtures, environmental preservation, energy storage, and food industry technologies. The work by Semenov et al. is of particular relevance regarding these applications. Because of this interest and in order to gain a better understanding of various processes in which the CO_2–water system plays an important role, precise knowledge of the singular points in its phase diagram is remarkably beneficial. Based on this premise, the authors formulated a new technique for the direct determination of the P and T coordinates of the lower quadruple point corresponding to the equilibrium between the gas–water solution, ice, and gas hydrate. In their study, they demonstrated the satisfactory reproducibility of their results, adequate reduction in measurement time, and a significant reduction in the combined standard uncertainty of the equilibrium temperature. This reduction in uncertainty is by an order of magnitude when compared with that of the previous data in the literature. Thus, the applied technique, which is described in great detail by the authors, is of great interest in validating similar techniques applicable to other systems.

Contribution 3 by Golikova et al. focuses, in particular, on the study of the excess enthalpies of the binary mixtures involved in the quaternary mixture acetic acid + n-butanol + n-butyl acetate + water. This four-component mixture is the most characteristic in the formation of butyl ester from the corresponding acid and alcohol. It should be noted that although it is a reactive mixture and the reaction processes are equilibrium type, for the reaction to proceed at an appreciable speed, it must be heated and homogeneous (e.g., strong acids) or heterogeneous catalysts must be present. In fact, butyl acetate has been classically manufactured via the Fischer esterification of butanol and acetic acid in the presence of sulfuric acid, which is used as a catalyst [36]. It should be noted that butyl acetate is an important chemical in high demand. The butyl acetate market represents about 2 billion USD annually. Because of its low volatility, its most common industrial use is in the production of lacquers and paints as a solvent. It is also applied in the manufacturing of artificial leather, plastic resins, and adhesives, as an anticorrosive agent, and for hardened coatings. In the pharmaceutical industry, it is used as a solvent or as an extraction agent. Additionally, it is applied in several cosmetic products such as lacquers, perfumes, or oils, in addition to nail polish removers. Thus, it is of interest to learn in detail all of the technical particularities that may be involved in the synthesis of the ester. In the contribution by Golikova et al., a detailed review of the mixing enthalpies of binary systems between the mentioned four components is carried out, highlighting the gaps or inconsistencies between the corresponding data for these mixtures. In order to address the main shortcoming mentioned above, the authors carried out experimental measurements, at 313.15 K and atmospheric pressure, of the excess enthalpies of three of the six possible binary mixtures between the four components. They include the binary mixtures acetic acid + n-butanol, acetic acid + n-butyl acetate, and n-butanol + n-butyl acetate. To this end, they used effective experimental equipment, namely a Setaram C80 isothermal mixing calorimeter. Their experimental results were correlated through the NRTL model and the Redlich–Kister equation. Using the obtained results and data from the literature, other thermodynamic magnitudes ($C^E_{p,m}$, S^E_m, $\Delta_{mix}S_m$, G^E_m, and $\Delta_{mix}G_m$) of the binary systems were estimated. The value and significance of the results obtained relate not only to their technical interest in relation to the industrial manufacturing of butyl acetate ester; there is also undouble theoretical interest derived from the close relationship between the magnitude studied experimentally (enthalpies of excess) and the interaction between the molecules that constitute the systems, among others. This type of interaction is a hot topic within studies focused on physical chemistry and other related sciences.

In Contribution 4 by Carbone et al., the authors discuss the surface tension of a series of binary mixtures with special characteristics. Surface tension, usually described as the tendency of liquid surfaces at rest shrink to the minimum surface area possible, is an important factor in industrial and technical processes; however, it is also of great theoretical significance in the scientific field. This is because surface tension is also directly related to intermolecular forces in the system. The binary mixtures considered by the authors consist primarily of 1,2 dibromoethane with a series of hydrocarbons (c-C6, n- and i-C7, n-C8, n-C14), coupled with 1-chloronaphthalene + 2,2,4-trimethylpentane, 1,4-dioxane +cyclohexane and methanol+methyl-tertbutyl-ether, with the latter selected as a reference. The primary aim of the work was to elucidate whether systems with W-shaped C^E_p vs. composition curves also show an anomalous dependence of surface tension on concentration. The authors rely on the hypothesis that the strong concentration fluctuations typical of systems with W-shaped C^E_p curves could also manifest themselves in the surface properties. It should be noted that, as reported by Saint-Victor and Patersoni [37], in the 1980s, Grolier, Wilhelm, and their collaborators found many systems with a "W-shape" for the C^E_p vs. composition curves. These systems were mainly constituted by an alkane plus a chemically dissimilar compound (1,4-dioxane, 2-butanone and 3-pentanone, 1,2-dichloroethane, 1,1,2,2-tetrachloroethane, 1,4-dichlorobutane, 1,6-dichlorohexane, and others). Although some authors initially suggested that this peculiar dependence could be a consequence of "a component existing as conformers of different polarity, the relative proportions of which are changed through mixing with the other component", the aforementioned authors, Saint-Victor and Paterson, although agreeing that such changes may be important, proposed another possibility. They pointed out that the W-shape may be due to deviation in local from bulk composition. Then, based on the quasi-chemical lattice theory of solutions of Guggenheim, a qualitative interpretation can be made. According to the theory, random mixing induces negative C_p^E, whereas local non-randomness (concentration fluctuations) contributes positive values of C_p^E, with the W-shape resulting as a consequence of a greater dominance of concentration fluctuations at intermediate compositions versus the tendency toward a random nature at both ends of the composition range [38]. The authors of Contribution 4 carried out surface tension measurements by using three different procedures and completed their laboratory study with depolarized light scattering experiments to determine the concentration fluctuations. Among the other results, an anomalous concentration dependence for the surface tension and the mixing surface tension for the systems under study was reported. In addition, the liquid/vapor interface is proposed to have a high thickness for the mixtures, as a consequence of the high values of the concentration–concentration correlation function associated with strong concentration fluctuations.

Contribution 5 by Gallego et al. refers to the study of the eutectic mixture of urea and sodium acetate trihydrate. Within the context of the aim of sustainable natural resource management, this mixture is proposed as a potential suitable solvent for the extraction of the valuable compound collagen from by-products of industrial fish and aquaculture processing. From this perspective, the solid–liquid equilibrium of the aforementioned eutectic mixture was studied. Subsequently, the solubility of the main amino acids of fish collagen in the proper eutectic at different temperatures, and also in the mixture resulting from the dilution of this mixture with water (50 wt %), was determined. The authors highlight in their work the importance of better management of the hitherto poorly valued by-products generated in the worldwide capture fisheries and aquaculture production, representing in some cases up to 70% of the total mass processed. For the purposes of economic calculations, it can be stated that the current global average consumption of fish and other forms of seafood per person is roughly 20 Kg/year, which represents an enormous amount of by-products. These by-products comprise heads, skin, fish ends, viscera, scales, bones, or cartilage. Such by-products have been traditionally used, following simple processing, for the production of fishmeal or animal feed, fish liver oil, glue, gelatin, or fertilizer, or they can be simply discarded as waste. The latter option implies a direct impact on the environment, in addition to significant economic losses. Therefore, adequate treatment of by-products and waste is essential for their revaluation, since they offer valuable sources of bioactive compounds. The authors provide new data on the composition of the eutectic mixture of sodium acetate trihydrate + urea. As they suggest in their work, its temperature close to ambient temperature and the fairly high values found for the solubilities of the most abundant amino acids in collagen and fish gelatin make this eutectic mixture promising as a solvent for corresponding industrial extraction processes. Furthermore, the authors found that the incorporation of water into the eutectic mixture increases the solubility and allows the operating temperature to be lowered, which expands its scope. The authors also suggest future areas of study, including molecular simulation, in order to gain a better understanding of how amino acids are solvated in the eutectic medium.

In Contribution 6, Nikolić et al. present their BioMThermDB 1.0 database, which is a freely available, web-based database that reports on the thermophysical and dynamic properties of an extensive series of proteins, predominantly globular proteins and antibodies, in aqueous solutions. Those properties, including the hydrodynamic radius, electrophoretic mobility, zeta potential, and the cloud-point temperature, are of crucial importance when determining the stability of this type of compound. Their database systematizes information dispersed in bibliographies and complements other existing databases in order to resolve questions about phase stability and molecular interactions in the aqueous solution or address the influence of changes in concentration, temperature, or pH on the properties of said solutions. In relation to the *hydrodynamic radius*, it can be stated that it is one of several parameters that describe molecular size, that is, how big or small a molecule is. It is defined as the radius of a sphere with the same hydrodynamic properties (frictional coefficient) as the molecule, and it is measurable, for example, through the use of dynamic light scattering [39]. This parameter provides, for example, a simple way to confirm the identity and oligomeric state of a protein in a solution. Another one of the parameters considered, *electrophoretic mobility*, can be highlighted as a key parameter characterizing the motion of a charged particle in an electric field in free solution. This electrophoretic mobility is defined as the speed of movement of the species under an electric field with the value of unity. It can be measured through the use of a moving-boundary method or, alternatively, via microelectrophoresis, that is, through the measurement of the time taken for single particles to travel a defined distance. Although it is widely acknowledged that this mobility depends primarily on the size and net charge of the molecule, a detailed relationship is still lacking for molecules with complexity in the shape and charge distribution of proteins [40]. As such, it should be stated that both proteins and other biomolecules may be separated according to their varying electrophoretic mobility, making electrophoresis a rapid, simple, and sensitive analytical tool for this purpose. For its part, the *zeta potential* (ζ) is defined as the electric potential on the sliding surface that separates the thin liquid layer attached to the solvated protein particle when this moves, with respect to the rest of the liquid mass. It describes the magnitude of electrostatic interactions of the particles in solution and, therefore, is a determinant of the stability of the dispersion. Usually, the zeta potential is measured indirectly through the electrophoretic mobility of the particles. Values of this potential are particularly useful in the design of drugs that maintain their function while suspended in solution. Regarding the *self-diffusion coefficient*, the IUPAC definition states that it is nothing but the diffusion coefficient when the chemical potential gradient equals zero. It is determined by the random, thermal motion of the molecules in the absence of a concentration gradient and is experimentally obtained by using isotopic markers or through the pulsed-field gradient-NMR method. The self-diffusion coefficient provides information not only on the size, shape, or molecular weight of a species but also on its aggregation or association with other species and changes in structure, or denaturation; however, quantitative interpretation of diffusion coefficient data can be difficult [41]. The *viscosity* of protein solution has also been determined to elicit specific information about these compounds. For example, the intrinsic viscosity (the ratio of the increase in the relative viscosity by the solute to its concentration in the limit of infinite dilution) has been used as a measure of "denaturation" extent. Conversely, the viscosity control of concentrated protein solutions is vital for the manufacturability and drug delivery routes of many therapeutic proteins developed by the pharmaceutical industry [42]. Lastly, the *cloud-point temperature*, defined as the temperature where the protein solution undergoes phase separation into two co-existing phases, is indicative of protein interactions and constitutes a useful tool for food product engineering [43]. For precise experimental determination, an array of optical detectors continuously monitors the sample to detect the first appearance of a cloud of the new phase. In summary, the BioMThermDB 1.0 database presents an overview of thermodynamic and thermophysical data on protein solutions that should aid scientists in solving problems such as predicting protein structures, studying protein–protein interactions, and use in protein family identification and other theoretical treatments. In addition, as indicated by the authors, the database can be used to help researchers in planning new experiments on these systems.

Contribution 7, by Pekar, refers to systems kinetically governed by the simultaneous existence of reaction processes and matter transport by diffusion. Chemical reactions involve the material transformation of chemical substances over time until reaching a final equilibrium state; diffusion, in contrast, is a process in which chemical species spread out as a result of existing concentration gradients. As indicated by the author of the work, the simultaneous occurrence of chemical reactions and diffusion is commonly present in engineering processes and in the biological processes of living systems such as metabolism or respiration. Key examples can also typically be found in geological and environmental systems. Although the analysis of the reaction and diffusion processes can be performed from a microscopic perspective, through atomic theory in the first case (collision theory and transition state theory) and through statistical treatment in the latter case (as the type carried out in the deduction of the Einstein–Smoluchowski equation), a macroscopic, phenomenological approach (as are that of formal kinetics and that of Fick equations) is also perfectly possible. In the present contribution, it is precisely this latter approach, the macroscopic one, which is exclusively addressed. The developed work specifically targets the named self-balancing diffusion, as defined by Truesdell [44]. Thus, based on general relations for exchanges of mass among the constituents of a mixture, it is considered that the interpretation of the mass changes that occur in the physicochemical processes does not strictly require the terms "atom" and "molecule", nor is a strict reference to "chemical" changes required. The only aspect that was necessary was to assume that "each constituent is made up by combination in fixed proportions from certain individually indestructible constituents". In turn, in the corresponding stoichiometric developments, Truesdell followed the results presented by Bowen [45] for their developments in the theory of fields of chemical reactions and their equilibria. For his part, in the present contribution, the author addresses in detail the applicability of the concept of reaction extent (the extent to which the reaction has proceeded) to those defined as self-balancing diffusion processes, and, in general terms, he shows how they could be detected in reality. The author also addresses the practical manifestations and consequences of these types of processes.

Contribution 8 by Volodin et al. has as its primary aim the goal of deepening knowledge at a microscopic level of the fascinating compound octamethylcyclotetrasiloxane, whose study, at that particular level, presents some gaps in the current literature. This chemical is the most important of the cyclic siloxanes, with production of well over one hundred million kilograms/year, and it is of significant industrial interest, especially in the production of polymers and lubricants or as a finishing agent or solvent. As with any other chemical, knowledge of this compound at a molecular level is important not only from the perspective of basic knowledge but also because of the close relationship between its structure and its macroscopic properties. The latter are what ultimately determine the applications of the product. Therefore, structure–property and structure–function relationships have long been considered important explanatory and crosscutting concepts in the disciplines of chemistry and biology [46] and also in engineering and materials science. Indeed, a detailed understanding of the relationship between structure, function, and reactivity is essential for the success of modern science. In the work by Alexander et al., experimental X-ray diffraction and thermochemical measurements are presented. Complemented by their theoretical calculations (density functional theory, ab initio molecular dynamics, and metadynamics), the authors were able to thoroughly elucidate the structure of the solid phases of the compound below its melting point up to 100 K. They also determined the values of the energy barriers associated with the phase and conformational changes occurring in their molecules. In relation to these changes in the 3D molecular structure mentioned above, it can be stated that, for theoretical chemists, conformational isomerism is a form of stereoisomerism in which the isomers can be interconverted simply by rotations around formally single bonds. If a conformation corresponds to a local minimum on the energy surface, then these stereoisomers are specifically referred to as conformational isomers or conformers. Conversion of one conformer to another involves overcoming an energy barrier. If this barrier is low, the rotations occur freely and quickly and then a sample of a given chemical consists of a mixture of several conformers. Conversely, if the height of the energy barrier is sufficiently high, rotation will thus be restricted, and a molecule may exist for a relatively long time period as a stable rotamer. As noted by Shklover and Struchkov [47], the rings in organocyclosiloxanes (RR′SiO)n with $n > 3$ (the same constitution as the compound considered in Contribution 8) are non-planar, and hence, it may be expected that they exist in different conformations. The single-crystal X-ray diffraction technique utilized by the authors is a powerful tool capable of providing detailed information on the type of crystal structure, including information such as distances and bond angles. In the present case, the X-ray results form a basis for discussion that was combined with thermodynamic parameters obtained through DSC experiments and modern computational methods (DFT, molecular dynamics and metadynamics calculations). This process allowed the authors, among other objectives, to establish energy parameters and the pathway of phase transition, in addition to the establishment of the conformational changes associated with the observed phase transition.

In the paper by Le Carpentier et al. (Contribution 9), the authors report on research work in the field of cardiac muscle tissue. The authors of the paper respond to an interest in improving the understanding of heart physiology from all points of view, including the most fundamental aspects, such as that concerning physical science. This particular field of work, heart physiology [48], is in continuous evolution, in search of more effective treatments for cardiovascular conditions. Specifically, through the above study, the authors analyze the effect of cardiac tissue anoxia on the molecular mechanism of myocardial function from a thermodynamic perspective. Cardiac muscle tissue, known as the myocardium, is the striated muscle tissue that forms most of the wall of the heart. It is specialized in its continuous and involuntary contraction, which allows blood to be pumped through blood vessels to supply oxygen and nutrients to the entire body. The myocardium contains cells called cardiomyocytes. They possess a cylindrical, elongated shape and connect to one another, forming myocardial fibers. These fibers are separated by collagen which supports the capillary network of the tissue. The junctions between myocytes consist of special formations called intercalated discs that allow the rapid propagation of electrical impulses, ensuring the synchronized contraction of the cells, which generate the heartbeat. Most of the cytoplasm of cardiomyocytes is occupied by myofibrils. Myofibrils are composed of aligned sarcomeres that constitute the contraction units. In turn, these sarcomeres are composed of thin filaments of actin and thicker filaments of myosin, extended between the so-called "Z discs". The muscular contraction of the myocardium occurs when the sarcomeres shorten owing to the fact that the filaments of these actin and myosin proteins slide over each other, making the Z discs move closer. Similar to many other aspects of biology [49], human cardiac activity has come to the attention of researchers from different scientific areas. As an example, Uehara and Sakane [50] utilized the thermodynamic approach in order to derive a relation between the referenced cardiac output (a global blood flow parameter of interest in hemodynamics) and the oxygen consumption rate. Shortly after their study, Dini et al. [51] examined the thermodynamics of metabolism and cardiac efficiency, including the entropy generation associated with the inefficiencies of the cardiac cycle. The experimental section of the work by Le Carpentier et al. is based on the observation of the effect of anoxia on the ventricular papillary muscles of adult male rats. Fundamentally, the shortening velocity of the aforementioned muscles was measured, and the associated isometric and isotonic tensions were determined. The subsequent theoretical treatment included the concepts of tribology corresponding to the friction of the myosin filaments, the rate of entropy production and excess entropy production associated with non-equilibrium, self-organization, and dissipative structures. The conclusions reached by the authors are of particular interest in relation to heart transplants.

In the review by Dehghan Niestanak and Unsworth (Contribution 10), the most relevant up-to-date knowledge on the interactions between blood proteins and specific uremic toxins (UTs) present in the human bloodstream when individuals are affected by certain conditions, are reviewed. When kidney function declines, reaching a value below a certain critical value, clinical alterations derived from the accumulation of UTs begin to manifest, which poses a serious risk to health. UTs have been defined as solutes normally excreted by the kidneys but retained in chronic kidney disease and interacting negatively with biological functions. They can be molecules derived from the intake of nutrients or products of metabolism itself. Roughly a quarter of the nearly 150 uremic retention solutes listed in 2012 by EUTox (a working group of the European Society of Artificial Organs) correspond to the so-called *Protein-Bound Uremic Toxins* (PBUTs). Although the molecular weight of most of these molecules is less than 500, they are difficult to remove as a result of their high affinity with plasma proteins. Of these molecules, the authors focused their attention on human serum albumin (HSA) in particular due to its relevance and abundance in the human bloodstream. The interactions between PBUTs and plasma proteins alter the conformation of the proteins, blocking other important unions with endogenous or exogenous substances that are necessary in order to maintain health. In addition, those undesirable unions with PBUTs reduce the effectiveness of palliative treatments. As a consequence of the aforementioned interactions, PBUTs are present in the bloodstream in two forms: a main bound fraction (bound to albumin and non-diffusible) and a small free fraction. From the numerous investigations focused on the elimination of UTs, it has been shown that the techniques to eliminate their free fraction are not very clinically effective due to their low proportion. For this reason, a considerable amount of research on the issue of patients with chronic kidney disease has been largely directed toward the use of so-called "chemical displacers" that could compete with PBUTs to bind to serum albumin [52]. The authors of the aforementioned review highlight the need to investigate in greater depth the molecular interactions that exist between the different species involved: PBUTs, serum albumin, and competitive binders. The authors discuss in detail the role of competitive binders in inhibiting uremic toxins. In summary, they carefully review the accuracy of critical information suitable for developing new pathways to remove undesirable uremic toxins from the blood compartment.

The review by Nie et al., corresponding to Contribution 11, represents a contribution to the scientific field of separation processes. These types of operations are frequently necessary in order to obtain pure substances or concentrates for domestic and industrial purposes or in research works. Chemical and engineering separations include a series of methods such as membrane techniques, distillation, crystallization, adsorption, absorption and stripping, and extraction. Specifically, liquid–liquid extraction is a technique involving two immiscible liquid phases. In this arrangement, one of the two phases, the solvent phase, extracts the solutes from the other liquid phase; however, this separation process must be followed by solvent recovery in addition to raffinate cleanup. Subsequently, because there are many options available to choose the fluid phases, for the raffinate phase cleaning method, and especially when we refer to industrial production, an adequate separation design is vital. This design must take into account factors such as efficiency, raw materials, energy cost, and also personnel and environmental safety. The work of the aforementioned authors refers to a specific type of liquid–liquid extraction, that is, the aqueous two-phase system (ATPS), and more specifically, the ionic liquid (IL)-based aqueous two-phase system (ATPS). This technique has attracted considerable attention in recent years as it includes the advantages of ILs and those of the ATPS. Such a combination represents an alternative that may be advantageous in certain cases compared to other types of ATPSs with other phase-forming chemicals [53] (polymer–polymer, polymer–salt, alcohol–salt, and micelles). In one instance, the ILs, usually defined as compounds that consist of an organic cation paired with an organic or inorganic anion with a melting point below 100 °C [54], constitute materials in which interest has grown exponentially in recent decades. They can compete successfully over conventional organic solvents, which are often more volatile and consequently more toxic, flammable, and environmentally harmful. Today, more than 137,066 results from the Web of Science Core Collection appear for the topic of *ionic liquid*. A significant portion of these works are devoted to separation studies with biphasic systems based on hydrophobic IL and water. However, it is necessary to warn that possible denaturation of biomacromolecules may occur due to the effect of IL. Faced with this issue, the aqueous two-phase system (ATPS) is composed of two liquid phases, which, although immiscible, have water as a major component in both phases, thus providing a mild environment adequate for the separation and purification of biological macromolecules, for example. However, as the authors of this contribution specify, despite previous studies highlighting possible important synergies of the corresponding combination of IL-based ATPSs, a broad study of their properties is required. Following a detailed analysis of the state of the art, the authors point out the urgent need for further basic physicochemical studies on the structure and properties of these types of systems and on phase separation mechanisms. Moreover, they highlight the need to establish rules that guide the selection of the most suitable ILs for both the effective separation of each type of analyte and that are also more suitable for the desired environmental preservation.

3. A Final Reflection

The eleven contributions comprising this Special Issue confirm the significant potential of thermodynamic tools in addressing the broad diversity of problems of both basic theoretical interest and practical applications. The contributions presented herein also offer broad perspectives that can serve to promote new research in fields of significant interest at present, such as process engineering, the field of human health, or environmental preservation.

Given the broad scope of its applications and the fact that, despite its maturity, there are still positive prospects for the development of the theory of thermodynamics, this subject must be maintained in the curricula of academic training centers and in research laboratories with full vigor. The current reformulations of scientific fields and their new interactions represent more than a mere threat to classical studies, a horizon of opportunities to which thermodynamics has much to contribute.

Author Contributions: Conceptualization, methodology, investigation, data curation, and visualization, A.M.M. and J.S.U.; resources, writing—original draft preparation, project administration, and funding acquisition, A.M.M.; writing—review and editing, and supervision J.S.U. All authors have read and agreed to the published version of the manuscript.

Funding: The authors thank the financial support provided by the Departamento de Ciencia, Tecnología y Universidad del Gobierno de Aragón-Fondo Social Europeo (Group E39_23R).

Conflicts of Interest: The authors declare no conflict of interest.

List of Contributions:

1. Gimeno, B.; Martinez, S.; Mainar, A.M.; Urieta, J.S.; Perez, P. Thermodynamic Behavior of (2-Propanol + 1,8-Cineole) Mixtures: Isothermal Vapor–Liquid Equilibria, Densities, Enthalpies of Mixing, and Modeling. *Int. J. Mol. Sci.* **2023**, *24*, 10380. https://doi.org/10.3390/ijms241210380.
2. Semenov, A.; Mendgaziev, R.; Stoporev, A.; Istomin, V.; Tulegenov, T.; Yarakhmedov, M.; Novikov, A.; Vinokurov, V. Direct Measurement of the Four-Phase Equilibrium Coexistence Vapor–Aqueous Solution–Ice–Gas Hydrate in Water–Carbon Dioxide System. *Int. J. Mol. Sci.* **2023**, *24*, 9321. https://doi.org/10.3390/ijms24119321.
3. Golikova, A.; Shasherina, A.; Anufrikov, Y.; Misikov, G.; Toikka, M.; Zvereva, I.; Toikka, A. Excess Enthalpies for Binary Mixtures of the Reactive System Acetic Acid + n-Butanol + n-Butyl Acetate + Water: Brief Data Review and Results at 313.15 K and Atmospheric Pressure. *Int. J. Mol. Sci.* **2023**, *24*, 5137. https://doi.org/10.3390/ijms24065137.
4. Carbone, C.; Guzmán, E.; Rubio, R.G. Anomalous Concentration Dependence of Surface Tension and Concentration-Concentration Correlation Functions of Binary Non-Electrolyte Solutions. *Int. J. Mol. Sci.* **2023**, *24*, 2276. https://doi.org/10.3390/ijms24032276.
5. Gallego, C.; Rodríguez, H.; Soto, A. Solubility of Amino Acids in the Eutectic Solvent Constituted by Sodium Acetate Trihydrate and Urea and in Its Mixture with Water. *Int. J. Mol. Sci.* **2023**, *24*, 1550. https://doi.org/10.3390/ijms24021550.
6. Nikolić, M.; Brudar, S.; Coutsias, E.; Dill, K.A.; Lukšič; M.; Simmerling, C.; Hribar-Lee, B. BioMThermDB 1.0: Thermophysical Database of Proteins in Solutions. *Int. J. Mol. Sci.* **2022**, *23*, 15371. https://doi.org/10.3390/ijms232315371.
7. Pekař, M. Reaction-Diffusion Systems: Self-Balancing Diffusion and the Use of the Extent of Reaction as a Descriptor of Reaction Kinetics. *Int. J. Mol. Sci.* **2022**, *23*, 10511. https://doi.org/10.3390/ijms231810511.
8. Volodin, A.D.; Smol'yakov, A.F.; Korlyukov, A.A. Various Approaches to Studying the Phase Transition in an Octamethylcyclotetrasiloxane Crystal: From X-ray Structural Analysis to Metadynamics. *Int. J. Mol. Sci.* **2022**, *23*, 9073. https://doi.org/10.3390/ijms23169073.

9. Lecarpentier, Y.; Claes, V.; Hébert, J.L.; Krokidis, X.; Schussler, O.; Vallée, A. Friction in Myocardial Anoxia Leads to Negative Excess Entropy Production, Self-Organization, and Dissipative Structures. *Int. J. Mol. Sci.* **2022**, *23*, 6967. https://doi.org/10.3390/ijms23136967
10. Dehghan Niestanak, V.; Unsworth, L.D. Detailing Protein-Bound Uremic Toxin Interaction Mechanisms with Human Serum Albumin in the Pursuit of Designing Competitive Binders. *Int. J. Mol. Sci.* **2023**, *24*, 7452. https://doi.org/10.3390/ijms24087452.
11. Nie, L.; Zheng, Z.; Lu, M.; Yao, S.; Guo, D. Phase Behavior of Ionic Liquid-Based Aqueous Two-Phase Systems. *Int. J. Mol. Sci.* **2022**, *23*, 12706. https://doi.org/10.3390/ijms232012706.

References

1. Ebeling, W. Thermodynamics—Past, Present and Future. In *Advances in Solid State Physics*; Kramer, B., Ed.; Springer: Berlin/Heidelberg, Germany, 2005; Volume 45, pp. 3–14. [CrossRef]
2. Fu, H.-Z.; Ho, Y.-S. Bibliometric Analysis of Thermodynamic Research: A Science Citation Index Expanded-Based Analysis (Chapter 21). In *Thermodynamics-Fundamentals and Its Application in Science*; Morales-Rodriguez, R., Ed.; IntechOpen Limited: London, UK, 2012; pp. 519–554. [CrossRef]
3. Prigogine, I.; Geheniau, J.; Gunzig, E.; Nardone, P. Thermodynamics and cosmology. *Gen. Relativ. Gravit.* **1989**, *21*, 767–776. [CrossRef]
4. Liu, Z.-K. Computational Thermodynamics and Its Applications. *Acta Mater.* **2020**, *200*, 745–792. [CrossRef]
5. Villacís, J. Thermodynamic Laws Applied To Economic Systems. *Am. J. Bus. Educ.* **2009**, *2*, 83–86.
6. Jakimowicz, A. The Role of Entropy in the Development of Economics. *Entropy* **2020**, *22*, 452. [CrossRef] [PubMed]
7. Weber, H.; Mahnke, R.; Liebe, C.; Kaupužs, J. Dynamics and Thermodynamic of Traffic Flow. In *Traffic and Granular Flow '07*; Appert-Rolland, C., Chevoir, F., Gondret, P., Lassarre, S., Lebacque, J.-P., Schreckenberg, M., Eds.; Springer Publishing: New York, NY, USA, 2009; pp. 427–433. [CrossRef]
8. Prausnitz, J.M. Molecular Thermodynamics for Chemical Process Design: The properties of fluid mixtures must be understood for economic manufacture of chemical products. *Science* **1979**, *205*, 759–766. [CrossRef] [PubMed]
9. Liu, Z.-K.; Wang, Y. Computational Thermodynamics of Materials. *MRS Bull.* **2017**, *42*, 162–163. [CrossRef]
10. Sieniutycz, S. *Thermodynamic Approaches in Engineering Systems*, 9th ed.; Elsevier: Amsterdam, The Netherlands, 2019.
11. Hall, K.R.; Iglesias-Silva, G.A. *Thermodynamics for Chemical Engineers*; Wiley BCH: Weinheim, Germany, 2022. [CrossRef]
12. Lucia, U. Bioengineering thermodynamics: An engineering science for thermodynamics of biosystems. *Int. J. Thermodyn.* **2015**, *18*, 254. [CrossRef]
13. de Cindio, B. Thermodynamic properties of food materials (Chapter 4). In *Engineering Principles of Unit Operations in Food Processing*; Jafari, S.M., Ed.; Elsevier: Sawston, UK, 2021; Volume 1: Unit Operations and Processing Equipment in the Food Industry; pp. 65–106. [CrossRef]
14. Kjelstrup, S.; Bedeaux, D. Thermodynamics of Electrochemical Systems. In *Springer Handbook of Electrochemical Energy*; Breitkopf, C., Swider-Lyons, K., Eds.; Springer: Berlin/Heidelberg, Germany, 2017; pp. 69–93. [CrossRef]
15. Struchtrup, H. *Thermodynamics and Energy Conversion*, 1st ed.; Springer: New York, NY, USA, 2014.
16. DeHoff, R. *Thermodynamics in Materials Science*, 2nd ed.; CRC Press: Boca Raton, FL, USA, 2006. [CrossRef]
17. Haynie, D.T. *Biological Thermodynamics, 2nd. ed.*; Cambridge University Press: New York, NY, USA, 2008.
18. Mukamel, E.A.; Glaser, A.M. Thermodynamics and the matter of life. *Science* **2021**, *371*, 38. [CrossRef]
19. Walsh, J.A. Obesity & the First Law of Thermodynamics. *Am. Biol. Teach.* **2013**, *75*, 413–415. [CrossRef]
20. Wiegert, R.G. Thermodynamic Considerations in Animal Nutrition. *Am. Zool.* **1968**, *8*, 71–81. [CrossRef]
21. Jordan, C.F. The Farm as a Thermodynamic System: Implications of the Maximum Power Principle. *BioPhys. Econ. Resour. Qual.* **2016**, *1*, 9. [CrossRef]
22. Henderson, P. Thermodynamics in Geology. *Mineral. Mag.* **1978**, *42*, 415. [CrossRef]
23. Li, L.; Tao, S. Towards Thermodynamic Architecture: Research on Systems-based Design Oriented by Renewable Energy. *IOP Conf. Ser. Earth Environ. Sci.* **2018**, *146*, 012060. [CrossRef]
24. Hofmeister, A.M.; Criss, R.E. A thermodynamic and mechanical model for formation of the Solar System via 3-dimensional collapse of the dusty pre-solar nebula. *Planet. Space Sci.* **2012**, *62*, 111–131. [CrossRef]
25. Wald, R.M. The Thermodynamics of Black Holes. *Living Rev. Relativ.* **2001**, *4*, 6. Available online: http://www.livingreviews.org/Articles/Volume4/2001-6wald/ (accessed on 10 July 2024). [CrossRef]
26. Jacobson, T. Thermodynamics of Spacetime: The Einstein Equation of State. *Gen. Relativ. Quantum Cosmol.* **1995**, *75*, 1260. [CrossRef] [PubMed]
27. Sánchez-Hernández, L.M. Emergent gravity: The Thermodynamic key of space-time? Ciencia ergo-sum. *Rev. Científica Multidiscip. Prospect.* **2020**, *27*. [CrossRef]
28. Dincer, I. Thermodynamics, Exergy and Environmental Impact. *Energy Sources* **2010**, *22*, 723–732. [CrossRef]
29. Sciubba, E. A Thermodynamic Measure of Sustainability. *Front. Sustain.* **2021**, *2*, 739395. [CrossRef]
30. Chapman, E.J.; Childers, D.L.; Vallino, J.J. How the Second Law of Thermodynamics Has Informed Ecosystem Ecology through Its History. *BioScience* **2015**, *66*, 27–39. [CrossRef]

31. Valero, A.; Uson, S.; Torres, C.; Valero, A. Application of Thermoeconomics to Industrial Ecology. *Entropy* **2010**, *12*, 591–612. [CrossRef]
32. Lollai, S.A. Quality Systems. A Thermodynamics-Related Interpretive Model. *Entropy* **2017**, *19*, 418. [CrossRef]
33. Weberdemenezes, E.; Dasilva, R.; Cataluna, R.; Ortega, R. Effect of ethers and ether/ethanol additives on the physicochemical properties of diesel fuel and on engine tests. *Fuel* **2006**, *85*, 815–822. [CrossRef]
34. Ahmadi, M.A.; Bahadori, A. Gas Hydrates (chapter 5). In *Fluid Phase Behavior for Conventional and Unconventional Oil and Gas Reservoirs*; Bahadori, A., Ed.; Gulf Professional Publishing: Houston, TX, USA, 2017; pp. 405–444. [CrossRef]
35. Jäger, A.; Vinš, V.; Gernert, J.; Span, R.; Hrubý, J. Phase equilibria with hydrate formation in H_2O+CO_2 mixtures modeled with reference equations of state. *Fluid Phase Equilib.* **2013**, *338*, 100–113. [CrossRef]
36. Suryawanshi, M.A.; Shinde, N.H.; Nagotkar, R.V. Kinetic Study of Esterification Reaction for the Synthesis of Butyl Acetate. *Int. J. Eng. Res. Technol.* **2014**, *3*, 246–2355.
37. Saint-Victor, M.-E.; Patterson, D. The w-shape concentration dependence of CEp and solution non-randomness: Ketones + normal and branched alkanes. *Fluid Phase Equilib.* **1987**, *35*, 237–252. [CrossRef]
38. Troncoso, J.; Cerdeiriña, C.A.; Carballo, E.; Romani, L. Quantitative analysis of the W-shaped excess heat capacities of binary liquid mixtures in the light of the local composition concept. *Fluid Phase Equilib.* **2005**, *235*, 201–210. [CrossRef]
39. Hardinc, S.E. Protein Hydrodynamics (Chapter 7). In *Protein: A Comprehensive Tleatise*; Geoffrey Allen, G., Ed.; JAI Press Inc.: Stamford, CT, USA, 1999; Volume 2, pp. 271–305.
40. Chae, K.S.; Lenhoff, A.M. Computation of the electrophoretic mobilty of proteins. *Biophys. J.* **1995**, *68*, 1120–1127. [CrossRef] [PubMed]
41. Tang, B.; Chong, K.; Massefski, W.; Evans, R. Quantitative Interpretation of Protein Diffusion Coefficients in Mixed Protiated–Deuteriated Aqueous Solvents. *J. Phys. Chem. B* **2022**, *126*, 5887–5895. [CrossRef]
42. Zhang, Z.; Liu, Y. Recent progresses of understanding the viscosity of concentrated protein solutions. *Curr. Opin. Chem. Eng.* **2017**, *16*, 48–55. [CrossRef]
43. Pincemaille, J.; Banc, A.; Chauveau, E.; Fromental, J.M.; Ramos, L.; Morel, H.; Menut, P. Methods for Screening Cloud Point Temperatures. *Food Biophys.* **2018**, *13*, 422–431. [CrossRef]
44. Truesdell, C. Thermodynamic of chemical reactions (Lecture 6). In *Rational Thermodynamics*, 2nd ed.; Truesdell, C., Ed.; Springer: New York, NY, USA, 1984; pp. 356–357.
45. Bowen, R.M. On the stoichiometry of chemically reacting mixtures. *Arch. Ration. Mech. Anal.* **1968**, *29*, 114–124. [CrossRef]
46. Tansey, J.T.; Baird, T.; Cox, M.M.; Fox, K.M.; Knight, J.; Sears, D.; Bell, E. Foundational concepts and underlying theories for majors in "biochemistry and molecular biology". *Biochem. Mol. Biol. Educ.* **2013**, *41*, 289–296. [CrossRef] [PubMed]
47. Shklover, V.E.; Struchkov, Y. The Structure of Organocyclosiloxanes. *Russ. Chem. Rev.* **1980**, *49*, 518–556. [CrossRef]
48. Iaizzo, P.A. *Handbook of Cardiac Anatomy, Physiology, and Devices*, 3rd ed.; Springer: New York, NY, USA, 2015. [CrossRef]
49. Martín, L.; González-Coloma, A.; Burillo, J.; Palavra, A.M.F.; Urieta, J.S.; Mainar, A.M. Microcalorimetric determination of the activity of supercritical extracts of wormwood (*Artemisia absinthium* L.) over Spodoptera littoralis. *J. Therm. Anal. Calorim.* **2012**, *111*, 1837–1844. [CrossRef]
50. Uehara, M.; Sakane, K.K. Thermodynamics of the Heart. In *Application of Thermodynamics to Biological and Materials Science*; Tadashi, M., Ed.; InTech Open: London, UK, 2011; pp. 227–242. Available online: http://www.intechopen.com/books/application-of-thermodynamics-to-biological-andmaterials-science/thermodynamics-of-the-heart (accessed on 10 July 2024).
51. Dini, F.L.; Guarini, G.; Morrone, D.; Marzilli, M. The second law of thermodynamics and the heart. *Future Cardiol.* **2012**, *8*, 697–706. [CrossRef]
52. Peters, T., Jr. *All about Albumin: Biochemistry, Genetics, and Medical Applications*; Academic Press: San Diego, CA, USA, 1996; pp. 76–133. [CrossRef]
53. Benavides, J.; Rito-Palomares, M.; Asenjo, J.A. Aqueous Two-Phase Systems. In *Comprehensive Biotechnology*, 2nd ed.; Moo-Young, M., Ed.; Elsevier: New York, NY, USA, 2011; pp. 697–713. [CrossRef]
54. Lei, Z.; Chen, B.; Koo, Y.-M.; MacFarlane, D.R. Introduction: Ionic Liquids. *Chem. Rev.* **2017**, *117*, 6633–6635. [CrossRef]

International Journal of
Molecular Sciences

MDPI

Article

Thermodynamic Behavior of (2-Propanol + 1,8-Cineole) Mixtures: Isothermal Vapor–Liquid Equilibria, Densities, Enthalpies of Mixing, and Modeling

Beatriz Gimeno [1], Santiago Martinez [1], Ana M. Mainar [1,2], José S. Urieta [1,2,*] and Pascual Perez [1,3,*]

1 Departamento de Química Física, Facultad de Ciencias, Universidad de Zaragoza, 50009 Zaragoza, Spain; ammainar@unizar.es (A.M.M.)
2 Group of Applied Thermodynamics and Surfaces (GATHERS), Aragon Institute for Engineering Research (I3A), 50018 Zaragoza, Spain
3 Instituto Agroalimentario de Aragón (I2A), Centro de Investigación y Tecnología Agroalimentaria de Aragón (CITA), Universidad de Zaragoza, 50009 Zaragoza, Spain
* Correspondence: urieta@unizar.es (J.S.U.); pascual@unizar.es (P.P.); Tel.: +34-876-555-469 (J.S.U.); +34-976-761-205 (P.P.)

Abstract: Vapor pressures and other thermodynamic properties of liquids, such as density and enthalpy of mixtures, are the key parameters in chemical engineering for designing new process units, and are also essential for understanding the physical chemistry, macroscopic and molecular behavior of fluid systems. In this work, vapor pressures between 278.15 and 323.15 K, densities and enthalpies of mixtures between 288.15 and 318.15 K for the binary mixture (2-propanol + 1,8-cineole) have been measured. From the vapor pressure data, activity coefficients and excess Gibbs energies were calculated via the Barker's method and the Wilson equation. Excess molar volumes and excess molar enthalpies were also obtained from the density and calorimetric measurements. Thermodynamic consistency test between excess molar Gibbs energies and excess molar enthalpies has been carried out using the Gibbs–Helmholtz equation. Robinson–Mathias, and Peng–Robinson–Stryjek–Vera together with volume translation of Peneloux equations of state (EoS) are considered, as well as the statistical associating fluid theory that offers a molecular vision quite suitable for systems having highly non-spherical or associated molecules. Of these three models, the first two fit the experimental vapor pressure results quite adequately; in contrast, only the last one approaches the volumetric behavior of the system. A brief comparison of the thermodynamic excess molar functions for binary mixtures of short-chain alcohol + 1,8-cineole (cyclic ether), or +di-n-propylether (lineal ether) is also included.

Keywords: 1,8-cineole; 2-propanol; vapor pressure; phase equilibria; excess thermodynamic functions; PRM; PRSV EoS; volume translation; SAFT

1. Introduction

Vapor pressures and other thermodynamic properties of pure liquid and mixtures are physical properties of great importance for not only being necessary for the daily challenges in chemical engineering but also from a theoretical point of view as they help to develop and improve representative models of the liquid state.

Within this context, the mixture (2-propanol + 1,8-cineole) is of particular interest due to the nature of the molecules it comprises, and because of its potential applications, as described below. In this work, an experimental study of vapor–liquid equilibria (VLE), densities, and enthalpies of mixing for the mixture (2-propanol + 1,8-cineole), together with some current models to fit them, are addressed.

1,3,3-Trimethyl-2-oxabicyclo[2.2.2]octane is a cyclic ether usually known as 1,8-cineole, or more commonly as eucalyptol because it is the main component of essential oil obtained from the eucalyptus, although it can be identified in the extracts of many others plants. The fairly recent paper of Cai et al. [1] includes a detailed review on sources, biological activities, and applications of this compound. 1,8-cineole is frequently used in food [2–4], cosmetics and care products [4,5], and has been used profusely in pharmaceutical trial [1,4]. It is effective as a mucolytic agent in inflammatory airway medical conditions, such as chronic obstructive pulmonary disease [6] and asthma [7]. Additionally, it has been claimed for many other biological properties, such as antimicrobial [5,8,9], antifungal [10], anxiolytic [11], and it has even been related to skin penetration enhancement of drugs [12]. More recently, other studies have been carried out on the biological activity of 1,8-cineole for the prevention of depressive-like behaviors [13], battle against ovarian and lung cancers [14,15], treatment of Alzheimer's disease [16,17] and the treatment of conditions, such as co-infections associated with SARS-CoV-2 (COVID-19) [18], among others.

Moreover, 1,8-cineole could be considered an environmentally friendly solvent [19,20] as it behaves as a low toxicity compound [21], and it is obtained mainly from renewable sources [1]. Regarding the extraction process of 1,8-cineole from vegetal material, one of the possible ways is to use supercritical CO_2 [22], and taking into account the low polarity of CO_2, it is convenient to add a polarity modifier in order to increase the yield of extraction. Usually, short-chain alcohols are used for that goal. Consequently, VLE and, in general, thermodynamic behavior of 2-propanol + 1,8-cineole mixture could be an interesting guide for the 1,8-cineole separation processes, even if conditions of extraction are somewhat different from those considered here.

Propan-2-ol, the simplest secondary alcohol, also called 2-propanol, isopropanol or isopropyl alcohol, is a well-known compound because of its use in the manufacturing of a wide variety of chemicals and products, and also due to its potential to be used as a multipurpose green solvent. In fact, although 2-propanol is currently obtained via synthesis (through hydration reaction of propene or by hydrogenating acetone), it appears among the first positions in the guides for the greenest solvents recommended according to safety, occupational health and environmental criteria [23]. Additionally, a sustainable synthesis of propanol from renewable glycerol has been proposed [24] in order to not only deliver opportunities to the biodiesel industry, but also to develop a green production route from the biomass-derived glycerol.

Another interesting feature in relation to their possible use as additives for diesel or gasoline is that not only bio-cineole [25] and 2-propanol [26] have been separately considered, but also types of ether/alcohol mixtures [27] such as the system analyzed in this work.

On the other hand, from a theoretical point of view, alcohol + ether mixtures are of great interest because ether molecules could compete with alcohol molecules in hydrogen bond formation through the lone electronic pairs of oxygen atom of ether [28,29].

In recent years, some papers have been published on the thermodynamic properties of both pure 1,8-cineole [30,31] and its mixtures with alcohols [32–37]. However, as far as we know, until the present work, no data on vapor pressure, P, excess molar enthalpies, H_m^E, and density, ρ, at the considered temperatures of this binary system could be found in the literature. Additionally, in this work, the activity coefficients, γ_i, and the excess molar Gibbs energies, G_m^E, were determined from experimental P data.

Our results on P are correlated by three equations of state (EoS): two of them are modifications of the Peng–Robinson (P-R) EoS [38], and a third equation is based on the statistical associating fluid theory (SAFT) [39].

EoS are useful approaches to describe the thermodynamic properties of fluid systems and phase equilibria. Since van der Waal, on the basis of the attractive forces between molecules and their spatial volumes, built his groundbreaking cubic equation [40], hundreds of EoS have been proposed along the years [41–43].

Peng and Robinson introduced a modification [38] of the attractive term of such equations in order to provide a more adequate prediction for densities of liquids. In this work, this modification is considered together with that proposed by Mathias [44] to improve the results when considering systems with polar substances, by the introduction of a polar parameter in the temperature depending on the attractive term of the EoS. On the other hand, the Stryjek and Vera modification [45] to the Peng–Robinson equation is considered. In this version, an adjustable parameter for pure components is introduced together with a modification of the polynomial for the acentric parameters.

In short, the EoS of Peng–Robinson–Mathias [44], PRM, and Peng–Robinson–Stryjek–Vera [45], PRSV, together with volume translation (VT) according to Peneloux [46] for both models are considered in this work.

Finally, as previously indicated, experimental results are compared with that of SAFT. The development and success of this model was due to the progress in the statistical mechanics methods, especially the perturbation theory and the valuable contribution of Wertheim [47,48] for chain and association effects. It provided a reasonably simple and accurate [49] general formulation which is accepted as the basis of the SAFT free energy model. This theoretical approach was derived by Chapman et al. [50,51], resulting in a well-grounded EoS, especially in the case of chain-like molecules and when effect of association is remarkable. In fact, the model may simulate the behavior of a wide range of systems having molecules ranging from nearly spherical non-associating to non-spherical associated ones, passing through the intermediate structural configurations. This third situation becomes the case of the system considered in this work, for which molecules are not very far from a spherical shape and one of the components is markedly associated while the other is not. Thus, for our intermediate system, it is interesting to compare the results of this particular SAFT molecular association model with those of the Van der Waals Modified Cubic EoS PRM, and PRSV previously mentioned. A number of user-friendly reviews on the derivation and applications of the SAFT model and the cubic EoS can be found in the literature [39,52–57].

It has been for the purpose of enhancing the vision of the models and the molecular interactions, that in addition to the measurements of P, the experimental determination of ρ and heats of mixture were carried out. In fact, the corresponding thermodynamic properties, V_m^E, and H_m^E, are very sensitive to the spatial effects derived from molecular geometry and the detail of intermolecular forces, which complements information provided by other thermophysical quantities [35,58,59]. Furthermore, the present work is also a part of a comprehensive study conducted by our group using 1.8-cineole as cyclic ether and a short-chain alcohol as the second component [32–37,60,61]. Likewise, in previous years, we also carried out a similar study with di-n-propylether as linear ether, and an alcohol as the second component of a series that also included the same short-chain alcohols [62,63]. Hence, we include a brief comparison of the different thermodynamic behavior of both sets of systems in the Discussion section.

2. Results and Discussion

2.1. Pure Components

The molar volumes of the pure components, V^0, used in the Barker analysis, together with the experimental vapor pressures, P^0, which are compared with the values in the literature are gathered in Table 1.

Experimental P^0 data of 2-propanol at ten temperatures between (278.15 K and 323.15 K) were fitted to an Antoine equation (Equation (1)):

$$\ln\left(P^0/\text{kPa}\right) = 17.6939 - [4114.55/(T/K - 39.969)] \quad (1)$$

The standard deviation of the experimental pressures with respect to that calculated are obtained according to Equation (2):

$$s = \left\{\sum_{i=1}^{N}(X - X_{calc})_i^2/(N-n)\right\}^{1/2} \quad (2)$$

X and X_{calc} correspond to the experimental pressures and calculated values, respectively, N is the number of experimental data and n is the number of adjusted parameters. Standard deviation results in a value of 25 Pa, 15 Pa being the maximum deviation at 298.15 K. The corresponding Antoine equation for 1,8-cineole was published previously [36]. Some new experimental vapor pressure data of this liquid have been published in recent years and they are shown in Figure 1 along with our data. The interval of temperature selected in Figure 1 allows for a better comparison between the data from different authors.

Table 1. Molar volumes, V^0, and vapor pressures, P^0, of pure liquids used in the Barker analysis [a], and data taken from the literature.

	2-Propanol			1,8-Cineole		
	$V^0 \times 10^6$/m^3·mol^{-1}	P^0/Pa		$V^0 \times 10^6$/m^3·mol^{-1}	P^0/Pa	
T/K	lit. [b,c]	exp. [b]	lit. [d]	lit. [b,e]	exp. [b]	lit.
278.15	75.35	1521	1514	164.6	71	54.80 [g]
283.15	75.70	2164	2162	165.3	93	80.12 [g]
283.51	-	-	-	-	-	83.55 [h]
288.15	76.12	3069	3044	166.1	130	133 [f]/115.17 [g]
293.15	76.52	4212	4228	166.9	178	163.72 [g]
293.49	-	-	-	-	-	170.01 [h]
298.15	76.93	5781	5797	167.7	253	229.50 [g]
303.15	77.35	7851	7852	168.5	346	317.05 [g]
303.34	-	-	-	-	-	323.77 [h]
308.15	77.78	10,515	10,511	169.2	477	431.34 [g]
313.15	78.23	13,934	13,913	170.0	627	-
313.23	-	-	-	-	-	591.71 [h]
318.15	78.66	18,242	18,220	170.9	828	-
323.15	79.15	23,623	23,618	171.7	1101	-

[a] Standard uncertainty u is u(T) = 0.01 K and the combined expanded uncertainty U_c is $U_c(P)$ = 0.1% with a 0.95 level of confidence (k = 2). [b] Used in the Barker analysis. [c] Ref. [64]. [d] Ref. [65]. [e] Ref. [35]. [f] Ref. [66]. [g] Ref. [31]. [h] Ref. [67].

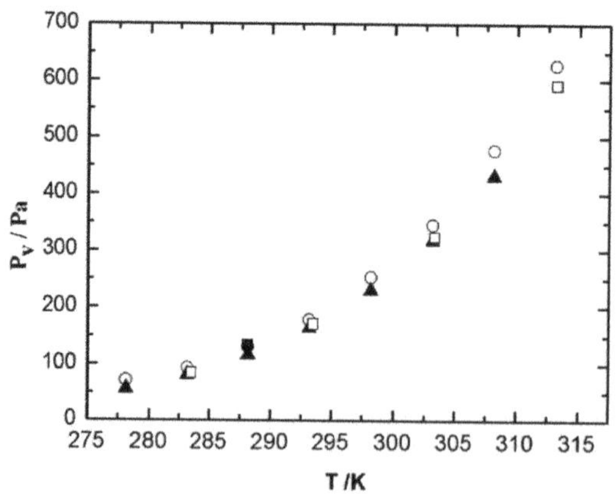

Figure 1. Vapor pressure of pure 1,8-cineole in the temperature range 278.15 K–313.15 K. ○, this work; ■, Stull et al. [66]; ▲, Štejfa et al. [31]; and □, Guetachew et al. [67].

The second virial coefficient at T = 325.0 K of 2-propanol ($B_{11} = -1810 \times 10^6$ m^3·mol^{-1} and 1,8-cineole ($B_{22} = -5490 \times 10^6$ m^3·mol^{-1}) were calculated from the Tsonopoulos correlation [68]. For mixtures, the cross virial coefficient, B_{12}, will be calculated according to the cubic Lorenz semiempirical combination rule [69], as mentioned in Equation (3),

$$B_{12} = \frac{1}{8}\left(B_{11}^{1/3} + B_{22}^{1/3}\right)^3 \quad (3)$$

2.2. Vapor Pressures and Derived Thermodynamic Parameters

Table 2 shows our vapor pressure measurements along with the activity coefficients, γ_1 and γ_2, and the excess molar Gibbs free energy, G_m^E, values fitted using Barker's method [70] to the Wilson correlation [71].

$$\frac{G_m^E}{RT} = -x_1 \ln(x_1 + \Lambda_{12}x_2) - x_2 \ln(x_2 + \Lambda_{21}x_1) \quad (4)$$

Table 2. Values of the vapor pressure P, deviations $\Delta P = P - P_{calc}$, activity coefficients, γ_1 and γ_2, and excess molar Gibbs energies, G_m^E [a], for the mixtures 2-propanol (1) + 1,8-cineole (2).

x_2	P/Pa	ΔP/Pa	γ_1	γ_2	G_m^E/J·mol^{-1}	x_2	P/Pa	ΔP/Pa	γ_1	γ_2	G_m^E/J·mol^{-1}
2-Propanol (1) + 1,8-Cineole (2)											
T/K = 278.15											
0.0309	1485	3	1.0012	2.9938	81	0.6307	944	6	1.5749	1.1958	649
0.1004	1425	22	1.0120	2.5666	244	0.7188	849	13	1.8228	1.1146	571
0.2031	1303	1	1.0488	2.1015	437	0.8182	661	−16	2.2277	1.0494	428
0.3294	1195	1	1.1308	1.7073	598	0.8670	576	7	2.4999	1.0271	335
0.4351	1107	−2	1.2377	1.4756	670	0.9382	347	−5	3.0297	1.0061	172
0.5233	1024	−13	1.3622	1.3297	686	0.9882	149	16	3.5396	1.0002	35
T/K = 283.15											
0.0310	2128	20	1.0012	2.9499	82	0.6308	1323	2	1.5661	1.1918	650
0.1006	2007	14	1.0119	2.5340	245	0.7189	1176	2	1.8084	1.1121	572
0.2032	1865	19	1.0482	2.0801	439	0.8182	940	−5	2.2020	1.0483	429
0.3295	1684	−6	1.1293	1.6938	601	0.8670	808	17	2.4652	1.0264	336
0.4352	1569	1	1.2347	1.4666	673	0.9382	471	−14	2.9741	1.0060	172
0.5234	1451	−13	1.3573	1.3233	688	0.9882	183	3	3.4600	1.0002	35
T/K = 288.15											
0.0313	3010	22	1.0012	2.9026	82	0.6309	1859	−1	1.5565	1.1875	651
0.1006	2838	13	1.0117	2.4997	246	0.7190	1655	6	1.7930	1.1095	572
0.2032	2622	7	1.0476	2.0570	440	0.8183	1329	6	2.1748	1.0471	428
0.3297	2398	7	1.1277	1.6793	602	0.8671	1112	7	2.4285	1.0257	335
0.4353	2205	−11	1.2314	1.4569	674	0.9382	657	−17	2.9160	1.0058	171
0.5235	2056	−10	1.3519	1.3165	689	0.9882	253	3	3.3775	1.0002	35
T/K = 293.15											
0.0316	4112	12	1.0012	2.8844	84	0.6310	2564	22	1.5514	1.1811	651
0.1007	3892	15	1.0119	2.4806	249	0.7191	2274	28	1.7811	1.1052	570
0.2033	3574	−15	1.0480	2.0388	444	0.8184	1800	8	2.1471	1.0450	426
0.3299	3264	−17	1.1283	1.6638	607	0.8672	1469	−22	2.3875	1.0245	333
0.4354	3010	−28	1.2315	1.4448	677	0.9383	896	−10	2.8430	1.0055	170
0.5236	2828	−2	1.3506	1.3070	691	0.9882	321	−16	3.2674	1.0002	34
T/K = 298.15											
0.0321	5654	29	1.0012	2.8630	87	0.6312	3492	12	1.5461	1.1752	651
0.1009	5320	−2	1.0121	2.4607	252	0.7193	3078	11	1.7695	1.1014	569
0.2035	4936	6	1.0484	2.0208	448	0.8185	2446	9	2.1213	1.0431	424
0.3302	4481	−25	1.1288	1.6490	611	0.8673	2003	−21	2.3501	1.0234	331
0.4356	4154	−18	1.2312	1.4336	680	0.9384	1221	−5	2.7782	1.0052	168
0.5238	3894	12	1.3491	1.2983	692	0.9882	475	10	3.1715	1.0002	34

Table 2. *Cont.*

x_2	P/Pa	ΔP/Pa	γ_1	γ_2	G^E_m/J·mol^{-1}	x_2	P/Pa	ΔP/Pa	γ_1	γ_2	G^E_m/J·mol^{-1}
2-Propanol (1) + 1,8-Cineole (2)											
T/K = 303.15											
0.0329	7621	−14	1.0013	2.8303	89	0.6314	4713	9	1.5397	1.1707	652
0.1013	7215	−10	1.0121	2.4362	255	0.7195	4153	16	1.7580	1.0986	569
0.2038	6699	8	1.0484	2.0025	451	0.8187	3313	38	2.0988	1.0418	423
0.3307	6105	−7	1.1285	1.6356	614	0.8675	2676	−37	2.3189	1.0226	330
0.4358	5640	−17	1.2298	1.4243	683	0.9384	1623	−17	2.7274	1.0050	168
0.5241	5248	−10	1.3462	1.2914	694	0.9883	633	7	3.0988	1.0002	34
T/K = 308.15											
0.0325	10,218	−11	1.0013	2.8076	89	0.6318	6249	−27	1.5338	1.1660	652
0.1017	9658	−16	1.0122	2.4119	257	0.7199	5537	30	1.7469	1.0956	569
0.2041	8951	−10	1.0485	1.9839	455	0.8189	4353	7	2.0764	1.0404	422
0.3315	8170	−11	1.1286	1.6216	618	0.8677	3589	−4	2.2877	1.0218	329
0.4362	7557	−11	1.2288	1.4145	685	0.9386	2140	−30	2.6763	1.0048	167
0.5245	7054	27	1.3437	1.2841	696	0.9883	871	28	3.0260	1.0002	34
T/K = 313.15											
0.0338	13,490	−48	1.0014	2.7686	93	0.6324	8269	10	1.5272	1.1610	651
0.1015	12,790	−27	1.0122	2.3878	258	0.7204	7246	21	1.7346	1.0924	567
0.2039	11,878	11	1.0481	1.9658	456	0.8193	5721	44	2.0524	1.0389	420
0.3310	10,824	−2	1.1274	1.6093	619	0.8680	4656	−24	2.2547	1.0210	326
0.4368	9977	−20	1.2274	1.4039	686	0.9387	2780	−36	2.6232	1.0046	165
0.5251	9254	−16	1.3408	1.2763	696	0.9883	1072	−26	2.9512	1.0002	33
T/K = 318.15											
0.0363	17,673	−13	1.0016	2.7299	100	0.6319	10780	15	1.5195	1.1565	650
0.1024	16,765	−2	1.0124	2.3622	262	0.7199	9481	84	1.7204	1.0896	565
0.2045	15,507	−19	1.0485	1.9459	460	0.8198	7322	−12	2.0291	1.0373	417
0.3322	14,127	−25	1.1281	1.5933	622	0.8684	5993	−37	2.2218	1.0200	323
0.4365	13,040	−31	1.2259	1.3944	687	0.9389	3592	−26	2.5692	1.0044	163
0.5247	12,103	−6	1.3373	1.2693	696	0.9883	1432	7	2.8747	1.0002	33
T/K = 323.15											
0.0410	22,785	−29	1.0020	2.6720	113	0.6328	13851	−11	1.5135	1.1516	649
0.1037	21,678	−11	1.0127	2.3316	266	0.7206	12092	22	1.7092	1.0867	563
0.2055	20,080	−7	1.0487	1.9245	463	0.8204	9415	32	2.0078	1.0358	414
0.3341	18,256	−32	1.1287	1.5768	625	0.8689	7677	−23	2.1926	1.0192	321
0.4375	16,849	−38	1.2249	1.3840	689	0.9392	4569	−50	2.5224	1.0042	162
0.5257	15,679	54	1.3349	1.2616	696	0.9883	1889	38	2.8096	1.0002	33

[a] Standard uncertainties, u, are u(x) = 0.0001, u(T) = 0.01 K, and the combined expanded uncertainty, U_c, is $U_c(P)$ = 0.1% with a 0.95 level of confidence (k = 2).

The activity coefficients are obtained through the appropriate differentiation of Equation (4).

$$\ln \gamma_1 = -\ln(x_1 + \Lambda_{12}x_2) + x_2\left[\frac{\Lambda_{12}}{x_1 + \Lambda_{12}x_2} - \frac{\Lambda_{21}}{\Lambda_{21}x_1 + x_2}\right] \quad (5)$$

$$\ln \gamma_2 = -\ln(x_2 + \Lambda_{21}x_1) - x_1\left[\frac{\Lambda_{12}}{x_1 + \Lambda_{12}x_2} - \frac{\Lambda_{21}}{\Lambda_{21}x_1 + x_2}\right] \quad (6)$$

with:

$$\Lambda_{ij} = \frac{V_j^0}{V_i^0}\exp\left(-\frac{\lambda_{ij} - \lambda_{ii}}{RT}\right) \quad (7)$$

where the subscripts 1 and 2 stand for 2-propanol and 1,8-cineole, respectively. V^0 is the molar volume and λ's are the interaction constants between the molecules designated in the subscripts.

The vapor pressure is then given by,

$$P_{calc} = x_1\gamma_1 P_1^o R_1 + x_2\gamma_2 P_2^o R_2, \tag{8}$$

and the non-ideality of the vapor phase is taken into account with the following corrections (Equations (9) and (10)):

$$R_1 = \exp\left\{\left[(V_1^o - B_{11})(P - P_1^o) - P\delta_{12}y_2^2\right]/RT\right\} \tag{9}$$

$$R_2 = \exp\left\{\left[(V_2^o - B_{22})(P - P_2^o) - P\delta_{12}y_1^2\right]/RT\right\} \tag{10}$$

where y_1 and y_2 are the vapor phase mole fractions of 1-propanol and 1,8-cineole, respectively and δ_{12} is defined by the following equation:

$$\delta_{12} = 2B_{12} - B_{11} - B_{22} \tag{11}$$

For every liquid mixture, the vapor pressure is measured at different temperatures from 278.15 to 323.15 K, so a slight modification of the true initial liquid mole fraction can be detected in Table 2, because of the variation in the amount and composition of the vapor phase.

In Table 3, the Wilson parameters, Λ_{12} and Λ_{21}, are collected, together with the standard deviations, defined by Equation (2).

Table 3. Wilson parameters, Λ_{12} and Λ_{21}, standard deviations, s(Pa), and Wilson coefficients, $\lambda_{12} - \lambda_{11}$, (J·mol^{-1}) and $\lambda_{12} - \lambda_{22}$ (J·mol^{-1}) of Equations (2) and (4)–(7).

	2-Propanol (1) + 1,8-Cineole (2)				
T/K	Λ_{12}	Λ_{21}	**s/Pa**	$\lambda_{12} - \lambda_{11}$	$\lambda_{12} - \lambda_{22}$
278.15	0.4255	0.5511	12	3783 (3815)	−429 (−460)
283.15	0.4341	0.5548	12	3803 (3800)	−452 (−452)
288.15	0.4434	0.5589	11	3818 (3784)	−476 (−443)
293.15	0.4633	0.5504	19	3776 (3768)	−445 (−434)
298.15	0.4815	0.5434	16	3743 (3752)	−420 (−425)
303.15	0.4942	0.5417	20	3739 (3737)	−417 (−416)
308.15	0.5082	0.5390	21	3725 (3721)	−408 (−407)
313.15	0.5227	0.5371	28	3710 (3705)	−402 (−399)
318.15	0.5401	0.5320	33	3682 (3689)	−383 (−390)
323.15	0.5543	0.5301	34	3666 (3674)	−375 (−381)

In parentheses, the values mentioned are those that linearly fitted.

In the same table, Wilson coefficients, $\lambda_{12} - \lambda_{11}$ and $\lambda_{12} - \lambda_{22}$, obtained from Equation (7), as well as values linearly fitted with temperature are also presented. Vapor pressure–liquid composition curves are shown in Figure 2. On the other hand, Figure 3 shows graphically the analytic calculations for G_m^E.

The negatively defined Wilson coefficients, λ_{ij}, stand for energies of interaction between the molecules type i and j. By combining our previous λ_{ij} values published for the 2-propanol (1) + di-n-propylether (2) system [62] and the λ_{ij} values for the 2-propanol (1) + 1,8-cineole (2) system from Table 3 at T = 298.15 K, we can obtain the differences:

$$\lambda_{12} - \lambda_{12'} = 851\ \text{J·mol}^{-1}$$
$$\lambda_{22} - \lambda_{2'2'} = 480\ \text{J·mol}^{-1}$$

where the notation used is 1 for 2-propanol, 2 for di-n-propylether and 2′ for 1,8-cineole.

These differences point to a stronger energy of interaction of 2-propanol-1,8-cineole than 2-propanol-di-n-propylether and also a stronger energy of interaction of 1,8-cineole-

1,8-cineole than di-n-propylether-di-n-propylether, taking into account that the calculated numerical values are relative and also the approximate nature of the Wilson model.

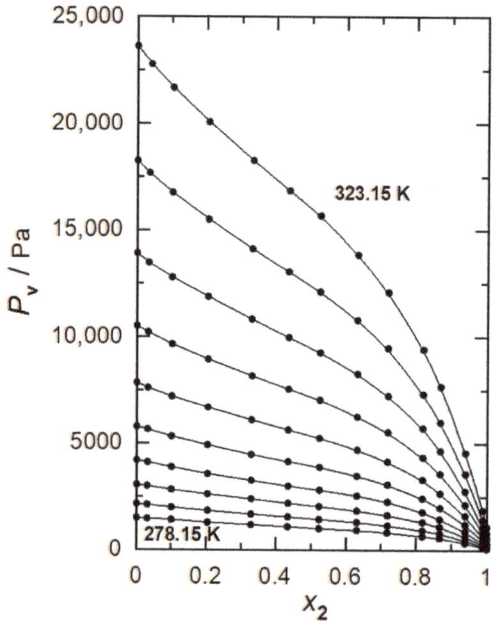

Figure 2. Vapor pressures plotted against liquid-phase composition of 1,8-cineole, at working temperatures between 278.15 K and 323.15 K at intervals of 5 K, for 2-propanol (1) + 1,8-cineole (2)): (•) experimental values, (—) polynomial curve fitting.

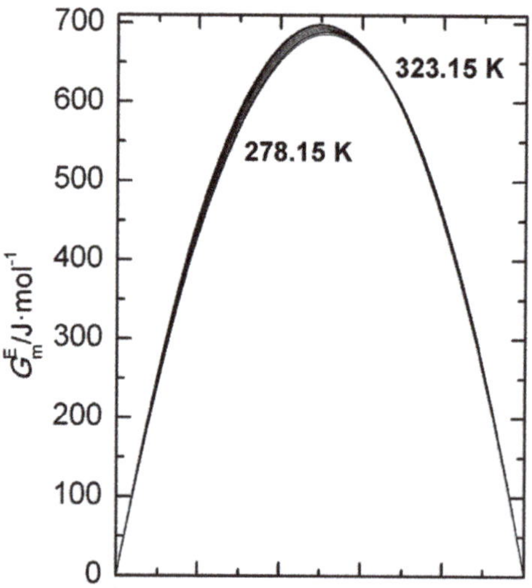

Figure 3. Excess molar Gibbs energies, G^E_m, calculated via the Barker's method, at temperatures between 278.15 K and 323.15 K at intervals of 5 K, for 2-propanol (1) + 1,8-cineole (2), plotted as a function of mole fraction of 1,8-cineole.

2.3. Excess Molar Enthalpies and Densities

Experimental excess molar enthalpies and densities at four temperatures are gathered in Tables 4 and 5, respectively.

Table 4. Excess molar enthalpies, H_m^E, for the 2-propanol + 1,8-cineole system at four temperatures [a].

x_2	H_m^E/Jmol^{-1}			
	288.15 K	**298.15 K**	**308.15 K**	**318.15 K**
0.000	0	0	0	0
0.050	65	71	77	88
0.102	146	160	175	196
0.206	291	325	360	401
0.303	396	448	495	554
0.409	474	542	609	676
0.510	517	599	663	749
0.606	518	602	683	762
0.703	483	558	644	719
0.804	401	461	530	593
0.897	245	283	334	368
0.952	124	143	172	195
1.0000	0	0	0	0

[a] Standard uncertainties, u, are u(x_2) = 0.001, u(T) = 0.01 K and the combined expanded uncertainty, U_c, is $U_c(H_m^E)$ = 2% with a 0.95 level of confidence (k = 2).

Table 5. Densities for the 2-propanol + 1,8-cineole system at four temperatures [a].

x_2	ρ/kg·m^{-3}			
	288.15 K	**298.15 K**	**308.15 K**	**318.15 K**
0.0000	789.34	780.99	772.37	763.40
0.1084	821.21	812.72	803.98	794.93
0.2025	842.65	834.07	825.27	816.20
0.2938	859.68	851.02	842.19	833.12
0.3996	875.90	867.22	858.36	849.56
0.5011	888.92	880.21	871.35	862.32
0.6004	899.70	890.97	882.13	873.15
0.6980	908.60	899.88	891.06	882.14
0.7953	916.30	907.60	898.81	889.96
0.9029	923.59	914.91	906.19	897.44
1.0000	928.96	920.34	911.73	903.13

[a] Standard uncertainties, u, are u(x_2) = 0.0001, u(T) = 0.01 K and the combined expanded uncertainty, U_c, is $U_c(\rho)$ = 0.04 kg·m^{-3} with a 0.95 level of confidence (k = 2).

Graphical representations of experimental densities as function of composition at the four temperatures considered appears in Figure 4.

Excess molar volumes were calculated using Equation (11):

$$V_m^E = x_1 M_1 \left(\frac{1}{\rho} - \frac{1}{\rho_1}\right) + x_2 M_2 \left(\frac{1}{\rho} - \frac{1}{\rho_2}\right) \quad (12)$$

where ρ stands for experimental density of the mixture and subscripts 1 and 2 for 2-propanol and 1,8-cineole, respectively.

The values of the excess molar properties, H_m^E and V_m^E, have been fitted via least squares to a Redlich–Kister polynomial:

$$Q_m^E = x_1 x_2 \sum_{j=1}^{k} A_j (x_1 - x_2)^j \quad (13)$$

where Q_m^E denotes H_m^E or V_m^E, x_1 and x_2 represent the mole fraction of 2-propanol and 1,8-cineole, respectively.

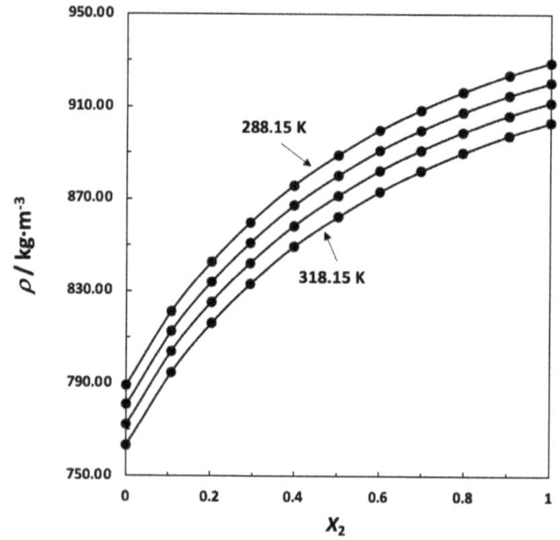

Figure 4. Densities, ρ, of mixtures 2-propanol (1) + 1,8-cineole (2) at four temperatures (288.15, 298.15, 308.15 and 318.15 K): (•) experimental values, (—) polynomial curve fitting.

The coefficients of Equation (13) are collected in Table 6 beside the standard deviation $s(Q_m^E)$ obtained from Equation (2).

Table 6. Coefficients and standard deviations, $s(Q_m^E)$, for least squares representation by Equation (2) of H_m^E (J·mol^{-1}) and V_m^E (cm^3·mol^{-1}) at the four temperatures studied.

	Q_m^E	A_0	A_1	A_2	A_2	$s(Q_m^E)$
288.15 K	H_m^E	2063	−617	136		6.3
	V_m^E	−2.017	−0.227	−0.770	−0.25	0.007
298.15 K	H_m^E	2380	−768	75		6.6
	V_m^E	−1.979	−0.247	−0.765	−0.300	0.006
308.15 K	H_m^E	2667	−981	197		7.8
	V_m^E	−1.912	−0.306	−0.737	−0.31	0.006
318.15 K	H_m^E	2978	−1107	204		7.0
	V_m^E	−1.847	−0.378	−0.642	−0.30	0.007

Graphical representations of both excess molar properties, H_m^E and V_m^E, are plotted in Figures 5 and 6, respectively. Both excess molar properties increase with temperature and in the case of excess molar enthalpies that increase at molar fraction around 0.5 is almost lineal, so we can calculate a value of $C_{P,m}^E \approx 7.6$ J·mol^{-1}K^{-1}.

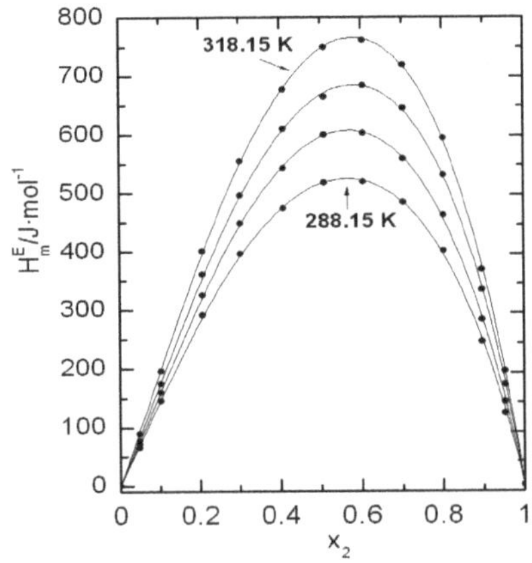

Figure 5. Excess molar enthalpies, H_m^E, at four temperatures for 2-propanol (1) + 1,8-cineole (2): (•) experimental values; (—) Equation (12).

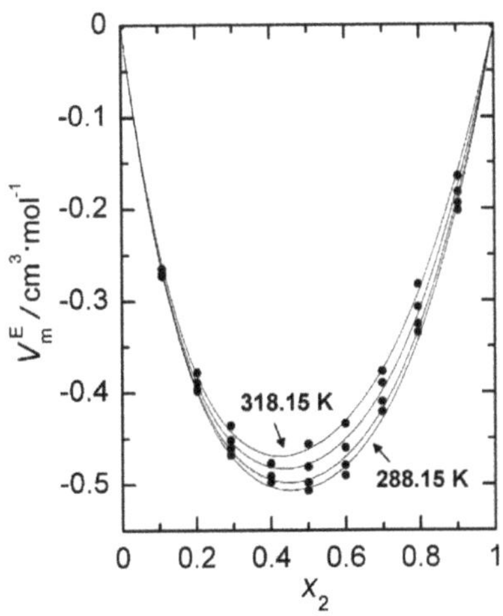

Figure 6. Excess molar volumes, V_m^E, at four temperatures for 2-propanol (1) + 1,8-cineole (2): (•) experimental values; (—), Equation (12).

In the absence of the independent values of the activity coefficients, we cannot use the Gibbs–Duhem relation to test the thermodynamic consistency of the vapor pressure measurements. However, we can test the consistency of the H_m^E and G_m^E values via the Gibbs–Helmholtz equation. The H_m^E values calculated at T = 298.15 K are shown as curves in Figure 7, together with our H_m^E experimental data. The match can be considered satisfactory although considerable uncertainty is implied by the quantitative evaluation of H_m^E from vapor pressures [72]. In the same figure and for the same temperature, TS_m^E curves, obtained from $TS_m^E = H_m^E - G_m^E$, are also plotted.

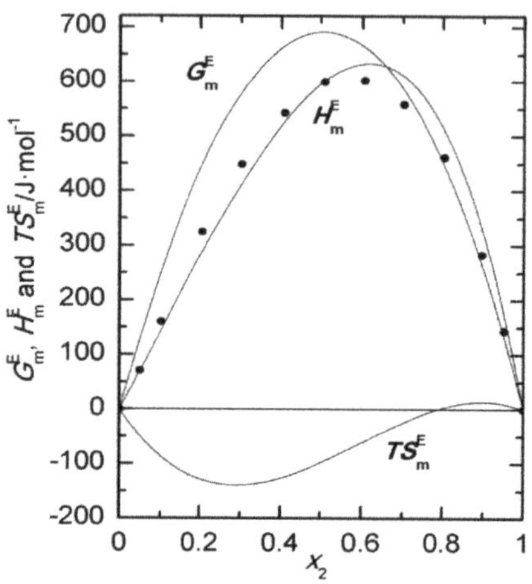

Figure 7. Thermal excess molar functions at 298.15 K for 2-propanol (1) + 1,8-cineole (2): (•) experimental H^E_m; (—) Gibbs–Helmholtz H^E_m, G^E_m and TS^E_m.

2.4. A Comparative Discussion of the Thermodynamic Excess Functions for Short-Chain Alcohol + 1,8-Cineol, or +di-n-Propylether

For comparative purposes, and in order to highlight the particularities of our binary mixture, the thermodynamic excess molar functions at T = 298.15 K and x = 0.5 of the short-chain alcohol + 1,8-cineole (cyclic ether) or +di-n-propylether (lineal ether) or +n-hexane (an inert solvent) are summarized in Table 7.

Table 7. Thermodynamic excess functions for {0.5 alkanol + 0.5 1,8-cineole or +0.5 di-n-propylether or +n-hexane} at 298.15 K.

Alkanol	1,8-Cineole [a] (μ = 1.544 D) [b]			Di-n-Propylether (μ = 1.2 D) [d]			n-Hexane [h]		
	H^E_m	TS^E_m	V^E_m	H^E_m	TS^E_m	V^E_m	H^E_m	TS^E_m	V^E_m
	J·mol^{-1}		cm^3·mol^{-1}	J·mol^{-1}		cm^3·mol^{-1}	J·mol^{-1}		cm^3·mol^{-1}
Ethanol	229	−501	−0.645	717 [e]	−254 [f]	−0.294 [e]	555	−850	0.410
1-Propanol	160	−384	−0.700	741 [g]	−80 [g]	−0.388 [g]	565	−680	0.180
2-Propanol	600 [c]	−400 [c]	−0.495 [c]	956 [g]	102 [g]	−0.027 [g]	787 [i]	−270 [j]	0.510 [k]
1-Butanol	213	−254	−0.664	743 [e]	4 [f]	−0.468 [e]	510	−630	0.080

[a] Ref. [33]. [b] Refs. [73,74]. [c] This work. [d] Refs. [75,76]. [e] Ref. [77]. [f] Ref. [63]. [g] Ref. [62]. [h] Ref. [78]. [i] Ref. [79]. [j] G^E—55 value calculated at 303.15 k from vapor pressures data taken from Barraza and Edwards [80]. [k] Ref. [81].

Among the reported values in Table 7, the low values of H^E_m and V^E_m for alcohol + 1,8-cineol mixtures stand out when comparing with those corresponding to the mixture of alcohol with di-n-propylether or with n-hexane. Such behavior can be qualitatively interpreted in terms of the type of interactions between the molecules that constitute the mixture, and the molecular shapes. Unlike what happens in the (alcohol + n-hexane) mixtures, where the most important contribution to the excess molar enthalpy is the breaking of hydrogen bonds in alcohol (endothermic contribution), in alcohol mixtures with both ethers we would also consider the breaking interactions of polar type in ether (endothermic contribution) and also the formation of alcohol—ether interactions (exothermic contribution). In the case of mixtures of alcohol with 1,8-cineole, we have to take into account that this molecule has a larger molar volume and a larger dipole moment than the corresponding di-n-propylether. For that reason, one would expect greater endothermic contributions to the excess molar enthalpy. However, experimental excess molar enthalpies are considerably lower in mixtures with 1,8-cineole than the corresponding ones with di-n-propylether. To justify this experimental behavior, we should consider that the alcohol-1,8-cineole (cyclic ether) interaction is stronger than the alcohol-di-n-propylether (linear ether) one. This justification is consistent with both the more negative values of the excess molar entropy and the excess molar volume of alcohol + 1,8-cineole mixtures, as shown in Table 7, as well as from the relative values of λ_{ij} interaction energies between 2-propanol-1,8-cineole and 2-propanol-di-n-propylether obtained from the vapor pressure data using Wilson's model, as indicated above.

The negative value of V_m^E and the low value of H_m^E for alcohol–cineol mixtures probably also has to do with the slightly curled shape of the ether molecule, which can favor the formation of alcohol–ether complexes, thanks to a good spatial coupling of the molecules. A sketch of the possible coupling between molecules of 1,8-cineole and 2-propanol is represented in Figure 8. The figure shows the maps of the electrostatic potential of both geometrically optimized molecules, obtained via the software BIOVIA COSMOtherm 2020; Version 20.0.0 (Revision 5273M). For a mutual orientation as that represented in the figure, both the interaction between the oxygen atom -O- of 1,8-cineole with the highly positive H- atom of the hydroxyl -OH group of alcohol, and the interaction between the highly negative -O- oxygen atom of the hydroxyl –OH group of alcohol with the -H atoms of $–CH_3$ groups of 1,8-cineole, would be favored.

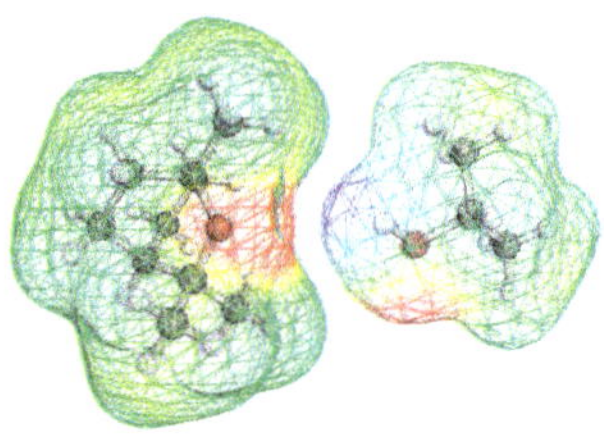

Figure 8. Favorable coupling between 1,8-cineole and 2-propanol molecules for the formation of a complex. The electronic charge density and the electrostatic potential surface were obtained using the software BIOVIA COSMOtherm 2020; Version 20.0.0 (Revision 5273M).

The greater strength of the alcohol–cyclic ether interaction with respect to that existing in the case of the linear ether could be also caused by an increase in electron density of the oxygen atom in the cyclic molecule (ring strain), as pointed out by other authors [82,83]. To confirm the validity of this hypothesis, we calculated the electron density around the oxygen atom in both the molecules, 1,8-cineole and di-n-propylether, obtaining the values of 5.12 and 4.98 electrons, respectively, which it is consistent with previous arguments. The calculations of electron density have been carried out at the B3LYP/6-31** level of theory [84–86] and the analysis of the electron-pairing was conducted using the Electron Localization Function (ELF) methodology as implemented in the Topmod program [87].

In the discussion above, we are assuming that the main contribution to the alcohol–ether interaction is the hydrogen bond formation between the lone pair of the oxygen atom in the ether and the hydrogen atom of the OH group in the alcohol. This has been established by numerous authors, among which we could cite a recent article by Patel et al. on binary systems 1,8-cineol + cresol [88].

Focusing on the comparison between 2-propanol and the other alcohols included in Table 8, we can see that mixtures including 2-propanol show greater positive value of H_m^E and appreciably less negative value of V_m^E than mixtures where 2-propanol is replaced by one of the other alcohols. Only excess molar entropy in mixture 1,8-cineole with 2-propanol are quite similar to that of mixture with 1-propanol. The extra increase in the excess molar functions has been attributed to cyclic multimers formation in the case of branched alcohols in low polarity solvents [89].

2.5. Equations of State (EoS)

Table 8 shows the properties of the pure compounds used in this work in order to describe both the phase equilibrium and the volumetric behavior of 1-propanol (1) + 1,8-cineole (2) mixtures via the PRM, PRSV and SAFT models.

Equations for calculation are described in detail in Appendix A1 (EoS Implemented in PE) of the work of Pfohl, Petkov and Brunner [90].

The cubic PRM and PRSV EoS parameters for 2-propanol were calculated from the correlation of vapor pressure and saturation properties. The SAFT parameters for 2-propanol were taken from the literature [39]. The cubic PRM and PRSV EoS parameters, and SAFT parameters corresponding to 1,8-cineole were calculated in a previous paper [35].

Table 8. Pure component properties and parameters used for the application of the studied equations of state.

	M_w/g·mol^{-1}		T_b/K	T_c/K	P_c/MPa	ω
1,8-cineole [a]	154.25		449.6	661.12	3.019	0.338
2-propanol [b]	60.096		355.4	508.3	4.760	0.665
PRM-VT	P_1		**c/b**		**Range *T*/K**	
1,8-cineole [c]	−0.003518		−0.086491		278–450	
2-propanol [d]	−0.256223		0.039731		278–493	
PRSV-VT	k_{ap}		**c/b**		**Range *T*/K**	
1,8-cineole [c]	0.007355		−0.086427		278–450	
2-propanol [d]	0.166735		0.039468		278–493	
SAFT	m	$v^{\circ\circ}$/L·mol^{-1}	$u^{\circ}\cdot k^{-1}$/K	κ	$\varepsilon\cdot k^{-1}$/K	**Range *T*/K**
1,8-cineole [c]	4.842	0.0178	263.43	-	-	278–450
2-propanol [e]	3.249	0.0120	202.94	0.0210	2670	293–493

[a] Ref. [32]. [b] Ref. [91]. [c] Ref. [35]. [d] This work. [e] Ref. [39].

The van der Waals one-fluid mixing rules [43] were used to determine the $P\rho T$ behavior of the mixtures. Classical quadratic combining rules for the cross-terms [43] were selected in all cases. A quadratic dependence between the interaction parameter, k_{ij}, and the temperature was found in the experimental range considered.

The k_{ij} interaction parameter has been set to our VLE data, showing a quadratic dependence on temperature. The fitted parameters for the equation,

$$k_{ij} = a + b\cdot T/\mathrm{K} + c\cdot T^2/\mathrm{K}^2 \quad (14)$$

appear in Table 9 together with the regression coefficients.

Table 9. Values of the coefficients a, b and c in Equation (14) and regression coefficient, R^2.

Model	a	b	c	R^2
PRM-VT	3.4140	-2.1981×10^{-2}	3.5557×10^{-5}	0.954
PRSV-VT	1.3550	-8.9000×10^{-3}	1.4854×10^{-5}	0.964
SAFT	0.5930	-4.0773×10^{-3}	6.9106×10^{-6}	0.976

Figure 9 shows the experimental VLE at three temperatures together with the obtained results using the selected EoS. It should be noted that bubble curves corresponding to the three models appear well separated for the temperatures of 278.15 and 323.15 K, but not at 298.15 K, where the bubble curves for PRM and PRSV models appear to be almost overlapping. A similar behavior is displayed in the dew curves for these two same models at 298.15 K. The dew curves corresponding to the PRM and SAFT models at 278.15 K and those of the PRM and PRSV ones at 323.15 K are also practically coincident. The best results for the correlations of the experimental data of the mixture under study were achieved with PRSV-VT and PRM-VT. The absolute average percentage deviation values (ADD) for these models were 9.27% and 10.99%, respectively. The ADD obtained for SAFT was 19.30%.

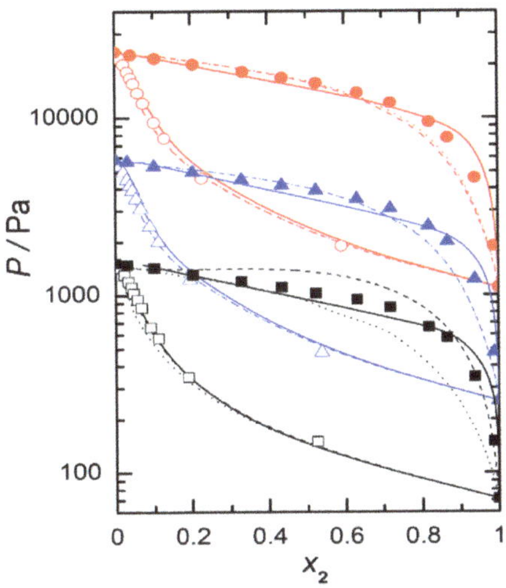

Figure 9. Isothermal vapor liquid equilibrium of the 2-propanol (1) + 1,8-cineole (2) system. Full symbols experimental data: (■), T = 278.15 K; (▲), T = 298.15 K; (•), and T = 323.15 K. Open symbols were obtained from the Wilson equation. Lines, EOS correlations: – – –, PRM-VT; ……, PRSV-VT; —, SAFT.

The major or minor capacity of the three EoS to reproduce the volumetric behavior of the system was also tested at 298.15 K. Figure 10 shows our experimental data for the excess molar volume at that temperature, together with the predictions of the three EoS tested.

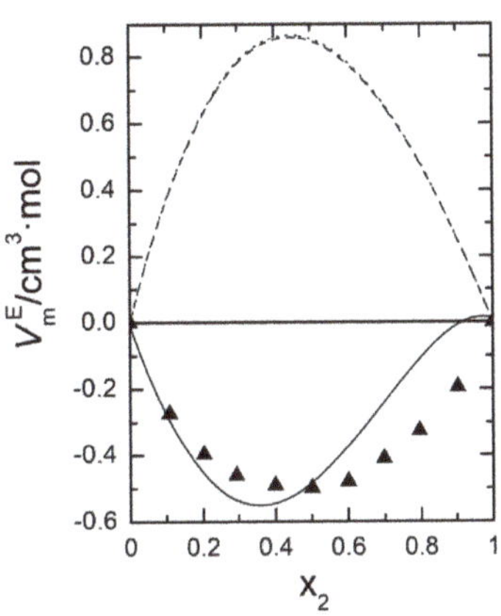

Figure 10. Volumetric behavior of the 2-propanol (1) + 1,8-cineole (2) system at 298.15 K. (▲), This work; – – –, PRM-VT; ……., PRSV-VT; —, SAFT.

As it can be observed, only the SAFT model correctly reproduces the sign of V_m^E, and even approaches the values of the volumetric behavior of the system quite well, something that the other two EoS cannot satisfy even when the refinement of the volume translation is used in them.

3. Materials and Methods

3.1. Chemicals

2-Propanol and 1,8-cineole were supplied by Aldrich, Seelze, Germany (mass fraction purity > 0.999 and >0.990, respectively). All the chemicals were low water content, stored over molecular sieve (4 Å), and used without further purification. The mass fraction purity was checked via gas chromatography and found to be 0.9999 for 2-propanol, and 0.9970 for 1,8-cineole.

3.2. Apparatus and Procedures

Vapor pressure measurements were carried out according a static method, using an apparatus similar to that of Marsh [92], but with the incorporation of some different details. The device and operating method have been thoroughly described previously [93]. Several important points should be noted here. The undesirable effects of condensation on the mercury meniscus were avoided by circulating the thermostat water at ±0.1 K to maintain the manometer temperature at 325 K. In the same way, most vapor phase was maintained at that temperature, T = 325 K. Liquid sample temperature was measured using Beckmann thermometers, calibrated against vapor-liquid equilibria (VLE) of pure benzene (Merck mole fraction > 0.999 and distilled twice), along with Ambrose's equation [94] relating temperature with pressure by means of a sum of Chebyshev polynomials up to the sixth degree. Thus, the accuracy in the temperature measurements was estimated to be ±0.01 K. The volume of the cell containing the liquid mixture was about 12 cm^3, and (8 to 10) cm^3 of liquid were used in each experiment. Previously, to be successively added by gravity into the cell immersed in liquid nitrogen, the liquids were degassed via magnetic stirring, with the (air plus vapor) phase being pumped away periodically. Masses of both components were determined by weighing with a precision electronic balance (0.0001 g). Caution was taken to prevent evaporation. Conversion to molar quantities was based on the relative atomic mass table issued by IUPAC, leading to an uncertainty in the mole fraction estimated to be less than ±0.0003. The vapor pressures, P, obtained from the difference in heights of the mercury meniscus in the two columns of the manometer, were calculated using the specific weight of that element for the value of gravitational acceleration at the laboratory where the measurements were performed. Manometer readings were performed using a Wild KM-305 cathetometer within ±0.01 mm, and pressure reproducibility was estimated to be better than 13 Pa. The uncertainty in the vapor pressure is estimated to be less than 0.1%.

Advantages and limitations of the static (isothermal) measurements method used in this work have been previously detailed by Smith and Menzies [95], together with the more general sources of error involved in steam pressure measurement, and particular reference to the individual sources of error in both the static and dynamic methods. Despite their greater laboriousness, compared to dynamic measurements, the main advantages of the static ones are not having to resort to physical or chemical analysis of the phases, directly providing the value of the Gibbs energy at each of the measured temperatures and avoiding inaccuracies derived from the accumulation of impurities throughout the measurement process. Regarding the more general sources of error involved in any of the measurement methods, both static and dynamic, those authors have highlighted deficiencies in (i) the stability, distribution and precise determination of temperature, (ii) the measurement and corrections for pressure measurements, and (iii) the presence of impurities in the chemical species used as components. In our work, we tried to minimize these deficiencies through methodological details such as those indicated above. Finally, and from the point of view of thermodynamic consistence compliance based on the Gibbs–Duhem equation, the static (isothermal) measurements are also more adequate than the dynamic (isobaric) ones.

A vibrating tube densimeter DMA 5000, Anton Paar GmbH, Graz, Austria was used for density measurements on the pure liquids and mixtures. The sample density is calculated from the vibration period with an uncertainty of ±0.04 kg·m^{-3}. The composition of the binary mixtures was determined by weighing the vapor pressure mixtures preparation in a similar way. From the experimental densities, the excess molar volumes were calculated. The uncertainty in excess molar volumes is estimated to be $\pm 2 \times 10^{-9}$ $m^3 \cdot mol^{-1}$.

A Thermometric 2277 Thermal Activity Monitor, American Laboratory Trading, San Diego, (CA, USA) together with two Shimadzu (model LC-10ADVP HPLC), Shimadzu Europe GmbH, Duisburg, Germany variable speed piston pumps, was used to determine excess enthalpies at the different temperatures. The pumps were programmed in order to be able to measure the excess molar enthalpies at the selected molar fractions of mixture and they were previously calibrated. The uncertainty in mole fraction is estimated to be ±0.001, and the uncertainty in H_m^E measurements is better than 2%, as verified by comparing the results for a standard system with those of the reference system [96]. Some additional details can be found in a previous paper [97].

In Table 10, experimental and literature densities, ρ, of pure liquids at the different temperatures are showed. The agreement between both sets of values could be considered as satisfactory.

Table 10. Experimental and literature densities of pure liquids [a].

	ρ/kg·m^{-3}			
	2-Propanol		**1,8-Cineole**	
T/K	**Experimental**	**Literature**	**Experimental**	**Literature**
288.15	789.34	-	928.96	928.78 [g]
298.15	780.99	780.98 [b] 780.82 [c] 781.26 [d]	920.34	920.29 [g]
308.15	772.37	772.51 [e] 772.6 [f]	911.73	-
318.15	763.40	-	903.13	-

[a] Standard uncertainty, u, is u(T) = 0.01 K and the combined expanded uncertainty, U_c, is $U_c(\rho)$ = 0.04 kg·m^{-3} with a 0.95 level of confidence (k = 2). [b] Ref. [98]. [c] Ref. [99]. [d] Ref. [64]. [e] Ref. [100]. [f] Ref. [101]. [g] Ref. [102].

4. Conclusions

New isothermal vapor–liquid equilibrium data at three temperatures, from 278.15 K to 323.15 K, and over the entire composition range for the binary mixture 2-propanol + 1,8-cineole are presented. Vapor pressures have been measured via a static method. From these data, activity coefficients and excess Gibbs energies have been calculated using Barker's method and the Wilson equation. The consistency of the obtained vapor pressure data is substantiated by the close values of excess molar enthalpies calculated from the vapor pressures by applying the Gibbs–Helmholtz equation to those experimental values obtained in the present work. Although the system is far from an ideal behavior, showing a large positive deviation from Raoult's law, azeotrope do not appear within the considered temperature interval.

On the other hand, excess molar volumes for this system at temperatures between 288.15 and 318.15 K have been calculated from the densities that were experimentally obtained.

A brief comparison is presented between the thermodynamic behavior of short-chain alcohol + 1,8-cineol binary mixtures and those where the cyclic ether has been replaced by a linear ether, di-n-propylether or inert solvent, n hexane. This comparison allowed us to attribute a higher OH–O interaction energy in the alcohol + cyclic ether mixture than that corresponding to the mixture with linear ether. This effect is probably due to an increase in the electronic density of the oxygen atom in the cyclic ether (which is associated with the ring strain) together with a steric effect derived from the possibility of a closer coupling between the alcohol molecule and the cyclic ether molecule.

Both experimental vapor-liquid equilibrium (VLE) and volumetric data were correlated using three different thermodynamic models, namely PRSV-VT, PRM-VT and SAFT. The best results for the correlations of the experimental VLE data are achieved when the first two models are used. This is probably due to the regression flexibility conferred in the cases of the modified PR models by considering the binary interaction parameter k_{ij} as an adjustable coefficient. On the contrary, only the SAFT model approaches the volumetric behavior of the real system better than the other EoS, even using the volume translation correction in them. The detail of the localized interactions between the areas of higher electron charge densities that SAFT considers could be responsible for that improvement. The approximations reported, as provided by the two considered versions of the Robinson–Stokes equation for VLE, and the one provided by SAFT for excess enthalpy, are within a very reasonable range.

Because of the special characteristics of the chemicals involved in this study, the results obtained in this work can be a good database in the development of advanced theoretical models such as molecular dynamics and direct simulation Monte Carlo. From an application point of view, these results, together with others of a more industrial type, can contribute to the development of biorefineries. More concretely, the values of parameters, as well as the reported experimental data, can be potentially suitable for simulation and for designing separation processes to obtain the interesting chemical 1,8-cineole.

Author Contributions: Conceptualization: P.P., J.S.U. and A.M.M.; Methodology: P.P., J.S.U. and A.M.M.; Investigation: B.G.; Data curation: P.P.; Resources and software: P.P., J.S.U., A.M.M. and S.M.; Formal analysis: P.P., S.M. and A.M.M.; Project administration and funding acquisition: A.M.M. and J.S.U.; Writing—original draft preparation: P.P. and B.G.; Writing—review and editing: P.P., J.S.U. and A.M.M.; Supervision: P.P., J.S.U. and A.M.M.; Visualization: S.M., P.P. and J.S.U. All authors have read and agreed to the published version of the manuscript.

Funding: The authors thank the financial support provided by the Departamento de Ciencia, Tecnología y Universidad del Gobierno de Aragón-Fondo Social Europeo (Group E39_23R).

Institutional Review Board Statement: Not applicable.

Informed Consent Statement: Not applicable.

Data Availability Statement: All data are contained within the article.

Acknowledgments: The authors thank Victor Polo (Departamento de Química Física, Facultad de Ciencias, Universidad de Zaragoza) for conducting the DFT calculations.

Conflicts of Interest: The authors declare no conflict of interest.

References

1. Cai, Z.M.; Peng, J.Q.; Chen, Y.; Tao, L.; Zhang, Y.Y.; Fu, L.Y.; Long, Q.D.; Shen, X.C. 1,8-Cineole: A review of source, biological activities, and application. *J. Asian Nat. Prod. Res.* **2020**, *23*, 938–995. [CrossRef] [PubMed]
2. De Vincenzi, M.; Mancini, E.; Dessi, M.R. Monographs on botanical flavouring substances used in foods. Part V. *Fitoterapia* **1996**, *67*, 241–251.
3. Scientific Committee on Food. *Opinion of the Scientific Committee on Food on Eucalyptol*; European Commission Health and Consumer Protection Directorate-General, Ed.; Directorate C—Scientific Opinions C2—Management of Scientific Committees II; Scientific Co-Operation and Networks: Bruxelles/Brussells, Belgium, 2002; p. 2.
4. Bhowal, M.; Gopal, M. Eucalyptol: Safety and Pharmacological Profile. *RGUHS J. Pharm. Sci.* **2015**, *5*, 125–131. [CrossRef]
5. Majed, M.M.; Shadi, F.G.; Karem, H.A.; Al-Azzam, S.I.; Wasfi, M.O. Antimicrobial Activity of Common Mouthwash Solutions on Multidrug-Resistance Bacterial Biofilms. *J. Clin. Med. Res.* **2013**, *5*, 389–394.
6. Worth, H.; Schacher, C.; Dethlefsen, U. Concomitant therapy with Cineole (*Eucalyptole*) reduces exacerbations in COPD: A placebo-controlled double-blind trial. *Respir. Res.* **2009**, *10*, 69. [CrossRef]
7. Juergens, U.R. Anti-inflammatory properties of the monoterpene 1.8-cineole: Current evidence for co-medication in inflammatory airway diseases. *Drug Res.* **2014**, *64*, 638–646. [CrossRef] [PubMed]
8. Van Vuuren, S.F.; Viljoen, A.M. Antimicrobial activity of limonene enantiomers and 1,8-cineole alone and in combination. *Flavour Fragr. J.* **2007**, *22*, 540–544. [CrossRef]
9. Merghni, A.; Noumi, E.; Hadded, O.; Dridi, N.; Panwar, H.; Ceylan, O.; Mastouri, M.; Snoussi, M. Assessment of the antibiofilm and antiquorum sensing activities of Eucalyptus globulus essential oil and its main component 1,8-cineole against methicillin-resistant Staphylococcus aureus strains. *Microb. Pathog.* **2018**, *118*, 74–80. [CrossRef]
10. Morcia, C.; Malnati, M.; Terzi, V. In vitro antifungal activity of terpinen-4-ol, eugenol, carvone, 1,8-cineole (eucalyptol) and thymol against mycotoxigenic plant pathogens. *Food Addit. Contam. Part A-Chem.* **2012**, *29*, 415–422. [CrossRef]

11. Kin, K.Y.; Seo, H.J.; Min, S.S.; Park, M.; Seol, G.H. The effect of 1,8-cineole inhalation on preoperative anxiety: A randomized clinical trial. Evid. Based Complement. *Altern. Med.* **2014**, *24*, 820126. [CrossRef]
12. Heard, C.M.; Kung, D.; Thomas, C.P. Skin penetration enhancement of mefenamic acid by ethanol and 1,8-cineole can be explained by the 'pull' effect. *Int. J. Pharm.* **2006**, *321*, 167–170. [CrossRef] [PubMed]
13. Dougnon, G.; Ito, M. Inhalation Administration of the Bicyclic Ethers 1,8- and 1,4-cineole Prevent Anxiety and Depressive-Like Behaviours in Mice. *Molecules* **2020**, *25*, 1884. [CrossRef] [PubMed]
14. Abdalla, A.N.; Shaheen, U.; Abdallah, Q.M.A.; Flamini, G.; Bkhaitan, M.M.; Abdelhady, M.I.S.; Ascrizzi, R.; Bader, A. Proapoptotic Activity of Achillea membranacea Essential Oil and Its Major Constituent 1,8-Cineole against A2780 Ovarian Cancer Cells. *Molecules* **2020**, *25*, 1582. [CrossRef] [PubMed]
15. Rodenak-Kladniew, B.; Castro, M.A.; Crespo, R.; Galle, M.; de Bravo, M.G. Anti-cancer mechanisms of linalool and 1,8-cineole in non-small cell lung cancer A549 cells. *Heliyon* **2020**, *6*, e05639. [CrossRef]
16. Paul, K.; Ganguly, U.; Chakrabarti, S.; Bhattacharjee, P. 1,8-Cineole-Rich Extract of Small Cardamom Seeds More Effective in Preventing Alzheimer's Disease than 1,8-Cineole Alone? *Neuromol. Med.* **2020**, *22*, 150–158. [CrossRef]
17. An, F.; Bai, Y.; Xuan, X.; Bian, M.; Zhang, G.; Wei, C. 8-Cineole Ameliorates Advanced Glycation End Products-Induced Alzheimer's Disease-like Pathology In Vitro and In Vivo. *Molecules* **2022**, *27*, 3913. [CrossRef]
18. Sharma, A.D.; Kaur, I. Targeting UDP-Glycosyltransferase, Glucosamine-6-Phosphate Synthase and Chitin Synthase by Using Bioactive 1,8 Cineole for "Aspergillosis" Fungal Disease Mutilating COVID-19 Patients: Insights from Molecular Docking, Pharmacokinetics and In-vitro Studies. *Chem. Afr.* **2022**, *5*, 149–160. [CrossRef]
19. Hamiche, S.; Bouzidi, N.; Daghbouche, Y.; Badis, A.; Garrigues, S.; de la Guardia, M.; El Hattab, M. Eucalyptol-based green extraction of brown alga Zonaria tournefortii. *Sustain. Chem. Pharm.* **2018**, *10*, 97–102. [CrossRef]
20. Campos, J.F.; Scherrmann, M.C.; Berteina-Raboin, S. Eucalyptol: A new solvent for the synthesis of heterocycles containing oxygen, sulfur and nitrogen. *Green Chem.* **2019**, *6*, 1531–1539. [CrossRef]
21. Eisenbrand, G.; Cohen, S.M.; Fukushima, S.; Gooderham, N.J.; Guengerich, F.P.; Hecht, S.S.; Taylor, S.V. FEMA GRAS assessment of natural flavor complexes: Eucalyptus oil and other cyclic ether-containing flavoring ingredients. *Food Chem. Toxicol.* **2021**, *155*, 112357. [CrossRef]
22. Sánchez-Vicente, Y.; Cabañas, A.; Renuncio, J.A.R.; Pando, C. Supercritical CO_2 as a green solvent for eucalyptus and citrus essential oils processing: Role of thermal effects upon mixing. *RSC Adv.* **2013**, *3*, 6065. [CrossRef]
23. Prat, D.; Hayler, J.; Wells, A. A survey of solvent selection guides. *Green Chem.* **2014**, *16*, 4546–4551. [CrossRef]
24. Samudrala, S.; Bhattacharya, S. Toward the Sustainable Synthesis of Propanols from Renewable Glycerol over MoO_3-Al_2O_3 Supported Palladium Catalysts. *Catalysts* **2018**, *8*, 385. [CrossRef]
25. Wang, K.; Strobel, G.; Yan, D.H. The Production of 1,8-Cineole, a Potential Biofuel, from an Endophytic Strain of *Annulohypoxylon* sp. FPYF3050 When Grown on Agricultural Residues. *J. Sustain. Bioenergy Syst.* **2017**, *7*, 65–84. [CrossRef]
26. Tasoren, E.; Aydogan, H.; Gokmen, M.S. Research of effect on gasoline-2-propanol blends on exhaust emission of gasoline engine with direct injection using Taguchi approach. *Eur. Mech. Sci.* **2021**, *5*, 177–182. [CrossRef]
27. Weberdemenezes, E.; Dasilva, R.; Cataluna, R.; Ortega, R. Effect of ethers and ether/ethanol additives on the physicochemical properties of diesel fuel and on engine tests. *Fuel* **2006**, *85*, 815–822. [CrossRef]
28. Raveendran, P.; Zimmermann, D.; Häber, T.; Suhm, M.A. Exploring a hydrogen-bond terminus: Spectroscopy of eucalyptol–alcohol clusters. *Phys. Chem. Chem. Phys.* **2000**, *2*, 3555–3563. [CrossRef]
29. Du, L.; Tang, S.; Hansen, A.S.; Frandsen, B.N.; Maroun, Z.; Kjaergaard, H.G. Subtle differences in the hydrogen bonding of alcohol to divalent oxygen and sulfur. *Chem. Phys. Lett.* **2017**, *667*, 146–153. [CrossRef]
30. Aparicio, S.; Alcalde, R.; Dávila, M.J.; García, B.; Leal, J.M. Properties of 1,8-Cineole: A Thermophysical and Theoretical Study. *J. Phys. Chem. B* **2007**, *11*, 3167–3177. [CrossRef]
31. Štejfa, V.; Fulem, M.; Růžička, K.; Červinka, C. Thermodynamic study of selected monoterpenes III. *J. Chem. Thermodyn.* **2014**, *79*, 280–289. [CrossRef]
32. Lasarte, J.M.; Martín, L.; Langa, E.; Urieta, J.S.; Mainar, A.M. Setup and Validation of a P ρ T Measuring Device. Volumetric Behavior of the Mixture 1,8-Cineole + Ethanol. *J. Chem. Eng. Data* **2008**, *53*, 1393–1400. [CrossRef]
33. Alfaro, P.; Langa, E.; Martínez-López, J.F.; Urieta, J.S.; Mainar, A.M. Thermophysical properties of the binary mixtures (1,8-cineole + 1-alkanol) at T = (298.15 and 313.15) K and at atmospheric pressure. *J. Chem. Thermodyn.* **2010**, *42*, 291–303. [CrossRef]
34. Torcal, M.; García-Abarrio, S.; Pardo, J.I.; Mainar, A.M.; Urieta, J.S. P, ρ, T Measurements and Isobaric Vapor-Liquid-Equilibria of the 1,3,3-Trimethyl-2-oxabicycle[2,2,2]octane + Propan-1-ol Mixture: Cubic and Statistical Associating Fluid Theory-Based Equation of State Analysis. *J. Chem. Eng. Data* **2010**, *55*, 5932–5940. [CrossRef]
35. Gimeno, B.; Torcal, M.; Mainar, A.M.; Pérez, P. Isothermal Vapor-Liquid Equilibrium of (1-butanol + 1,8-cineole) at Ten Temperatures between 278.15 K and 323.15 K. *J. Chem. Eng. Data* **2011**, *56*, 2443–2448. [CrossRef]
36. Gimeno, B.; Torcal, M.; Mainar, A.M.; Urieta, J.S.; Pérez, P. Total vapour pressure and excess Gibbs energy of ethanol with 1,8-cineole at temperatures between 278.15 K and 323.15 K. *Fluid Phase Equilib.* **2011**, *309*, 83–88. [CrossRef]
37. Gimeno, B.; Martínez, S.; Urieta, J.S.; Pérez, P. Vapor Pressures and Activity Coefficients of (1-propanol + 1,8-cineole) at 10 Temperatures between 278.15 K and 323.15 K. *J. Chem. Eng. Data* **2012**, *57*, 3026–3031. [CrossRef]
38. Peng, D.; Robinson, D.B. A New Two-Constant Equation of State. *Ind. Eng. Chem. Fundam.* **1976**, *15*, 59–64. [CrossRef]

39. Huang, S.H.; Radosz, M. Equation of State for Small, Large, Polydisperse and Associating Molecules. *Ind. Eng. Chem. Res.* **1990**, *29*, 2284–2294. [CrossRef]
40. van der Waals, J.D. Over de Continuiteit van den Gas en Vloeistoftoestand. Ph.D. Thesis, Leiden University, Leiden, The Netherlands, 1873.
41. Papadopoulos, A.I.; Tsivintzelis, I.; Seferlis, P.; Linke, P. Computer-aided molecular design: Fundamentals, methods, and applications. In *Elsevier Reference Module in Chemistry, Molecular Sciences and Chemical Engineering*; Reedijk, J., Ed.; Elsevier: Waltham, MA, USA, 2018; pp. 1–76. [CrossRef]
42. Lopez-Echeverry, J.S.; Reif-Acherman, S.; Araujo-Lopez, E. Peng-Robinson equation of state: 40 years through cubics. *Fluid Phase Equilib.* **2017**, *447*, 39–71. [CrossRef]
43. Valderrama, J.O. The state of the cubic equations of state. *Ind. Eng. Chem. Res.* **2003**, *42*, 1603–1618. [CrossRef]
44. Mathias, P.M. A Versatile Phase Equilibrium Equation of State. *Ind. Eng. Chem. Process Des. Dev.* **1983**, *22*, 385–391. [CrossRef]
45. Stryjek, R.; Vera, J.H. PRSV: An Improved Peng-Robinson Equation of State for Pure Compounds and Mixtures. *Can. J. Chem. Eng.* **1986**, *64*, 323–333. [CrossRef]
46. Peneloux, A.; Rauzy, E.; Freze, R. A Consistent Correction for Redlich-Kwong-Soave Volumes. *Fluid Phase Equilib.* **1982**, *8*, 7–23. [CrossRef]
47. Wertheim, M.S. Fluids with highly directional attractive forces: I. Statistical thermodynamics. *J. Stat. Phys.* **1984**, *35*, 19–34. [CrossRef]
48. Wertheim, M.S. Fluids with highly directional attractive forces: IV. Equilibrium polymerization. *J. Stat. Phys.* **1986**, *42*, 477–942. [CrossRef]
49. Müller, E.A.; Gubbins, K.E. Molecular-Based Equations of State for Associating Fluids: A Review of SAFT and Related Approaches. *Ind. Eng. Chem. Res.* **2001**, *40*, 2193–2211. [CrossRef]
50. Chapman, W.G.; Gubbins, K.E.; Jackson, G.; Radosz, M. SAFT Equationof- State Solution Model for Associating Fluids. *Fluid Phase Equilib.* **1989**, *52*, 31–38. [CrossRef]
51. Chapman, W.G.; Gubbins, K.E.; Jackson, G.; Radosz, M. New Reference Equation of State for Associating Liquids. *Ind. Eng. Chem. Res.* **1990**, *29*, 1709–1721. [CrossRef]
52. Gross, J.; Sadowski, G. Application of the Perturbed-Chain SAFT Equation of State to Associating Systems. *Ind. Eng. Chem. Res.* **2002**, *41*, 5510–5515. [CrossRef]
53. Diamantonis, N.I.; Boulougouris, G.C.; Mansoor, E.; Tsangaris, D.M.; Economou, I.G. Evaluation of Cubic, SAFT, and PC-SAFT Equations of State for the Vapor–Liquid Equilibrium Modeling of CO_2 Mixtures with Other Gases. *Ind. Eng. Chem. Res.* **2013**, *52*, 3933–3942. [CrossRef]
54. Velasquez, J.A.; Hernandez, J.P.; Forero, L.A.; Cardona, L.F. Prediction of phase equilibria, density, speed of sound and viscosity of 2-alkoxyethanols mixtures: A comparison study between SAFT type EoSs and a modified PR EoS. *Fluid Phase Equilib.* **2022**, *563*, 113570. [CrossRef]
55. Pliego, J.R. Building the liquid-vapour equilibrium curve through a cubic equation of state: Use of excel for teaching physical-chemistry. *Quim. Nova* **2016**, *39*, 641–645. [CrossRef]
56. Kontogeorgis, G.M.; Privat, R.; Jaubert, J.-N. Taking Another Look at the van der Waals Equation of State—Almost 150 Years Later. *J. Chem. Eng. Data* **2019**, *64*, 4619–4637. [CrossRef]
57. Kamesh, R.; Kumari, A.; Rani, K.Y. Measurements, Correlations, and Modified UNIFAC Predictions of Isobaric Vapor-Liquid Equilibrium Data for the Binary System of Dimethyl Carbonate plus Anisole at Different Pressures. *J. Chem. Eng. Data* **2021**, *66*, 3788–3801. [CrossRef]
58. Sarkoohaki, B.; Almasi, M.; Karimkhani, M. Theoretical and experimental study of physicochemical behavior of binary mixtures: SAFT and PC-SAFT models. *J. Chem. Sci.* **2019**, *13*, 53. [CrossRef]
59. Fattahi, M.; Iloukhani, H. Excess molar volume, viscosity, and refractive index study for the ternary mixture {2-methyl-2-butanol + tetrahydrofuran + propylamine} at different temperatures. Application of the ERAS-model and Peng–Robinson–Stryjek–Vera equation of state. *J. Chem. Thermodyn.* **2010**, *42*, 1335–1345. [CrossRef]
60. Martínez-López, J.F.; Schneider, S.; Salavera, D.; Mainar, A.M.; Urieta, J.S.; Pardo, J.I. Molar heat capacities of the mixture {1,8-cineole + ethanol} at several temperatures and atmospheric pressure. *J. Chem. Thermodyn.* **2016**, *92*, 146–151. [CrossRef]
61. Torcal, M.; Langa, E.; Pardo, J.I.; Mainar, A.M.; Urieta, J.S. Isobaric VLE of the mixture {1,8-cineole + ethanol}. EOS analysis and COSMO-RS modeling. *J. Chem. Thermodyn.* **2016**, *97*, 88–92. [CrossRef]
62. Garriga, R.; Sánchez, F.; Pérez, P.; Gracia, M. Isothermal vapour-liquid equilibrium at eight temperatures and excess functions at 298.15 K of di-n-propylether with 1-propanol or 2-propanol. *Fluid Phase Equilib.* **1997**, *138*, 131–144. [CrossRef]
63. Garriga, R.; Pérez, P.; Gracia, M. Vapour Pressures at Eight Temperatures of Mixtures of Di-n-Propylether + Ethanol or 1-ButanolThermodynamic Description of Mixtures of Di-n-Propylether + Alkanol According to the ERAS Model. *Ber. Bunsen-Ges. Phys. Chem.* **1997**, *101*, 1466–1473. [CrossRef]
64. Thermodynamics Research Center. *TRC-Thermodynamic Tables-Non-Hydrocarbons*; The Texas A&M University System: College Station, TX, USA, 1966; p. d-5030.
65. Ambrose, D.; Ghiassee, N.B. Vapour pressures and critical temperatures of some alkanoic acids: C1 to C10. *J. Chem. Thermodynamic.* **1987**, *19*, 505–519. [CrossRef]
66. Stull, D.R. Vapour Pressure or Pure Substances. Organic Compound. *Ind. Eng. Chem.* **1947**, *39*, 517–540. [CrossRef]

67. Guetachew, T.; Mokbel, I.; Batiu, I.; Cisse, Z.; Jose, J. Vapor pressures and sublimation pressures of eight constituets of essential oils at pressures in the range from 0.3 to 83,000 Pa. *ELDATA Int. Electron. J. Phys. Chem. Dat.* **1999**, *5*, 43–53.
68. Tsonopoulos, C. Empirical Correlation of Second Virial Coeficients. *AIChE J.* **1974**, *20*, 263–273. [CrossRef]
69. Erickson, W.D.; Roach, J.F. *Composition and Thermodynamic Properties of Reacting Gas Mixtures under High Pressures Using the Lewis and Randall Rule*; Nasa Technical Note TN D-2591; National Aeronautics and Space Administration: Washington, DC, USA, 1965; p. 18.
70. Barker, J.A. Determination of Activity Coefficients from Total Pressure Measurements. *Aust. J. Chem.* **1953**, *6*, 207–210. [CrossRef]
71. Wilson, G.M. Vapor-Liquid Equilibrium. 11. A New Expression for the Excess Free Energy of Mixing. *J. Am. Chem. Soc.* **1964**, *86*, 127–130. [CrossRef]
72. Rowlinson, J.S. Excess thermodynamic functions. In *Liquids and Liquid Mixtures*, 3rd ed.; Swinton, F.L., Ed.; Butterworth: London, UK, 1982; Chapter 5; p. 138.
73. Le Fevre, R.J.W.; Sundaram, A.; Pierens, R.K. Molecular Polarisability: The Anisotropy of the Carbon-Oxygen Link. *J. Chem. Soc.* **1963**, *2*, 479–488. [CrossRef]
74. Mc Clellan, A.L. (Ed.) *Tables of Experimental Dipole Moments*; Rahara Enterprises: El Cerrito, CA, USA, 1974; Volume 2, p. 433.
75. Meyer, L.; Büchner, A. The dipolar moment of the n-propyl ether. *Physik. Z.* **1932**, *33*, 390–391.
76. Mc Clellan, A.L. (Ed.) *Tables of Experimental Dipole Moments*; Freeman: San Francisco, CA, USA, 1963; p. 220.
77. Iñarrea, J.; Valero, J.; Pérez, P.; Gracia, M.; Gutiérrez Losa, C. H_m^E and V_m^E of some (butanone or dipropylether + an alkanol) mixtures. *J. Chem. Thermodyn.* **1988**, *20*, 93–199. [CrossRef]
78. Brown, I.; Fock, W.; Smith, F. The thermodynamics properties of solutions of normal and branched alcohols in benzene and n-hexane. *J. Chem. Thermodyn.* **1969**, *1*, 273–291. [CrossRef]
79. Christensen, C.; Gmehling, J.; Rasmussen, P.; Weidlich, U. Part 1: Heats of Mixing Data Collection. In *DECHEMA Chemistry Data Series*; DECHEMA: Frankfurt, Germany, 1984; Volume III, p. 594.
80. Barraza, R.; Edwards, J. Thermodynamics of the isopropanol/n-hexane and isopropanol/n-heptane systems. Part III. Gas/liquid equilibrium. *Monatsh. Chem.* **1981**, *112*, 925–933. [CrossRef]
81. Kashyap, P.; Rani, M.; Gahlyan, S.; Tiwari, D.P.; Maken, S. Volumetric, acoustic and optical properties of binary mixtures of 2-propanol with n-alkanes (C 6–C 10) from 293.15 K to 303.15 K. *J. Mol. Liq.* **2018**, *268*, 303–314. [CrossRef]
82. Letcher, T.M.; Govender, U.P. CH_4O and C_4H_8O. Primary linear-alkyl monoalcohols + Cyclic monoethers: Heat of Mixing and Solution. *J. Chem. Eng. Data* **1995**, *40*, 1097–1100. [CrossRef]
83. González, J.A.; de la Fuente, I.G.; Cobos, J.C. Disquac analysis of binary-liquid organic mixtures containing cyclic or linear alkanols and cycloalkanes or n-alkanes. *Thermochim. Acta* **1996**, *278*, 57–69. [CrossRef]
84. Lee, C.; Yang, W.; Parr, R.G. Development of the Colle-Salvetti correlation-energy formula into a functional of the electron density. *Phys. Rev. B* **1988**, *37*, 785–788. [CrossRef]
85. Becke, A.D. A new mixing of Hartree–Fock and local density-functional theories. *J. Chem. Phys.* **1993**, *98*, 1372–1377. [CrossRef]
86. Becke, A.D. Density functional thermochemistry. III. The role of exact exchange. *J. Chem. Phys.* **1993**, *98*, 5648–5652. [CrossRef]
87. Noury, S.; Krokidis, X.; Fuster, F.; Silvi, B. Computational tools for the electron localization function topological analysis. *Comput. Chem.* **1999**, *23*, 597–604. [CrossRef]
88. Patel, P.; Bhalodia, J.; Sharma, S.S.; Chandra, P. Refractive index, speed of sound, FT-IR and computational study of intermolecular interaction between binary mixtures of 1,8-cineole with o-, m- and p-cresol at 303.15, 308.15 and 313.15 K. *J. Mol. Liq.* **2016**, *222*, 1192–1211. [CrossRef]
89. Jimenez, J.; Valero, J.; Gracia, M.; Gutiérrez Losa, C. Hm^E of (an n-alkane+ a butanol isomer). *J. Chem. Thermodyn.* **1988**, *20*, 931–936. [CrossRef]
90. Pfohl, O.; Petkov, S.; Brunner, G. *PE 2000. A Poweful Tool to Correlate Phase Equilibria*; Herbert Utz: Munich, Germany, 2000; pp. 103–105, 115.
91. Poling, B.E.; Prausnitz, J.M.; O'Connell, J.P. *The Properties of Gases and Liquids*, 5th ed.; Mc Graw-Hill: New York, NY, USA, 2007.
92. Marsh, K.N. Thermodynamics of Octamethylcyclotetrasiloxane Mixtures. *Trans. Faraday Soc.* **1968**, *64*, 883–893. [CrossRef]
93. Garriga, R.; Martínez, S.; Pérez, P.; Gracia, M. Isothermal (vapour + liquid) equilibrium at several temperatures of (1-chlorobutane + 1-butanol, or 2-methyl-2-propanol). *J. Chem. Thermodyn.* **2001**, *33*, 523–534. [CrossRef]
94. Ambrose, D. Recommended reference materialsfor realization of physicochemical properties. *IUPAC Pure Appl. Chem.* **1977**, *49*, 1437–1464.
95. Smith, A.; Menzies, A.W.C. Studiesin Vapor Pressure: III. A Static Method for Determining the Vapor Pressures of Solids and Liquids. *J. Am. Chem. Soc.* **1910**, *32*, 1412–1434. [CrossRef]
96. Head, A.J.; Sabbah, R. Enthalpy. In *Recommended Reference Materials for the Realization of Physicochemical Properties*; Marsh, K.N., Ed.; Blackwell Scientific Publications: Oxford, UK, 1987; Chapter 9; pp. 282–283.
97. Lafuente, C.; Artigas, H.; Lopez, M.C.; Royo, F.M.; Urieta, J.S. Excess Molar Enthalpies for Isomeric Chlorobutanes with Isomeric Butanols. *Phys. Chem. Liq.* **2001**, *39*, 665–673. [CrossRef]
98. Zafarani-Moattar, M.T.; Shekaaria, H. Density and speed of sound of lithium bromide with organic solvents: Measurement and correlation. *J. Chem. Thermodyn.* **2007**, *39*, 1649–1660. [CrossRef]
99. Haase, R.; Tillmann, W. Properties of the liquid-system water plus 2-propanol. *Z. Phys. Chem.* **1994**, *187*, 223–231. [CrossRef]

100. Dubey, G.P.; Sharma, M. Studies of mixing properties of binary systems of 2-propanol with hexadecane and squalane at T = (298.15, 303.15, and 308.15). *J. Chem. Thermodyn.* **2009**, *41*, 115–122. [CrossRef]
101. Aminabhavi, T.M.; Aralaguppi, M.I.; Harogoppad, S.B.; Balundgi, R.H. Densities, viscosities, refractive indices, and speeds of sound for methyl acetoacetate + aliphatic alcohols (C1–C8). *J. Chem. Eng. Data* **1993**, *38*, 31–39. [CrossRef]
102. Barata, P.A.; Serrano, M.L. Densities and Viscosities of Thymol + 1,8-Cineole. *J. Chem. Eng. Data* **1994**, *39*, 298–301. [CrossRef]

International Journal of
Molecular Sciences

MDPI

Article

Direct Measurement of the Four-Phase Equilibrium Coexistence Vapor–Aqueous Solution–Ice–Gas Hydrate in Water–Carbon Dioxide System

Anton Semenov [1,*], Rais Mendgaziev [1], Andrey Stoporev [1,2,*], Vladimir Istomin [1,3], Timur Tulegenov [1], Murtazali Yarakhmedov [1], Andrei Novikov [1] and Vladimir Vinokurov [1]

1 Department of Physical and Colloid Chemistry, Gubkin University, 65, Leninsky Prospekt, Building 1, 119991 Moscow, Russia; meda810@mail.ru (R.M.); vlistomin@yandex.ru (V.I.); tulegenov08@list.ru (T.T.); murtazali99@bk.ru (M.Y.); novikov.a@gubkin.ru (A.N.); vladimir@vinokurov.me (V.V.)

2 Department of Petroleum Engineering, Kazan Federal University, Kremlevskaya Str. 18, 420008 Kazan, Russia

3 Skolkovo Institute of Science and Technology (Skoltech), Nobelya Str. 3, 121205 Moscow, Russia

* Correspondence: semenov.a@gubkin.ru (A.S.); andrey.stoporev@kpfu.ru (A.S.)

Abstract: Precise data on the non-variant equilibrium of the four phases (vapor–aqueous solution–ice–gas hydrate) in *P*–*T* coordinates are highly desired for developing accurate thermodynamic models and can be used as reference points (similar to the triple point of water). Using the two-component hydrate-forming system CO_2–H_2O, we have proposed and validated a new express procedure for determining the temperature and pressure of the lower quadruple point Q_1. The essence of the method is the direct measurement of these parameters after the successive formation of the gas hydrate and ice phases in the initial two-phase gas–water solution system under intense agitation of the fluids. After relaxation, the system occurs in the same equilibrium state (T = 271.60 K, P = 1.044 MPa), regardless of the initial parameters and the order of crystallization of the CO_2 hydrate and ice phases. Considering the combined standard uncertainties (±0.023 K, ±0.021 MPa), the determined *P* and *T* values agree with the results of other authors obtained by a more sophisticated indirect method. Validating the developed approach for systems with other hydrate-forming gases is of great interest.

Keywords: gas hydrates; carbon dioxide; phase equilibria; ice; quadruple point; measurement

1. Introduction

The interaction between water and low molecular weight substances (carbon dioxide, lower hydrocarbons, nitrogen, hydrogen sulfide, etc.) at low temperatures/high pressure leads to the formation of crystalline inclusion compounds related to clathrate hydrates. In these compounds, the hydrate-forming molecules (guest) occupy the cavities of the crystal lattice built up by water molecules (host) [1]. The physicochemical properties of clathrate hydrates and their applications are discussed in recent reviews [2,3]. Recently, the study of the phase equilibria of carbon dioxide hydrate has received special attention due to its significant fundamental and applied relevance. From an applied standpoint, the formation of carbon dioxide hydrates is promising for seawater desalination [4–6], wastewater treatment [7], the prevention of CO_2 emissions to the atmosphere through its sequestration and capture [8–11], the separation of gas mixtures [12–15], and in food industry technologies [16]. Carbon dioxide hydrate is a promising phase change material for cold energy storage [17,18]. A deeper understanding of the phase behavior in the CO_2–H_2O system is crucial for explaining processes in the lithosphere of Mars and other astronomical objects [19–21].

Experimental data on the phase behavior of carbon dioxide hydrates are necessary to predict the stability of the phases under certain conditions, which are the basis for the development of hydrate-based technologies. Data on different types of phase equilibria in the CO_2–H_2O system can be found in the literature. We will use the following symbols to denote the equilibrium phases: V—carbon dioxide-rich vapor phase; L_W–water-rich liquid phase; L_W*—supercooled water-rich liquid phase; L_{CO2}—carbon dioxide-rich liquid phase; I—hexagonal ice; H—carbon dioxide hydrate. Based on the Gibbs rule of phases in the case of two independent variables (temperature and pressure), the number of degrees of freedom for any three-phase equilibrium in a two-component system is one. Such an equilibrium must correspond to a line in the phase diagram in the *P*–*T* coordinates. The V-L_W-H monovariant three-phase equilibrium is the most common and has the greatest practical importance. The V-L_W-H equilibrium temperatures and pressures for the CO_2–H_2O system have been experimentally determined by Ng and Robinson [22], Adisasmito et al. [23], Dholabhai et al. [24], Fan and Guo [25], Yang et al. [26], Wendland et al. [27], Mohammadi et al. [28], Yasuda and Ohmura [29], Melnikov et al. [30], Sami et al. [31], Nema et al. [32], Adeniyi et al. [33], Tariq et al. [34], Semenov et al. [35,36], and Cruz-Castro et al. [37]. Sun et al. [38] investigated the effect of a magnetic field on the V-L_W-H equilibrium for carbon dioxide hydrate. The authors showed that the impact of the magnetic field with an induction of 0.39 T leads to a shift of the hydrate phase boundary by up to 3 K to a higher temperature region. Affecting the equilibrium conditions of hydrate formation with a magnetic field may develop a new method for accelerating hydrate formation and stabilizing the hydrate phase. However, the cited results need to be verified by other authors. Anderson [39] calculated the enthalpy of dissociation ΔH_d of carbon dioxide hydrate into water and gas, and the hydrate number *n* using the Clapeyron equation from the V-L_W-H equilibrium data. Over the whole range of the V-L_W-H phase coexistence, the ΔH_d value ranged from 63.6 $\pm$ 1.8 to 57.7 $\pm$ 1.8 kJ/mol, and the hydrate number varied from 6.6 $\pm$ 0.3 to 5.6 $\pm$ 0.3.

At temperatures below the H_2O melting point, ice is a more thermodynamically stable phase than water, so the main type of equilibrium for such conditions is a monovariant V-I-H. For the considered CO_2–H_2O binary system, the temperatures and pressure of the coexistence of gas, ice, and hydrates have been measured by Miller and Smythe [40], Adamson and Jones [41], Wendland et al. [27], Yasuda and Ohmura [29], Fray et al. [42], Mohammadi and Richon [43], and Nagashima et al. [44]. V-I-H and V-L_W-H equilibria for carbon dioxide hydrates with different isotopic compositions of $^{12}CO_2$ and $^{13}CO_2$ were studied by Kimura et al. [45]. It was found that the difference in the position of the equilibrium curves for $^{12}CO_2$ and $^{13}CO_2$ hydrates is small and is 0.007–0.012 MPa in the range of 269–278 K. At the same time, the $^{12}CO_2$ hydrate has a slightly lower equilibrium pressure than $^{13}CO_2$. Takeya et al. applied direct space methods combined with the Rietveld analysis to study the crystal structure of host–guest materials, including carbon dioxide hydrates [46]. The relevant studies of the phase equilibria of gas hydrates at temperatures below the freezing point of water are discussed in reviews [47,48].

At temperatures below the melting point of H_2O, a metastable three-phase equilibrium of V-L_W*-H with water supercooled relative to the ice phase is possible. This equilibrium is a metastable continuation of the V-L_W-H line into the lower temperature region. Experimental data on the V-L_W*-H equilibrium for the CO_2–H_2O system have been obtained by Wendland et al. [27], Melnikov et al. [30], and Nema et al. [32]. The dissociation of hydrates at temperatures < 273 K can lead to the formation of an ice layer on the surface of the hydrate particle, dramatically reducing the rate of the subsequent decomposition of the hydrate (the phenomenon of self-preservation of gas hydrates). Several studies have shown that gas hydrates initially decompose into gas and supercooled water, which can then turn into ice. The formation and long-term existence of supercooled water and gas hydrates in a metastable state has been documented by visual observations of hydrate dissociation at temperatures below the H_2O melting point not only for carbon dioxide [30] but also for other gases such as propane, methane [49–51], and ethane [52]. The formation of supercooled water during hydrate dissociation has been confirmed using Raman spectroscopy for the ethane hydrate [52] and nuclear magnetic resonance relaxation spectroscopy for the difluorodichloromethane hydrate [53].

The ice melting curve in a carbon dioxide atmosphere (V-L_w-I equilibrium) has been studied by Melnikov et al. [54]. The authors found that for the CO_2–H_2O system, the three-phase V-L_w-I equilibrium line has a slope dT/dP that is 20 times greater (in modulo) than the L_w-I equilibrium line of the one-component system with pure water. This is due to the high solubility of carbon dioxide in water, which lowers the equilibrium temperature of ice melting compared to the system without CO_2. Melnikov et al. [54] also observed a metastable three-phase equilibrium V-L_w*-I (at a pressure > 1.04 MPa) when the liquid water-rich phase is supercooled relative to the carbon dioxide hydrate phase. Although, according to the Schreinemaker's method [55], an L_w-H-I equilibrium should also be observed in this region, to the best of our knowledge, there is no such experimental data. We will be grateful to the reviewers and readers if they provide such data.

The peculiarity of carbon dioxide as a hydrate-forming gas is that at temperatures and pressures typical of the hydrate formation, CO_2 is not a supercritical fluid and, therefore, can exist in either gas or liquid form. This complicates the phase behavior of the CO_2–H_2O system compared to similar systems containing non-condensing gases under the conditions studied. Reference data on the vapor–liquid equilibrium of pure carbon dioxide over a wide range of temperatures and pressures have been obtained by Duschek et al. [56]. Experimental data on V-L_{CO2}-H and V-L_{CO2}-L_W three-phase equilibria for the CO_2–H_2O system have been reported by Larson [57], Vlahakis et al. [58], Fan and Guo [25], and Wendland et al. [27]. Our comparative analysis showed that the experimental points of the two-phase equilibrium of the one-component system with CO_2 [56] coincide with the data for the three-phase equilibrium V-L_{CO2}-H and V-L_{CO2}-L_W in the two-component CO_2–H_2O system [57,58] within the measurement uncertainties. At pressures above the V-L_{CO2}-H and V-L_{CO2}-L_W lines, the thermodynamic stability of carbon dioxide hydrate is determined by the L_{CO2}-L_w-H three-phase equilibrium. Experimental data for the latter were obtained by Takenouchi and Kennedy [59], Ng and Robinson [22], Fan and Guo [25], Yang et al. [26], and Adeniyi et al. [33]. In [60], the authors proposed a thermodynamic model to describe the phase behavior of gas hydrates in the system H_2O and CO_2 in a wide range of temperatures 150–295 K and pressures up to 500 MPa.

Figure 1 shows a fragment of the phase diagram for the two-component CO_2–H_2O system, constructed according to the above literature data, for the pressure and temperature range of 0.08–8 MPa and 220–300 K (with a logarithmic pressure scale). Figure S1 shows the same phase diagram with a linear pressure scale. At the lower quadruple point Q_1, the lines of the monovariant equilibria V-L_w-H, V-I-H, and V-L_w-I intersect. Four phases V-L_w-I-H coexist at this point, so this equilibrium is non-variant because the number of degrees of freedom must equal zero according to the Gibbs phase rule for a two-component system. Therefore, the four phases V-L_w-I-H can only be in equilibrium at the fixed pressure and temperature values. Precise temperature and pressure data at the quadruple point of the CO_2–H_2O system are valuable for developing accurate thermodynamic models. They can also be used as a reference point (analogous to the triple point of water).

The Q_1 point coordinates are usually determined by experimentally measuring two types of equilibria, V-L_w-H and V-I-H [29], V-L_w(L_w*)-H and V-I-H [32], or V-L_w(L_w*)-I and V-I(L_w)-H [54], approximating the experimental points for each equilibrium by a function (usually, $\ln P = A + B \cdot T^{-1}$), and finding the intersection of the two functions. This procedure requires a considerable amount of time and other resources.

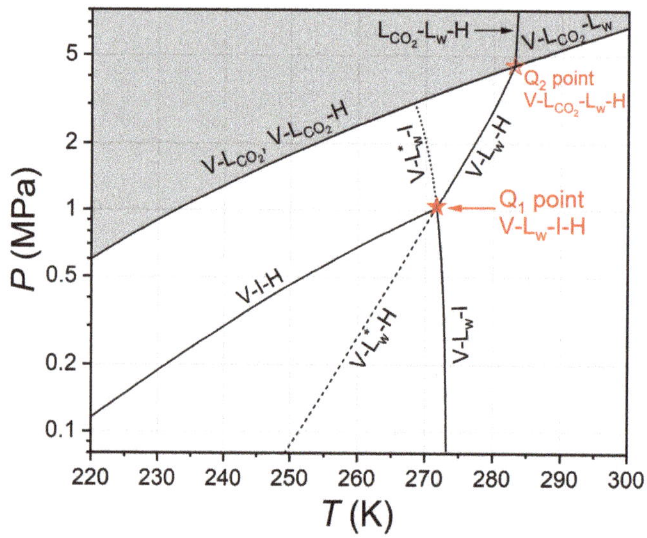

Figure 1. Phase diagram for the H_2O–CO_2 system based on literature data for the temperature and pressure range of 220–300 K and 0.08–8 MPa; gray and white colors correspond to the regions of existence of liquid and gaseous carbon dioxide, respectively; solid line V-L_w-H is the three-phase gas–aqueous solution–gas hydrate equilibrium (fit based on [35,36]); dashed line V-L_w*-H is the metastable three-phase gas–supercooled aqueous solution–gas hydrate equilibrium (fit based on [30,32]); solid line V-I-H is the three-phase gas–ice–gas hydrate equilibrium (fit based on [29,44]); solid line V-L_w-I is the three-phase gas–liquid aqueous solution–ice equilibrium (fit based on [54]); dotted line V-L_w*-I is the metastable three-phase gas–supercooled aqueous solution–ice equilibrium (fit based on [54]); solid line V-L_{CO2} is the two-phase equilibrium of gaseous and liquid carbon dioxide (fit based on [56]) overlapping with solid line V-L_{CO2}-(H or L_w) of the three-phase equilibrium of gaseous and liquid carbon dioxide and gas hydrate (or aqueous solution, fit based on [57,58]); solid line L_{CO2}-L_w-H is the three-phase liquid carbon dioxide-rich phase–aqueous solution–gas hydrate equilibrium (fit based on [33,59]); the red stars represent the non-variant four-phase equilibria: gas–aqueous solution–ice–gas hydrate (Q_1 point (271.60 K, 1.044 MPa), data of this work) and gas–liquid carbon dioxide-rich phase–aqueous solution–gas hydrate (Q_2 point, (283.13 K, 4.494 MPa), intersection of fits based on [35,36,57,58]).

In this paper, we report the results of our experimental observations, which indicate the possibility of a well-reproducible measurement of pressure and temperature at the lower quadruple point Q_1 in a single experiment, using the example of a CO_2–H_2O system with the successive crystallization of gas hydrates and ice (or ice and then gas hydrate) under intense stirring and good heat transfer in the system. Our results suggest reducing the time required to determine the equilibrium conditions of the four-phase equilibrium coexistence of the vapor–aqueous solution–ice–gas hydrate and significantly reducing the measurement uncertainty.

2. Results

To confirm the validity of the pressure and temperature readings of the GHA350 setup used for the CO_2 hydrate equilibrium measurements, we examined two reference systems with a priori known equilibrium P and T parameters. As such comparison references, we chose the freezing point of water at atmospheric pressure and the equilibrium pressure/temperature of V-L_w-H coexistence in the CO_2–H_2O system.

2.1. Measurement Results of Reference Systems

2.1.1. Ice Freezing Point at Ambient Pressure

Figure 2 shows the temperature curve obtained by measuring the ice freezing point of distilled water with a Pt100 resistance thermometer from a GHA350 autoclave under intensive agitation at 600 rpm. The temperature in the system was lowered at a rate of 3 K/h (linear section starting at 30 min). Crystallization of the supercooled water started at 271.18 K, after which the temperature rapidly increased (10 s) to 273.14 K and remained at a plateau. The mean and standard deviation of the temperature at the ice freezing plateau were 273.144 ± 0.007 K. The average value differs from the reference ice melting temperature of 273.1525 K (at 101,325 Pa) [61] by no more than 0.0085 K. It is also clear from the data in Figure 1 that the measured temperature fluctuations are in the range of no more than 0.03 K (from 273.13 to 273.16 K).

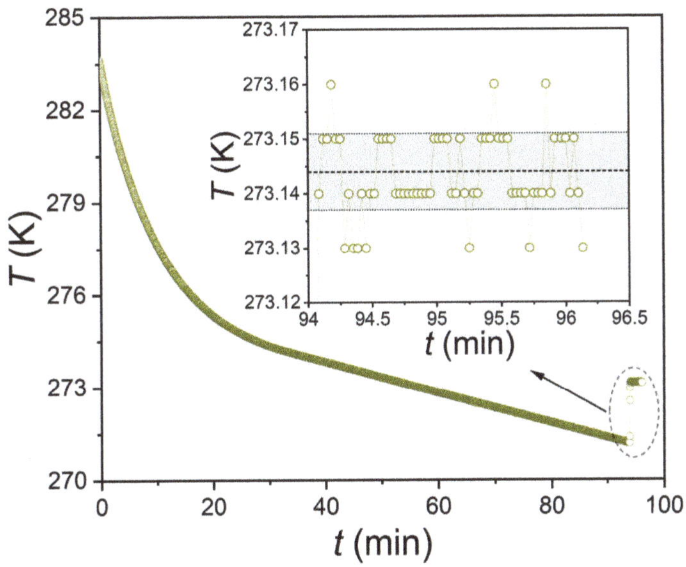

Figure 2. Temperature evolution in GHA350 autoclave during measurement of ice freezing point at ambient pressure.

2.1.2. Three-Phase Equilibrium Conditions V-L_W-H of CO_2–H_2O System

Figure 3a shows the experimental values of the equilibrium conditions for the coexistence of the gaseous CO_2-rich phase, the aqueous solution, and the CO_2 hydrate measured by us [35,36], as well as the literature data [23,24,28–32]. As seen, there is good agreement between our values and the literature.

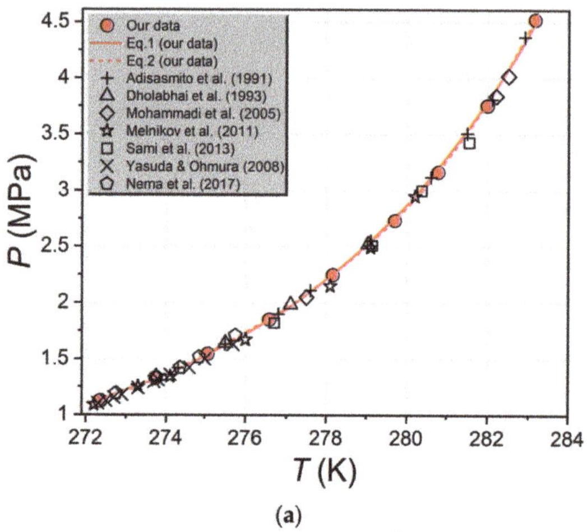

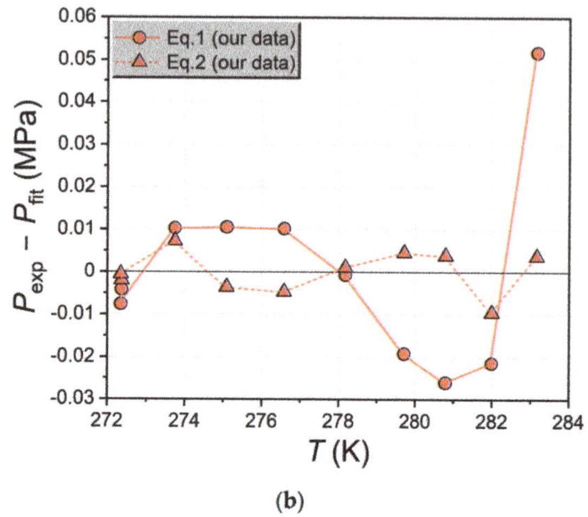

Figure 3. (**a**) Measured pressures and temperatures of the monovariant three-phase equilibrium V-L_w-H of H_2O–CO_2 system; red circles—our data [35,36]; black symbols—literature data [23,24,28–32]; solid red line—fit from our data by three-parameter Equation (1) (see ref. [36]); red dashed line—fit from our data by a fourth-degree polynomial (see Equation (2)); (**b**) fitting residuals for both models as a function of equilibrium temperature.

To derive the correlation between the pressure and temperature at the V-L_w-H equilibrium line for the H_2O–CO_2 system, we used Equations (1) [36] and (2):

$$P = \exp(A + \frac{B}{T} + C \cdot \ln T), \quad (1)$$

$$P = A_0 + A_1 T + A_2 T^2 + A_3 T^3 + A_4 T^4, \quad (2)$$

where P and T are the equilibrium pressure and temperature, respectively; A, B, C, and A_0–A_4 are the fitted parameters. The approximation results are shown in Figure 3a as solid and dashed lines. The points in Figure 3b illustrate the difference between the experimental and calculated equilibrium pressure as a function of temperature for each of the two models. The numerical values of the coefficients are given in Tables S1 and S2 (Supplementary). As expected, the polynomial Equation (2) is more flexible compared to model 1, and the residuals plot (Figure 3b) shows that the difference between the experimental and calculated values is several times smaller in the case of Equation (2), especially at higher pressures. From the data in Tables S1 and S2, we can see that Equation (1) has an average absolute deviation (AAD) of 0.016 MPa (0.63%), while for Equation (2), this value is 0.004 MPa (0.19%). Thus, the accuracy of the description of the experimental data increases by a factor of 3.5 in the transition from model 1 to 2.

To quantify the differences between the results of our measurements of the V-L_w-H equilibrium conditions for the carbon dioxide hydrate and the literature data, we calculated the relative difference using Equation (3):

$$\delta = \frac{P_{exp} - P_{fit}}{P_{fit}} \times 100\%, \quad (3)$$

where P_{exp} is the experimental value of the equilibrium pressure, P_{fit} is the calculated value of the equilibrium pressure obtained by substituting the experimental equilibrium temperature T_{exp} into approximations 1 and 2. Both approximations were derived only from our experimental data [35,36]. The calculated relative differences of equilibrium pressure from Equation 3 are shown in Figures 4 and 5.

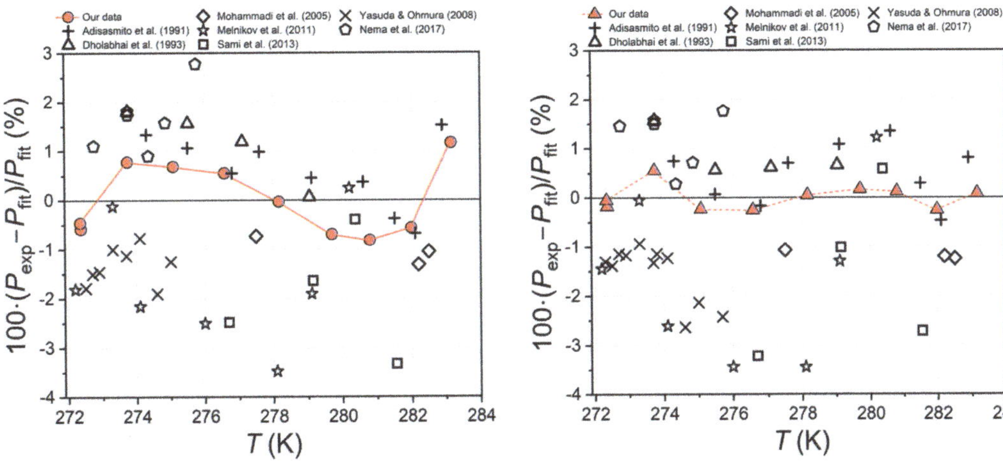

(a) **(b)**

Figure 4. (**a**) The relative difference between the experimental [23,24,28–32,35,36] and calculated (approximation by Equation (1) using data from [35,36]) equilibrium pressure of carbon dioxide hydrate; (**b**) similar relative difference for the fit with Equation (2).

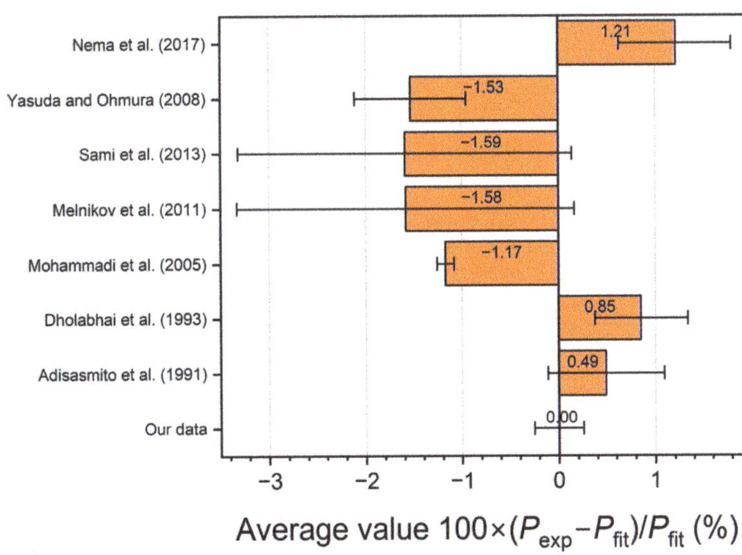

Figure 5. Average relative pressure difference in three-phase equilibrium V-L_W-H in system CO_2–H_2O calculated by Equation (3) for polynomial approximation 2 (data in Figure 4b); experimental data are from [23,24,28–32,35,36]; error bars are standard deviations of average.

The comparison in Figures 4 and 5 shows that our measured equilibrium pressures for the carbon dioxide hydrate [35,36] are in best agreement with the results of Adisasmito et al. [23] and Dholabhai et al. [24]. The equilibrium pressures of Adisasmito et al. [23] are overestimated by 0.49% on average compared to our values [35,36], and the points of Dholabhai et al. [24] are overestimated by 0.85%. The experimental points of Nema et al. [32] (δ = 1.21%) are also shifted to slightly higher pressures. The experimental points of Mohammadi et al. [28], Yasuda and Ohmura [29], Melnikov et al. [30], and Sami et al. [31] (δ = −1.17%, −1.53%, −1.58%, and −1.59%, respectively) are underestimates of the equilibrium pressure relative to our data.

Summarizing the results obtained for the second reference system (V-L_W-H equilibrium in the CO_2–H_2O system), we can conclude a good agreement between our measured equilibrium temperatures and pressures [35,36] and the literature data [23,24,28,32]. There is also a satisfactory agreement of our results with the data [29–31].

2.2. Measurement of Pressure and Temperature in the Lower Quadruple Point Q_1 of CO_2–H_2O System

Figure 6 shows the experimental curves of the pressure and temperature versus the time obtained in the study of the hydrate equilibria in the CO_2–H_2O system [35,36]. The curves are not shown for the entire time range but only for the hydrate formation stage during cooling. The experiments were carried out using a GHA350 autoclave equipped with a stirring system that allows high-intensity stirring of the fluids. The results of our analyses [36,62] show that for distilled water (without accounting for changes in density and viscosity due to gas dissolution) at a stirrer speed of 600 rpm and a temperature of 273–283 K, the Reynolds number is more than 20,000, i.e., the character of the flow in the GHA350 autoclave during operation of the stirring system is fully turbulent [63]. High-intensity fluid agitation eliminates the diffusion limitations of the hydrate formation reaction and contributes to the rapid heat transfer of the hydrate and ice crystallization. We performed six runs, all of which observed the sequential formation of carbon dioxide hydrate and ice (or vice versa, ice and then hydrate). Runs 1–6 differed in their initial conditions (P, T). Consider the curves of run 1 in more detail.

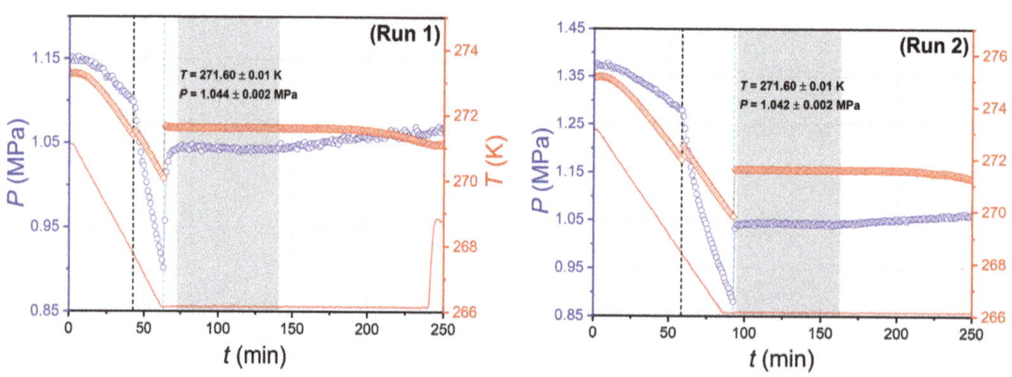

(a) (b)

Figure 6. *Cont.*

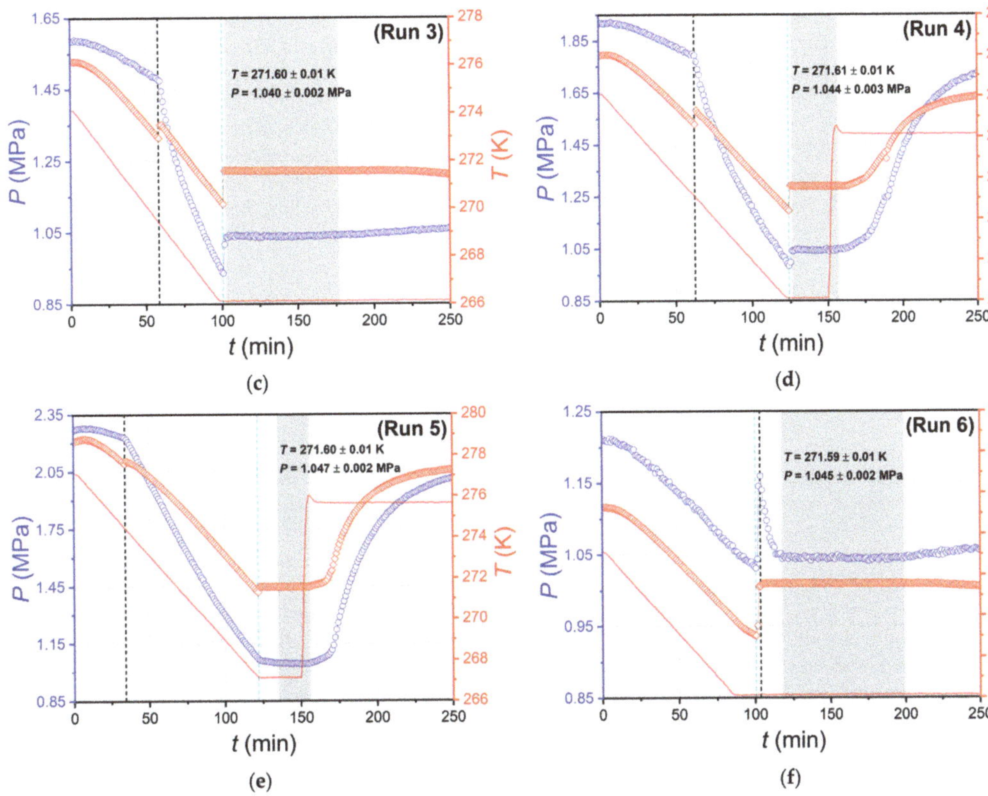

Figure 6. (**a**) Evolution of pressure and temperature over time in the GHA350 autoclave (blue and red symbols) in run 1; the onsets of CO_2 hydrate formation and ice crystallization are shown by vertical black and cyan dashed lines, respectively; the solid red line reflects temperature change in the coolant circulating in the GHA350 autoclave jacket; the gray fill shows the time with steady temperature and pressure readings in the autoclave after the occurrence of hydrate and ice in the system and the next reaching a plateau of T and P; the variation of temperature and pressure at the plateau is displayed as the mean and standard deviation of T and P; (**b–f**) similar results from runs 2–6 at different initial pressure and temperature in the system.

The initial pressure and temperature of the fluids in the autoclave in run 1 were 1.15 MPa and 273.26 K, corresponding to conditions beyond the CO_2 hydrate stability zone. Hydrate formation was induced by cooling the system at a 5 K/h rate. CO_2 hydrate growth started at 42.3 min (time marked with a black dashed line) and was followed by a short temperature peak (of 0.13 K) due to the release of heat from the carbon dioxide hydrate formation reaction and a concurrent kink in the pressure curve due to the onset of gas uptake. Hydrate growth then proceeded at a decreasing temperature (due to the difference between the temperature of the fluids in the autoclave and the temperature of the coolant circulating in the autoclave jacket). The temperature in the autoclave declined until ice crystallization occurred, which started at 63.4 min (time marked with a cyan dashed line). Due to ice freezing, the autoclave temperature increased by 1.55 K to 271.61 K in $\approx$ 1 min and remained at the plateau for over 1 h. The onset of ice crystallization also increased the pressure, which is associated with a reduction in the free volume of the system [64]. However, the pressure relaxation in the system was slower than the temperature relaxation. The pressure plateau after the onset of ice crystallization took about 6 min, which is six times longer than the time for the temperature to reach the plateau. In the range of

72.5–140.9 min with the simultaneous presence of four phases in the system (vapor, water solution, hydrate, and ice), it occurs while intensive stirring, pressure, and temperature are constant (271.60 ± 0.01 K, 1.044 ± 0.002 MPa), which can be interpreted as an equilibrium state. The temperature and pressure values at the plateau agree very well with the literature data on the coordinates of the lower quadruple point in the CO_2–H_2O system [29,32,54], which were determined indirectly. After 140.9 min, a trend of slow increases in the pressure and decreases in the temperature is observed, which we attribute to the accumulation of carbon dioxide hydrates and ice in the system. At a certain amount of solids, the stirrer can no longer effectively mix the fluids in the system, resulting in a gradual shift of the system state away from equilibrium for the four phases V-L_w-I-H.

Runs 2–5 (Figure 6b–e) differ from run 1 in the higher initial pressure in the system (more CO_2 in the autoclave). However, the cooling curves for runs 2–5 are qualitatively the same as run 1. The carbon dioxide hydrate begins to crystallize first during cooling, followed by ice. During the transition from run 1 to run 5, the regularity of the increase in time between the onset of hydrate and ice crystallization (21.1 min for run 1 and 87.6 min for run 5) is evident. Ice freezing starts at approximately the same temperature (269.8–271.3 K) in the considered experiments. Since both the initial T and P increase from run 1 to run 5, an expected consequence is an increase in the cooling time of the ice nucleation temperature. In the case of runs 2–5, as a result of the successive appearance of hydrate and ice in the system, the temperature and pressure at the plateau (after relaxation) take the same values within the measurement errors as in run 1, coinciding with the numerical values of parameters of the lower quadruple point of the CO_2–H_2O system [29,32,54].

For run 6, the initial parameters were close to those of run 1, but the shape of the pressure and temperature curves was qualitatively different from runs 1–5. The difference for run 6 was that the ice phase started to crystallize at 100.6 min, resulting in a rapid temperature rise of 2.6 K (from 269 K to 271.6 K) for 2 min. The carbon dioxide hydrate began to form 3 min after the appearance of ice. The occurrence and growth of ice and then CO_2 hydrate caused the pressure to increase to a peak value of 1.159 MPa and the subsequent relaxation of P to 1.045 MPa. The second thermal effect on the temperature curve caused by the appearance of the CO_2 hydrate phase was absent in the case of run 6. From the curve for run 1, the exothermic effect of hydrate crystallization is an order of magnitude smaller in amplitude than that of ice formation, in agreement with our previous results [65]. Therefore, the superposition of the two exotherms in run 6 resulted in the less intense temperature signal of hydrate formation not being detected. It took 14 min for the pressure to relax from a peak of 1.159 MPa to 1.045 MPa, after which the parameters remained constant (271.59 ± 0.01 K, 1.045 ± 0.002) for 82 min with four phases in the system: gaseous CO_2, aqueous solution, ice, and carbon dioxide hydrate.

Thus, regardless of the initial parameters (T and P) and the order in which the CO_2 hydrate and ice phases occur during cooling, the system appears in the same equilibrium state after relaxation with values of temperature 271.60 ± 0.01 K and pressure 1.044 ± 0.002, corresponding to the non-variant equilibrium of four phases V-L_w-I-H in the CO_2–H_2O system.

The resulting $P(T)$-trajectories for runs 1–6 are shown in Figure 7a–f. Each panel of this figure also offers a point corresponding to the mean temperature and pressure value at the plateau according to the results of six independent measurements.

Table 1. The measurement results of runs 1–6 of the temperature and pressure of the four-phase equilibrium gaseous carbon dioxide–aqueous solution–ice–carbon dioxide hydrate.

Run	*T*, K	*P*, MPa	Standard Deviation *T*, K	Standard Deviation *P*, MPa
1	271.60	1.044	0.008	0.002
2	271.60	1.042	0.007	0.002
3	271.60	1.040	0.009	0.002
4	271.61	1.044	0.007	0.003
5	271.60	1.047	0.009	0.002
6	271.59	1.045	0.009	0.002
Average	271.60	1.044	0.006	0.002

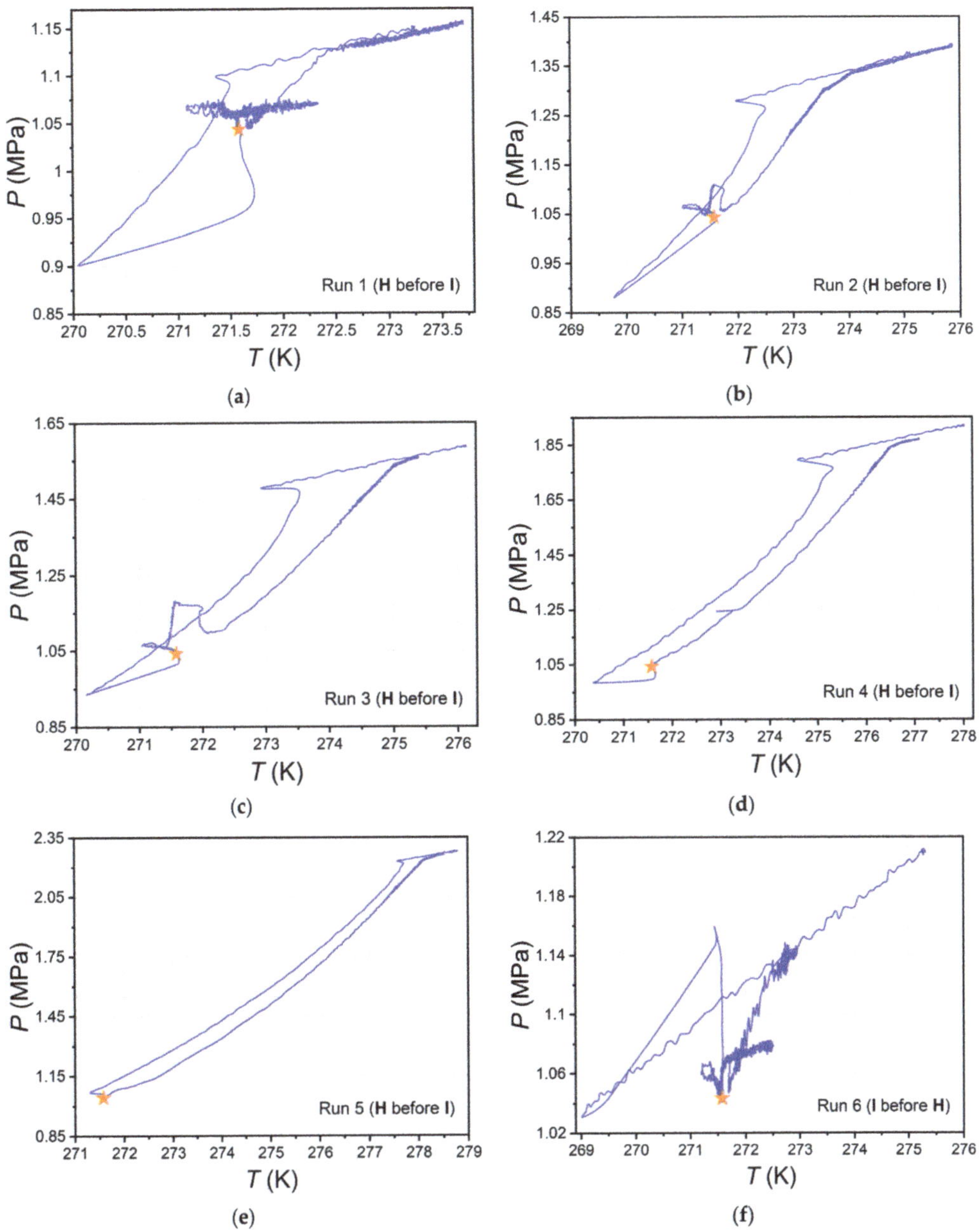

Figure 7. (**a**–**f**) Pressure–temperature trajectories from runs 1–6 (solid blue line); orange star depicts *P* and *T* at lower quadruple point (four phases in equilibrium V-L_w-I-H) for H_2O–CO_2 system (see our data in Table 1).

One can see from the data in Figure 7 that when varying the initial parameters *P* and *T* in quite a wide range, the experimental trajectory in each case crosses the equilibrium point of the four phases V-L_w-I-H in the CO_2–H_2O system. This observation confirms the reproducibility and universal nature of the observed phenomenon.

3. Discussion

The results of the measurements of the equilibrium parameters at the quadruple point for the H_2O–CO_2 system are shown graphically in Figure 8; the numerical values of the pressure and temperature are in Table 1. One can conclude that the results from runs 1–6 have good repeatability.

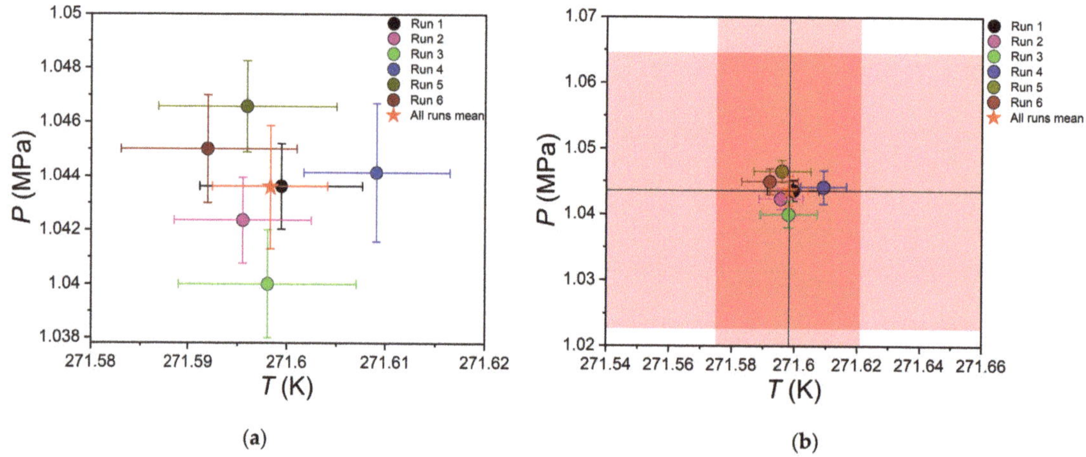

Figure 8. (**a**) Measured values of temperature and pressure at the plateau after the successive occurrence of hydrate and ice (runs 1–5) or ice and hydrate (run 6) in the H_2O–CO_2 system. Panel (**a**) also demonstrates the average temperature and pressure of the non-variant equilibrium (red star) calculated from data of all single measurements (runs 1–6); (**b**) similar quantities plotted considering the combined standard uncertainties in temperature (±0.023 K) and pressure (±0.021 MPa) for the average (runs 1–6). The uncertainties u_c in temperature and pressure are shown as a light-red fill in the corresponding domain.

The value of T and P for each run corresponds to the mean value of the parameter at the plateau (averaged over time). The average (last row in Table 1) is calculated by averaging the parameters over the six points obtained. Based on the results in Table 1, the metrological characteristics of the measuring instruments and standards used for calibration, we calculated the combined standard and expanded uncertainties of the equilibrium pressure and temperature according to JCGM 100:2008 [66] using Equations (4) and (5):

$$u_c = \sqrt{u_1^2 + u_2^2 + u_3^2}, \quad (4)$$

$$U = k \cdot u_c, \quad (5)$$

where u_c is the combined standard uncertainty of the equilibrium temperature or pressure, u_1 is the maximum error of the standards (0.011 K and 0.017 MPa) used to calibrate the pressure and temperature sensors of the GHA350 rig. A 1524 reference thermometer coupled with a PRT 5616-12 reference sensor (both Fluke, Everett, WA, USA) was utilized to calibrate the Pt100 temperature sensor. The pressure sensor was calibrated with a 717 5000G (Fluke, Everett, WA, USA). u_2 is the repeatability of the GHA350 sensors (0.019 K and 0.012 MPa). The repeatability of the Pt100 was evaluated by measuring the freezing point of water at atmospheric pressure three times (one measurement per day). The repeatability of the GHA350 pressure sensor was determined by comparing the difference in readings with the reference gauge 700G30 (Fluke, Everett, WA, USA) at 1.1–2.5 MPa (after completion of runs 1–6). u_3 is the standard deviation of the equilibrium pressure or temperature from runs 1–6 (0.006 K and 0.002 MPa; last row in Table 1). U is the expanded uncertainty of the equilibrium temperature or pressure, and k is the coverage factor, which was taken as two. The combined standard uncertainty of equilibrium temperature and pressure of Q_1 point is 0.023 K and 0.021 MPa, and the expanded uncertainty (at 95% confidence level) is 0.045 K and 0.042 MPa.

A comparison of the results with the literature data is in Table 2. Taking into account the combined standard uncertainties (±0.023 K, ±0.021 MPa) in the temperature and pressure (see Figure 8b and Table 2), the measured parameters at the quadruple point agree with the literature data from Yasuda and Ohmura (2008) [29], Nema et al. (2017) [32], and Mel'nikov et al. (2014) [54]. It should be emphasized again that in this work, the temperature and pressure of the four-phase V-L_W-I-H equilibrium were measured directly during the successive crystallization of hydrate and ice (or in reverse order) in the system. While in the cited papers [29,32,54], the coordinates of the Q_1 point were determined indirectly by a more time-consuming method by measuring the equilibrium points of two different three-phase equilibria (see Table 2), approximating the obtained points by two equations, and determining their intersection point. Direct measurement of the four-phase equilibrium conditions reduced the combined standard uncertainty of the equilibrium temperature by an order of magnitude compared to literature data [29,32,54].

Table 2. Comparison of the temperature and pressure of the four-phase equilibrium gaseous carbon dioxide–aqueous solution–ice–carbon dioxide hydrate determined in this study with the literature data [29,32,54].

T, K	P, MPa	$u_c(T)$, K	$u_c(P)$, MPa	Determination Method	Reference
271.60	1.044	0.023	0.021	Direct measurement	Mean of runs 1–6 (this work)
271.7	1.03	0.20	0.02	Intersection of V-L_W-H and V-I-H lines	Yasuda and Ohmura 2008 [29]
271.6	1.04	0.2	0.02	Intersection of V-L_W(L_W*)-H and V-I-H lines	Nema et al. 2017 [32]
271.65	1.03	n/a	n/a	Intersection of V-L_W(L_W*)-I and V-I(L_W)-H lines	Mel'nikov et al. 2014 [54]

Let us discuss the possible reasons for directly measuring the P and T parameters of the four-phase equilibrium for the CO_2–H_2O system. First, the high turbulence of the GHA350 stirring system at 600 rpm ($Re > 2 \times 10^4$ [36,62,67]), which significantly intensifies the heat and mass transfer processes, contributes to the rapid reaching of the equilibrium state in the system. Second, in the simultaneous presence of the CO_2 hydrate and ice phases in the system, the processes of their formation compete with each other [68]. When ice crystallizes in the system, the free volume decreases, increasing pressure [64]. Hydrate formation leads to the opposite effect of a pressure decrease due to the gas uptake and reduction in its amount in the free volume. Thus, the system's equilibrium state (P and T at the plateau) with the simultaneous crystallization of hydrate and ice in the system can be explained by the rates of these processes that may be comparable. In this case, the system is self-regulating, i.e., the effect of the pressure increase due to ice crystallization is compensated by the pressure decrease due to gas hydrate formation. The ice freezing process usually occurs at a higher rate than the formation of the sI methane hydrate or sII C_1-C_3 hydrate [65]. However, CO_2 has a higher solubility in water compared to hydrocarbons, so the kinetics of carbon dioxide hydrate formation are faster and more favorable [69–72]. Therefore, the difference between the ice and gas hydrate formation rates becomes smaller in the case of carbon dioxide.

Verification of the proposed technique on other similar systems will allow one to use it for the direct measurement of the conditions of the non-variant four-phase V-L_W-I-H equilibrium of various hydrate-forming systems to obtain precise data for the construction and refinement of phase diagrams. In future work, we plan to investigate how the nature of the hydrate-forming gas and the type of hydrate structure affect the feasibility of directly measuring the four-phase equilibrium using the described technique. The nature of the gas and the type of hydrate structure (sI or sII) influence the kinetics of hydrate formation. Therefore, it is of interest to compare the CO_2–H_2O system (high gas solubility in water, sI hydrate) with others such as N_2-H_2O (lower gas solubility in water, sII hydrate) and CH_4-H_2O (lower gas solubility in water, sI hydrate). Special attention should also be paid to the study of the proposed technique for systems with kinetic promoters of hydrate formation [73,74], which allows the rate of hydrate formation to be significantly increased. This can reduce the difference in the rate of hydrate and ice growth processes. The direct measurement of the four-phase equilibrium in hydrate-forming systems according to the procedure described in the manuscript significantly improves the accuracy of the determination of the coordinates of the lower quadruple point. It reduces the time and material costs of research compared to conventional methods.

4. Materials and Methods

The pressurized carbon dioxide with at least 99.99 vol% of the main component was purchased from NIIKM (Moscow, Russia). The Simplicity UV laboratory water purification system produced deionized water with a resistivity of 18.2 MΩ·cm at 298.15 K (Merck Millipore, Burlington, MA, USA).

The studies were performed with a GHA350 apparatus (PSL Systemtechnik, Osterode am Harz, Germany) [75], which comprises a 600 mL Hastelloy autoclave rated at a peak operating pressure of 35 MPa. The autoclave is coupled to a Hei-TORQUE 400 precision drive (Heidolph, Schwabach, Germany), a Minipower magnetic coupling (Premex, Lyss, Switzerland), and a four-blade propeller (diameter of 6.1 cm). The temperature control system involves an outer jacket with an ethanol coolant whose circulation is provided by a Ministat 240 (Huber, Offenburg, Germany). Ministat 240 maintains coolant temperature stability within ±0.02 K. The autoclave is fitted with calibrated temperature transducer (Pt100) with a resolution of ±0.01 K.

A P3251 sensor (Tecsis, Offenbach am Main, Germany), in combination with a 700G30 electronic reference manometer (Fluke, Everett, WA, USA), is used to measure the pressure in the autoclave with a resolution of 0.001 MPa. The Pt100 resistance thermometer and the GHA350 pressure sensor were calibrated prior to the experiments using the reference instruments and the methodology described previously [76]. The sensors, the Hei-TORQUE 400 precision stirrer, and the Ministat 240 thermostat are connected to a PC running WinGHA software (version 4.0.10.790, PSL Systemtechnik, Osterode am Harz, Germany). After executing the WinGHA script, the setup operation is controlled, and the measured parameters are automatically collected. When the script is stopped, the data saved in the file with the .dat extension can be processed.

The autoclave, previously cleaned by washing with distilled water and blowing with compressed air, was filled with 300 mL of 18.2 MΩ·cm water. The empty volume was blown out three times with carbon dioxide to remove residual air. At 295 K, the autoclave was pressurized with gaseous CO_2 and stirred at 600 rpm to accelerate the dissolution of the carbon dioxide in the water. Stirring was continued until the end of the experiment. The specified stirring speed was standard for us when measuring gas hydrate equilibria with the GHA350 [77]. After pressure stabilization, the autoclave temperature was lowered to a level where the P and T parameters were outside the hydrate stability zone but close to the V-L_w-H curve for the CO_2–H_2O system [23,24,35,36]. The system was then held for 1 h to allow mutual saturation of the gas and liquid phases, after which the coolant temperature was reduced by 5–10 K at a rate of 5 K/h. After the ramp cooling, the thermostat maintained the coolant temperature at a constant level. Successive formation of CO_2 hydrate and ice was detected by thermal effects and pressure changes.

5. Conclusions

We have proposed a new technique for the direct determination of the P and T coordinates of the lower quadruple point (non-variant four-phase equilibrium gas–water solution–ice–gas hydrate) on the example of the two-component system carbon dioxide–water. The technique directly measures the temperature and pressure after the successive appearance of gas hydrates and ice phases in the initial two-phase gas–water solution system under intensive fluid agitation. Irrespective of the initial parameters (T and P) and the order of hydrate and ice occurrence, the system is in the same equilibrium state after relaxation $T = 271.60 \pm 0.01$ K (mean and standard deviation) and $P = 1.044 \pm 0.002$ MPa. A good reproducibility of the measurement results from six independent experiments was demonstrated. The obtained values of the P and T parameters at the lower quadruple point for the CO_2–H_2O system, considering the combined standard uncertainty of the temperature and pressure measurements (± 0.023 K, ± 0.021 MPa), are in perfect agreement with the literature data [29,32,54] determined by a more complicated and indirect method. The direct measurement allows us to reduce the combined standard uncertainty of the equilibrium temperature by an order of magnitude compared to the literature data. It is of great interest to validate the developed technique for other hydrate-forming systems.

Supplementary Materials: The following supporting information can be downloaded at https://www.mdpi.com/article/10.3390/ijms24119321/s1.

Author Contributions: Conceptualization, A.S. (Anton Semenov); methodology, A.S. (Anton Semenov); validation, A.S. (Anton Semenov) and R.M.; formal analysis, A.S. (Anton Semenov); investigation, R.M., T.T. and M.Y.; resources, V.V.; data curation, R.M. and A.S. (Anton Semenov); writing—original draft preparation, A.S. (Anton Semenov), A.S. (Andrey Stoporev) and A.N.; writing—review and editing, A.S. (Anton Semenov), A.S. (Andrey Stoporev) and A.N.; visualization, A.S. (Anton Semenov); supervision, A.S. (Anton Semenov), V.I.; project administration, A.S. (Anton Semenov); funding acquisition, A.S. (Anton Semenov) All authors have read and agreed to the published version of the manuscript.

Funding: This research was funded by the Russian Science Foundation, grant number 20-79-10377.

Institutional Review Board Statement: Not applicable.

Informed Consent Statement: Not applicable.

Data Availability Statement: Raw data obtained by measuring three-phase V-L_w-H and four-phase equilibrium coexistence V-L_w-I-H in H_2O–CO_2 system are available at https://data.mendeley.com/datasets/5xjdgvjb84/2 (accessed on 24 January 2023) or on request from the corresponding authors.

Conflicts of Interest: The authors declare no conflict of interest.

References

1. Powell, H.M. 15. The Structure of Molecular Compounds. Part IV. Clathrate Compounds. *J. Chem. Soc.* **1948**, *16*, 61–73. [CrossRef] [PubMed]
2. Manakov, A.Y.; Stoporev, A.S. Physical Chemistry and Technological Applications of Gas Hydrates: Topical Aspects. *Russ. Chem. Rev.* **2021**, *90*, 566–600. [CrossRef]
3. Hassanpouryouzband, A.; Joonaki, E.; Vasheghani Farahani, M.; Takeya, S.; Ruppel, C.; Yang, J.; English, N.J.; Schicks, J.M.; Edlmann, K.; Mehrabian, H.; et al. Gas Hydrates in Sustainable Chemistry. *Chem. Soc. Rev.* **2020**, *49*, 5225–5309. [CrossRef] [PubMed]
4. Kang, K.C.; Linga, P.; Park, K.; Choi, S.-J.; Lee, J.D. Seawater Desalination by Gas Hydrate Process and Removal Characteristics of Dissolved Ions (Na^+, K^+, Mg^{2+}, Ca^{2+}, B^{3+}, Cl^-, $SO_4{}^{2-}$). *Desalination* **2014**, *353*, 84–90. [CrossRef]
5. Nakane, R.; Gima, E.; Ohmura, R.; Senaha, I.; Yasuda, K. Phase Equilibrium Condition Measurements in Carbon Dioxide Hydrate Forming System Coexisting with Sodium Chloride Aqueous Solutions. *J. Chem. Thermodyn.* **2019**, *130*, 192–197. [CrossRef]
6. Nakane, R.; Shimosato, Y.; Gima, E.; Ohmura, R.; Senaha, I.; Yasuda, K. Phase Equilibrium Condition Measurements in Carbon Dioxide Hydrate Forming System Coexisting with Seawater. *J. Chem. Thermodyn.* **2021**, *152*, 106276. [CrossRef]
7. Gambelli, A.M.; Rossi, F. Review on the Usage of Small-Chain Hydrocarbons (C_2—C_4) as Aid Gases for Improving the Efficiency of Hydrate-Based Technologies. *Energies* **2023**, *16*, 3576. [CrossRef]
8. Babu, P.; Linga, P.; Kumar, R.; Englezos, P. A Review of the Hydrate Based Gas Separation (HBGS) Process for Carbon Dioxide Pre-Combustion Capture. *Energy* **2015**, *85*, 261–279. [CrossRef]
9. Liu, Y.; Zhang, L.; Yang, L.; Dong, H.; Zhao, J.; Song, Y. Behaviors of CO_2 Hydrate Formation in the Presence of Acid-Dissolvable Organic Matters. *Environ. Sci. Technol.* **2021**, *55*, 6206–6213. [CrossRef]

10. Song, Y.; Wang, S.; Cheng, Z.; Huang, M.; Zhang, Y.; Zheng, J.; Jiang, L.; Liu, Y. Dependence of the Hydrate-Based CO_2 Storage Process on the Hydrate Reservoir Environment in High-Efficiency Storage Methods. *Chem. Eng. J.* **2021**, *415*, 128937. [CrossRef]
11. Nesterov, A.N.; Reshetnikov, A.M. New Combination of Thermodynamic and Kinetic Promoters to Enhance Carbon Dioxide Hydrate Formation under Static Conditions. *Chem. Eng. J.* **2019**, *378*, 122165. [CrossRef]
12. Filarsky, F.; Schmuck, C.; Schultz, H.J. Development of a Biogas Production and Purification Process Using Promoted Gas Hydrate Formation—A Feasibility Study. *Chem. Eng. Res. Des.* **2018**, *134*, 257–267. [CrossRef]
13. Kida, M.; Goda, H.; Sakagami, H.; Minami, H. CO_2 Capture from CH_4–CO_2 Mixture by Gas–Solid Contact with Tetrahydrofuran Clathrate Hydrate. *Chem. Phys.* **2020**, *538*, 110863. [CrossRef]
14. Kudryavtseva, M.S.; Petukhov, A.N.; Shablykin, D.N.; Stepanova, E.A.; Vorotyntsev, V.M. Calculating Coefficients of the Gas Hydrate Distribution of CO_2 and H_2S When Removing Them from a Methane-Containing Gas Mixture. *Russ. J. Phys. Chem. A* **2023**, *97*, 561–564. [CrossRef]
15. Skiba, S.; Chashchin, D.; Semenov, A.; Yarakhmedov, M.; Vinokurov, V.; Sagidullin, A.; Manakov, A.; Stoporev, A. Hydrate-Based Separation of the CO_2 + H_2 Mixtures. Phase Equilibria with Isopropanol Aqueous Solutions and Hydrogen Solubility in CO_2 Hydrate. *Int. J. Hydrog. Energy* **2021**, *46*, 32904–32913. [CrossRef]
16. Nkosi, N.; Nkazi, D.; Tumba, K. A Review of Thermodynamic and Kinetic Studies Relevant to Gas Hydrate-Based Fruit Juice Concentration. *J. Food Eng.* **2023**, *341*, 111323. [CrossRef]
17. Wang, X.; Zhang, F.; Lipiński, W. Carbon Dioxide Hydrates for Cold Thermal Energy Storage: A Review. *Sol. Energy* **2020**, *211*, 11–30. [CrossRef]
18. Bukhalkin, D.D.; Semenov, A.P.; Novikov, A.A.; Mendgaziev, R.I.; Stoporev, A.S.; Gushchin, P.A.; Shchukin, D.G. Phase Change Materials in Energy: Current State of Research and Potential Applications. *Chem. Technol. Fuels Oils* **2020**, *55*, 733–741. [CrossRef]
19. Longhi, J. Phase Equilibrium in the System CO_2-H_2O: Application to Mars. *J. Geophys. Res.* **2006**, *111*, E06011. [CrossRef]
20. Mousis, O.; Chassefière, E.; Lasue, J.; Chevrier, V.; Elwood Madden, M.E.; Lakhlifi, A.; Lunine, J.I.; Montmessin, F.; Picaud, S.; Schmidt, F.; et al. Volatile Trapping in Martian Clathrates. *Space Sci. Rev.* **2013**, *174*, 213–250. [CrossRef]
21. Trainer, M.G.; Tolbert, M.A.; McKay, C.P.; Toon, O.B. Enhanced CO_2 Trapping in Water Ice via Atmospheric Deposition with Relevance to Mars. *Icarus* **2010**, *206*, 707–715. [CrossRef]
22. Ng, H.J.; Robinson, D.B. Hydrate Formation in Systems Containing Methane, Ethane, Propane, Carbon Dioxide or Hydrogen Sulfide in the Presence of Methanol. *Fluid Phase Equilib.* **1985**, *21*, 145–155. [CrossRef]
23. Adisasmito, S.; Frank, R.J.; Sloan, E.D. Hydrates of Carbon Dioxide and Methane Mixtures. *J. Chem. Eng. Data* **1991**, *36*, 68–71. [CrossRef]
24. Dholabhai, P.D.; Kalogerakis, N.; Bishnoi, P.R. Equilibrium Conditions for Carbon Dioxide Hydrate Formation in Aqueous Electrolyte Solutions. *J. Chem. Eng. Data* **1993**, *38*, 650–654. [CrossRef]
25. Fan, S.-S.; Guo, T.-M. Hydrate Formation of CO_2 -Rich Binary and Quaternary Gas Mixtures in Aqueous Sodium Chloride Solutions. *J. Chem. Eng. Data* **1999**, *44*, 829–832. [CrossRef]
26. Yang, S.O.; Yang, I.M.; Kim, Y.S.; Lee, C.S. Measurement and Prediction of Phase Equilibria for Water+CO_2 in Hydrate Forming Conditions. *Fluid Phase Equilib.* **2000**, *175*, 75–89. [CrossRef]
27. Wendland, M.; Hasse, H.; Maurer, G. Experimental Pressure−Temperature Data on Three- and Four-Phase Equilibria of Fluid, Hydrate, and Ice Phases in the System Carbon Dioxide−Water. *J. Chem. Eng. Data* **1999**, *44*, 901–906. [CrossRef]
28. Mohammadi, A.H.; Anderson, R.; Tohidi, B. Carbon Monoxide Clathrate Hydrates: Equilibrium Data and Thermodynamic Modeling. *AIChE J.* **2005**, *51*, 2825–2833. [CrossRef]
29. Yasuda, K.; Ohmura, R. Phase Equilibrium for Clathrate Hydrates Formed with Methane, Ethane, Propane, or Carbon Dioxide at Temperatures below the Freezing Point of Water. *J. Chem. Eng. Data* **2008**, *53*, 2182–2188. [CrossRef]
30. Melnikov, V.P.; Nesterov, A.N.; Reshetnikov, A.M.; Istomin, V.A. Metastable States during Dissociation of Carbon Dioxide Hydrates below 273K. *Chem. Eng. Sci.* **2011**, *66*, 73–77. [CrossRef]
31. Sami, N.A.; Das, K.; Sangwai, J.S.; Balasubramanian, N. Phase Equilibria of Methane and Carbon Dioxide Clathrate Hydrates in the Presence of (Methanol+$MgCl_2$) and (Ethylene Glycol+$MgCl_2$) Aqueous Solutions. *J. Chem. Thermodyn.* **2013**, *65*, 198–203. [CrossRef]
32. Nema, Y.; Ohmura, R.; Senaha, I.; Yasuda, K. Quadruple Point Determination in Carbon Dioxide Hydrate Forming System. *Fluid Phase Equilib.* **2017**, *441*, 49–53. [CrossRef]
33. Adeniyi, K.I.; Deering, C.E.; Grynia, E.; Marriott, R.A. Water Content and Hydrate Dissociation Conditions for Carbon Dioxide Rich Fluid. *Int. J. Greenh. Gas Control* **2020**, *101*, 103139. [CrossRef]
34. Tariq, M.; Soromenho, M.R.C.; Rebelo, L.P.N.; Esperança, J.M.S.S. Insights into CO_2 Hydrates Formation and Dissociation at Isochoric Conditions Using a Rocking Cell Apparatus. *Chem. Eng. Sci.* **2022**, *249*, 117319. [CrossRef]
35. Semenov, A.P.; Mendgaziev, R.I.; Stoporev, A.S.; Istomin, V.A.; Sergeeva, D.V.; Tulegenov, T.B.; Vinokurov, V.A. Dimethyl Sulfoxide as a Novel Thermodynamic Inhibitor of Carbon Dioxide Hydrate Formation. *Chem. Eng. Sci.* **2022**, *255*, 117670. [CrossRef]
36. Semenov, A.P.; Mendgaziev, R.I.; Stoporev, A.S.; Istomin, V.A.; Sergeeva, D.V.; Tulegenov, T.B.; Vinokurov, V.A. Dataset for the Dimethyl Sulfoxide as a Novel Thermodynamic Inhibitor of Carbon Dioxide Hydrate Formation. *Data Br.* **2022**, *42*, 108289. [CrossRef]

37. Cruz-Castro, L.; Ramirez-Jaramillo, E.; Albiter-Hernández, A. Investigation of Gas Hydrate Phase Equilibria in Bulk and in a Large Particle Size Natural Quartz Sand for Methane, Carbon Dioxide and Natural Gas. *Pet. Sci. Technol.* **2023**, 2192243. [CrossRef]
38. Sun, S.; Li, Y.; Gu, L.; Yang, Z.; Zhao, J. Experimental Study on Carbon Dioxide Hydrate Formation in the Presence of Static Magnetic Field. *J. Chem. Thermodyn.* **2022**, *170*, 106764. [CrossRef]
39. Anderson, G.K. Enthalpy of Dissociation and Hydration Number of Carbon Dioxide Hydrate from the Clapeyron Equation. *J. Chem. Thermodyn.* **2003**, *35*, 1171–1183. [CrossRef]
40. Miller, S.L.; Smythe, W.D. Carbon Dioxide Clathrate in the Martian Ice Cap. *Science* **1970**, *170*, 531–533. [CrossRef]
41. Adamson, A.W.; Jones, B. Physical Adsorption of Vapors on Ice. IV. Carbon Dioxide. *J. Colloid Interface Sci.* **1971**, *37*, 831–835. [CrossRef]
42. Fray, N.; Marboeuf, U.; Brissaud, O.; Schmitt, B. Equilibrium Data of Methane, Carbon Dioxide, and Xenon Clathrate Hydrates below the Freezing Point of Water. Applications to Astrophysical Environments. *J. Chem. Eng. Data* **2010**, *55*, 5101–5108. [CrossRef]
43. Mohammadi, A.H.; Richon, D. Ice–Clathrate Hydrate–Gas Phase Equilibria for Argon + Water and Carbon Dioxide + Water Systems. *Ind. Eng. Chem. Res.* **2011**, *50*, 11452–11454. [CrossRef]
44. Nagashima, H.D.; Fukushima, N.; Ohmura, R. Phase Equilibrium Condition Measurements in Carbon Dioxide Clathrate Hydrate Forming System from 199.1 K to 247.1 K. *Fluid Phase Equilib.* **2016**, *413*, 53–56. [CrossRef]
45. Kimura, H.; Fuseya, G.; Takeya, S.; Hachikubo, A. Carbon Isotope Fractionation during the Formation of CO_2 Hydrate and Equilibrium Pressures of $^{12}CO_2$ and $^{13}CO_2$ Hydrates. *Molecules* **2021**, *26*, 4215. [CrossRef]
46. Takeya, S.; Udachin, K.A.; Moudrakovski, I.L.; Susilo, R.; Ripmeester, J.A. Direct Space Methods for Powder X-Ray Diffraction for Guest−Host Materials: Applications to Cage Occupancies and Guest Distributions in Clathrate Hydrates. *J. Am. Chem. Soc.* **2010**, *132*, 524–531. [CrossRef]
47. Nagashima, H.D.; Miyagi, T.; Yasuda, K.; Ohmura, R. Clathrate Hydrates at Temperatures below the Freezing Point of Water: A Review. *Fluid Phase Equilib.* **2020**, *517*, 112610. [CrossRef]
48. Heidaryan, E.; Robustillo Fuentes, M.D.; Pessôa Filho, P.d.A. Equilibrium of Methane and Carbon Dioxide Hydrates Below the Freezing Point of Water: Literature Review and Modeling. *J. Low Temp. Phys.* **2019**, *194*, 27–45. [CrossRef]
49. Mel'nikov, V.P.; Nesterov, A.N.; Reshetnikov, A.M. Formation of Supercooled Water upon Dissociation of Propane Hydrates at $T < 270$ K. *Dokl. Phys. Chem.* **2007**, *417*, 304–307. [CrossRef]
50. Melnikov, V.P.; Nesterov, A.N.; Reshetnikov, A.M.; Zavodovsky, A.G. Evidence of Liquid Water Formation during Methane Hydrates Dissociation below the Ice Point. *Chem. Eng. Sci.* **2009**, *64*, 1160–1166. [CrossRef]
51. Melnikov, V.P.; Nesterov, A.N.; Reshetnikov, A.M.; Istomin, V.A.; Kwon, V.G. Stability and Growth of Gas Hydrates below the Ice–Hydrate–Gas Equilibrium Line on the P–T Phase Diagram. *Chem. Eng. Sci.* **2010**, *65*, 906–914. [CrossRef]
52. Ohno, H.; Oyabu, I.; Iizuka, Y.; Hondoh, T.; Narita, H.; Nagao, J. Dissociation Behavior of C_2H_6 Hydrate at Temperatures below the Ice Point: Melting to Liquid Water Followed by Ice Nucleation. *J. Phys. Chem. A* **2011**, *115*, 8889–8894. [CrossRef]
53. Melnikov, V.P.; Nesterov, A.N.; Podenko, L.S.; Reshetnikov, A.M.; Shalamov, V.V. NMR Evidence of Supercooled Water Formation during Gas Hydrate Dissociation below the Melting Point of Ice. *Chem. Eng. Sci.* **2012**, *71*, 573–577. [CrossRef]
54. Mel'nikov, V.P.; Nesterov, A.N.; Podenko, L.S.; Reshetnikov, A.M. Influence of Carbon Dioxide on Melting of Underground Ice. *Dokl. Earth Sci.* **2014**, *459*, 1353–1355. [CrossRef]
55. Zen, E. *Construction of Pressure-Temperature Diagrams for Multicomponent Systems after the Method of Schreinemakers—A Geometric Approach*; Department of the Interior, US Geological Survey: Reston, VA, USA, 1966.
56. Duschek, W.; Kleinrahm, R.; Wagner, W. Measurement and Correlation of the (Pressure, Density, Temperature) Relation of Carbon Dioxide II. Saturated-Liquid and Saturated-Vapour Densities and the Vapour Pressure along the Entire Coexistence Curve. *J. Chem. Thermodyn.* **1990**, *22*, 841–864. [CrossRef]
57. Larson, S.D. *Phase Studies of the Two-Component Carbon Dioxide-Water System, Involving the Carbon Dioxide Hydrate*; University of Illinois: Urbana, IL, USA, 1955.
58. Vlahakis, J.G.; Chen, H.S.; Suwandi, M.S.; Barduhn, A.J. The Growth Rate of Ice Crystals: Properties of Carbon Dioxide Hydrates, A Review of Properties of 51 Gas Hydrates. *Res. Dev. Rep.* **1972**, *830*.
59. Takenouchi, S.; Kennedy, G.C. Dissociation Pressures of the Phase $CO_2 \cdot 5\ 3/4H_2O$. *J. Geol.* **1965**, *73*, 383–390. [CrossRef]
60. Jäger, A.; Vinš, V.; Gernert, J.; Span, R.; Hrubý, J. Phase Equilibria with Hydrate Formation in H_2O+CO_2 Mixtures Modeled with Reference Equations of State. *Fluid Phase Equilib.* **2013**, *338*, 100–113. [CrossRef]
61. *Handbook of Chemistry and Physics*, 95th ed.; Haynes, W.M.; Lide, D.R.; Bruno, T.J. (Eds.) CRC Press: Boca Raton, FL, USA, 2014.
62. Semenov, A.P.; Stoporev, A.S.; Mendgaziev, R.I.; Gushchin, P.A.; Khlebnikov, V.N.; Yakushev, V.S.; Istomin, V.A.; Sergeeva, D.V.; Vinokurov, V.A. Synergistic Effect of Salts and Methanol in Thermodynamic Inhibition of SII Gas Hydrates. *J. Chem. Thermodyn.* **2019**, *137*, 119–130. [CrossRef]
63. Merkel, F.S.; Schmuck, C.; Schultz, H.J. Investigation of the Influence of Hydroxyl Groups on Gas Hydrate Formation at Pipeline-Like Conditions. *Energy Fuels* **2016**, *30*, 9141–9149. [CrossRef]
64. Yarakhmedov, M.B.; Semenov, A.P.; Stoporev, A.S. Effect of Lower Alcohols on the Formation of Methane Hydrate at Temperatures Below the Ice Melting Point. *Chem. Technol. Fuels Oils* **2023**, *58*, 962–966. [CrossRef]

65. Stoporev, A.S.; Semenov, A.P.; Medvedev, V.I.; Sizikov, A.A.A.; Gushchin, P.A.; Vinokurov, V.A.; Manakov, A.Y. Visual Observation of Gas Hydrates Nucleation and Growth at a Water—Organic Liquid Interface. *J. Cryst. Growth* **2018**, *485*, 54–68. [CrossRef]
66. Jcgm, J.C.G.M. Evaluation of Measurement Data—Guide to the Expression of Uncertainty in Measurement. *Int. Organ. Stand. Geneva ISBN* **2008**, *50*, 134.
67. Semenov, A.P.; Mendgaziev, R.I.; Stoporev, A.S.; Istomin, V.A.; Sergeeva, D.V.; Ogienko, A.G.; Vinokurov, V.A. The Pursuit of a More Powerful Thermodynamic Hydrate Inhibitor than Methanol. Dimethyl Sulfoxide as a Case Study. *Chem. Eng. J.* **2021**, *423*, 130227. [CrossRef]
68. Li, Y.; Maria Gambelli, A.; Chen, J.; Yin, Z.; Rossi, F.; Tronconi, E.; Mei, S. Experimental Study on the Competition between Carbon Dioxide Hydrate and Ice below the Freezing Point. *Chem. Eng. Sci.* **2023**, *268*, 118426. [CrossRef]
69. Lim, V.W.S.; Barwood, M.T.J.; Metaxas, P.J.; Johns, M.L.; Aman, Z.M.; May, E.F. Nucleation Rates of Carbon Dioxide Hydrate. *Chem. Eng. J.* **2022**, *443*, 136359. [CrossRef]
70. Maeda, N. Nucleation Curve of Carbon Dioxide Hydrate from a Linear Cooling Ramp Method. *J. Phys. Chem. A* **2019**, *123*, 7911–7919. [CrossRef]
71. Wei, Y.; Nobuo, M. Nucleation Curves of Carbon Dioxide Hydrate in the Absence of a Solid Wall. *Energy Fuels* **2023**, *37*, 3760–3774. [CrossRef]
72. Drachuk, A.O.; Molokitina, N.S.; Kibkalo, A.A.; Podenko, L.S. Production of Carbon Dioxide Hydrates Using Frozen Aqueous Solutions of Polyvinyl Alcohol. *Russ. J. Appl. Chem.* **2022**, *95*, 506–511. [CrossRef]
73. Ricaurte, M.; Torré, J.-P.; Diaz, J.; Dicharry, C. In Situ Injection of THF to Trigger Gas Hydrate Crystallization: Application to the Evaluation of a Kinetic Hydrate Promoter. *Chem. Eng. Res. Des.* **2014**, *92*, 1674–1680. [CrossRef]
74. Dicharry, C.; Diaz, J.; Torré, J.-P.; Ricaurte, M. Influence of the Carbon Chain Length of a Sulfate-Based Surfactant on the Formation of CO_2, CH_4 and CO_2–CH_4 Gas Hydrates. *Chem. Eng. Sci.* **2016**, *152*, 736–745. [CrossRef]
75. Semenov, A.P.; Medvedev, V.I.; Gushchin, P.A.; Kotelev, M.S.; Yakushev, V.S.; Stoporev, A.S.; Sizikov, A.A.; Ogienko, A.G.; Vinokurov, V.A. Phase Equilibrium for Clathrate Hydrate Formed in Methane + Water + Ethylene Carbonate System. *Fluid Phase Equilib.* **2017**, *432*, 1–9. [CrossRef]
76. Gong, Y.; Mendgaziev, R.I.; Hu, W.; Li, Y.; Li, Z.; Stoporev, A.S.; Manakov, A.Y.; Vinokurov, V.A.; Li, T.; Semenov, A.P. Urea as a Green Thermodynamic Inhibitor of SII Gas Hydrates. *Chem. Eng. J.* **2022**, *429*, 132386. [CrossRef]
77. Semenov, A.P.; Mendgaziev, R.I.; Tulegenov, T.B.; Stoporev, A.S. Analysis of the Techniques for Measuring the Equilibrium Conditions of Gas Hydrates Formation. *Chem. Technol. Fuels Oils* **2022**, *58*, 628–636. [CrossRef]

International Journal of
Molecular Sciences

MDPI

Article

Excess Enthalpies for Binary Mixtures of the Reactive System Acetic Acid + n-Butanol + n-Butyl Acetate + Water: Brief Data Review and Results at 313.15 K and Atmospheric Pressure

Alexandra Golikova *, Anna Shasherina, Yuri Anufrikov, Georgii Misikov, Maria Toikka, Irina Zvereva and Alexander Toikka

Institute of Chemistry, St. Petersburg State University, Universitetskiy Prospect 26, Peterhof, Saint Petersburg 198504, Russia
* Correspondence: a.d.golikova@spbu.ru

Abstract: The data on molar excess enthalpies, H_m^E, for the binary mixtures acetic acid + n-butanol, acetic acid + n-butyl acetate and n-butanol + n-butyl acetate at 313.15 K and atmospheric pressure were obtained with use of the C80 isothermal mixing calorimeter (Setaram). The correlation of the data was carried out using the NRTL model and Redlich–Kister equation. A comparative analysis with the literature data on all available binary subsystems of the quaternary system was carried out. Other thermodynamic properties ($C_{p,m}^E$, S_m^E, $\Delta_{mix}S_m$, G_m^E and $\Delta_{mix}G_m$) of the binary systems were estimated using literature data and well-known formulas of classical thermodynamics.

Keywords: heat of mixing; calorimetry; n-butyl acetate; thermochemistry; Redlich–Kister equation; NRTL

1. Introduction

Features of molecular interaction in a solution can be described using thermodynamic characteristics, which primarily include the molar excess enthalpy. The excess enthalpies are the direct information about the energetic effects occurring between the molecules present in the mixtures. Such research is necessary not only for improving the solution theories and the development of appropriate databases, but also for describing the nature of the processes.

The data sets on enthalpy include important thermodynamic properties in many engineering applications and for the development of chemical engineering processes. Thus, excess thermodynamic properties in general quantify the deviations from ideality of the thermodynamic functions of mixtures.

Esters are promising substances for such studies. These substances are associated with a broad market in the chemical industry. One of the most common solvents of the paint and coatings industry, as well as a dehydrant in a number of industrial applications [1,2], is n-butyl acetate. In addition, n-butyl acetate is perfectly suitable as a solvent for the environment [3]. It is used in fragrances in cosmetic product, pharmaceutical and food industries [2]. One of the important properties of n-butyl acetate is its ability to act as an additive to gasoline and diesel fuel, and as biofuel [4,5].

Due to the possibility of using n-butyl acetate in various industries, systems with n-butyl acetate are among the most studied. Quite common are solubility, chemical equilibrium, vapor–liquid and liquid–liquid equilibrium and critical states studies for n-butyl acetate system synthesis [6–10] at deferent temperatures, e.g., 308.15, 318.15, 328.15 K, mostly at atmospheric pressure (excluding vapor–liquid experiments). In our work, we decided to pay attention to the thermal characteristics of this system. Despite the seemingly relatively large amount of data on excess enthalpies for the system acetic acid + n-butanol + n-butyl acetate + water, after a careful review of the literature, it turned out that the available results are not enough for a complete thermodynamic picture of the behavior of the system as a whole. For the convenience of perceiving the data on molar excess enthalpy available in the literature for the quaternary system investigated, including binary subsystems, we have collected them in Table 1.

Table 1. Literary data on molar excess enthalpy data for binary subsystems and quaternary system acetic acid + n-butanol + n-butyl acetate + water.

System	Temperature T/K	Comments	Literature
Acetic acid + n-butanol	298.15	Experimental data	[11]
	298.15, 318.15	Experimental data, graphs only	[12]
	313.15	Experimental data (NRTL)	[13]
Acetic acid + n-butyl acetate	313.15	Experimental data	[13]
Acetic acid + water	290.15, 293.15, 298.15, 303.15, 313.15, 323.15	Experimental data (fitted to the equation)	[14]
	293.15, 313.15	Calculated data (UNIQUAC model)	[15]
	293.15, 313.15	Calculated data (association model, +NRTL)	[16]
	293.15	Experimental data	[17]
	296.15–298.15	Experimental data (Redlich and Kister equation)	[18]
	298.15	Experimental data (fitted to the equation)	[19]
	298.15	Experimental data	[20]
	313.15	Experimental data (NRTL)	[13]
	313.15	Calculated (UNIFAC)	[21]
n-Butanol + n-butyl acetate	298.15	Experimental data (fitted to the equation)	[22]
	298.15	Experimental data (Redlich–Kister and SSF equations, UNIQUAC and NRTL models)	[23]
	298.15, 313.15	Experimental data (fitted to the equation)	[24]
	303.15	Experimental data (fitted to the equation)	[25]
	313.15	Experimental data (NRTL)	[13]
	353.15	Calculated data (Douglas–Avakian method)	[26]
n-Butanol + water	298.15	Experimental data (fitted to the equation)	[27]
	303.15	Experimental data	[28]
	303.15	Experimental data	[29]
	303.15, 328.15	Experimental data	[30]
	313.15	Experimental data (NRTL)	[13]
n-Butyl acetate + water	298.15	Experimental data (fitted to the equation)	[31]
	313.15	Experimental data (NRTL)	[13]
Acetic acid + n-butanol + n-butyl acetate + water	313.15	Experimental data (NRTL)	[13]

For the acetic acid + n-butanol system, experimental data in the temperature range 298.15–318.15 are presented by the authors of the articles [11–13]. Authors [12] have published the results as the function of excess enthalpies (H^E) on temperature and composition, and they fit the experimental data to the power series. In [13], the NRTL parameters were calculated for all experimental results, and it should be noted that temperatures indicated in the text differ from the ones in tables. Experimental data on excess enthalpies for the system acetic acid + n-butyl acetate were published only by the authors of [13]. There is a sufficient amount of data in the literature concerning the system acetic acid + water. Experimental results are presented in [17]. The authors of [14,18,19] fitted experimental data with different equations. The corresponding graphs are plotted in [14] for all temperatures except 298.15 K (the latter is listed in the table), but results for this temperature can be found in [32]. Data on excess enthalpy and computed the molar excess entropy values are listed in [20]. In [15,16,21], the data were obtained using methods to predict the molar excess enthalpies. In [13,22–25], the information on experimental molar excess enthalpy for the n-butanol + n-butyl acetate system was collected and results have also been correlated using various equations and models. The authors of [26] calculated the molar excess enthalpies and presented them in diagrams. The experimental molar excess enthalpies of mixing for the system n-butanol + water are represented with dots in graphs only in [27]. These data were also obtained by the authors of [28–30]. There is a fairly limited amount of data on molar excess enthalpy for the n-butyl acetate + water system in the literature. Experimental and calculated results are presented by the authors of [13,31]. Data on molar excess enthalpy for the acetic acid + n-butanol + n-butyl acetate + water system are presented in [13]. There are no data on the excess enthalpies for all ternary subsystems in the literature.

As a result of this study, we provide new experimental data on the excess enthalpies of mixing for binary subsystems of the acetic acid + n-butanol + n-butyl acetate + water system. Obtained data were correlated using the local composition NRTL model that is well known to be a good approach for correlating experimentally measured thermodynamic properties of various systems with different natures of deviation from the ideal solution. Due to the fact the NRTL model is a thermodynamically consistent model of local composition, it is widely used not only for the case of thermodynamic analysis of organic mixtures, but also within the chemical engineering field. We conducted a broad comparative analysis of the results available in the literature on this topic, as a result of which it was found that the data have a strong discrepancy among themselves. Additionally, thermodynamic functions, such as molar excess entropy, molar excess heat capacity, molar excess Gibbs energy, molar entropy of mixing and molar Gibbs energy of mixing, were also evaluated.

2. Results and Discussion

The new results for the binary systems acetic acid + n-butanol, acetic acid + n-butyl acetate and n-butanol + n-butyl acetate are given in Tables 2–4, respectively, and plotted in Figures 1–3.

Table 2. Molar excess enthalpies of the acetic acid + n-butanol system at 313.15 K [a] (J mol^{-1}), x—mole fraction of acetic acid.

x(AcOH)	H_m^E/J mol^{-1}	x(AcOH)	H_m^E/J mol^{-1}
0.0520	90.0	0.6092	475.2
0.1046	170.9	0.7082	443.8
0.2067	290.1	0.8063	369.4
0.3109	382.3	0.8563	307.7
0.4107	434.6	0.9032	229.3
0.5078	468.9	0.9521	128.0

[a] Standard uncertainties of temperature $u(T)$ = 0.05 K, mole fraction $u(x)$ = 0.0001 and molar excess enthalpies is $U_r(H_m^E)$ = 0.03 (95% level of confidence).

Table 3. Excess enthalpies of the acetic acid + n-butyl acetate system at 313.15 K [a] (J mol^{-1}), *x*—mole fraction of acetic acid.

x(AcOH)	H_m^E/J mol^{-1}	x(AcOH)	H_m^E/J mol^{-1}
0.0505	39.2	0.5957	48.7
0.1011	76.6	0.7023	21.0
0.2014	100.2	0.7995	0.3
0.2994	100.7	0.8010	−0.6
0.4004	94.3	0.8992	−11.5
0.5006	70.6	0.9496	−11.2

[a] Standard uncertainties of temperature $u(T) = 0.05$ K, mole fraction $u(x) = 0.0001$ and molar excess enthalpies is $U_r(H_m^E) = 0.03$ (95% level of confidence).

Table 4. Excess enthalpies of the n-butanol + n-butyl acetate system at 313.15 K [a] (J mol^{-1}), *x*—mole fraction of acetic acid.

x(BuOH)	H_m^E/J mol^{-1}	x(BuOH)	H_m^E/J mol^{-1}
0.0527	376.9	0.5984	1465.8
0.0980	652.5	0.6982	1283.8
0.1972	1104.0	0.8044	962.5
0.2965	1392.9	0.9006	549.7
0.4006	1512.3	0.9499	291.1
0.4967	1566.3		

[a] Standard uncertainties of temperature $u(T) = 0.05$ K, mole fraction $u(x) = 0.0001$ and molar excess enthalpies is $U_r(H_m^E) = 0.03$ (95% level of confidence).

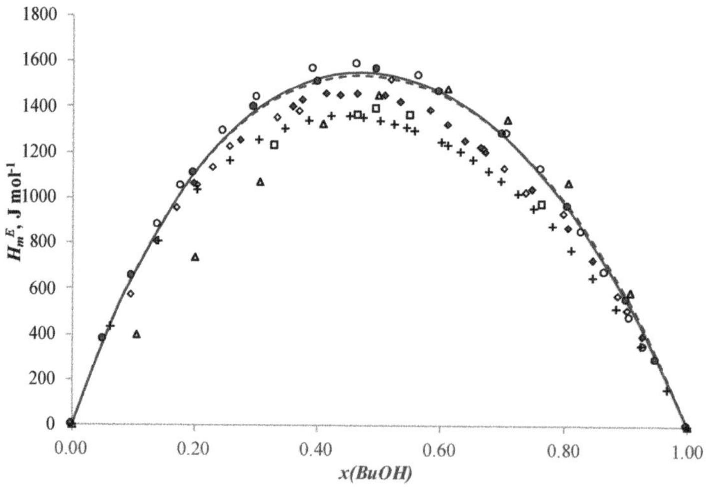

Figure 1. Molar excess enthalpies for binary system n-butanol + n-butyl acetate, (J mol^{-1}): the experimental solid circles (•) at 313.15 K, redrawn from Ref. [23] open up triangles (△) at 298.15 K, redrawn from Ref. [24] open rhomb (◊) at 298.15 K, redrawn from Ref. [22] solid rhomb (♦) at 298.15 K, redrawn from Ref. [25] plus (+) at 303.15 K, redrawn from Ref. [24] open rectangle (□) at 313.15 K, Ref. [13] open circles (○) at 313.15 K, calculated by Redlich-Kister equation (—) and NRTL model (- - -), *x*—mole fraction of n-butanol.

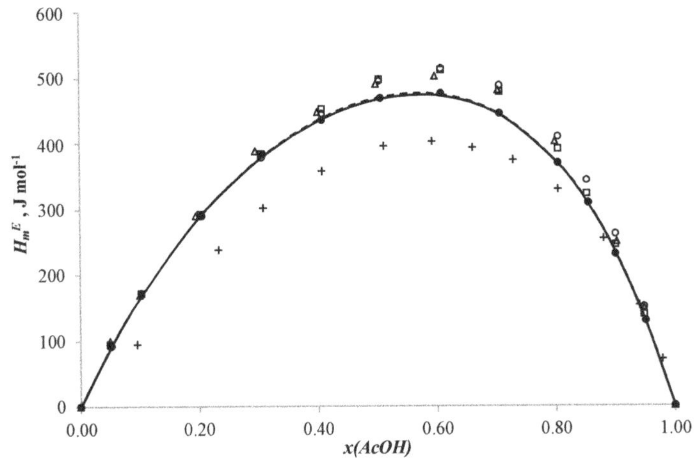

Figure 2. Molar excess enthalpies for binary system acetic acid + n-butanol, (J mol^{-1}): the experimental solid circles (•) at 313.15 K, redrawn from Ref. [12] open circles (○) at 298.15 K, redrawn from Ref. [11] plus (+) at 298.15 K, redrawn from Ref. [13] open up triangles (△) at 313.15 K, redrawn from Ref. [12] open rectangle (□) at 318.15 K, calculated by Redlich-Kister equation (—) and NRTL model (- - -), x—mole fraction of acetic acid.

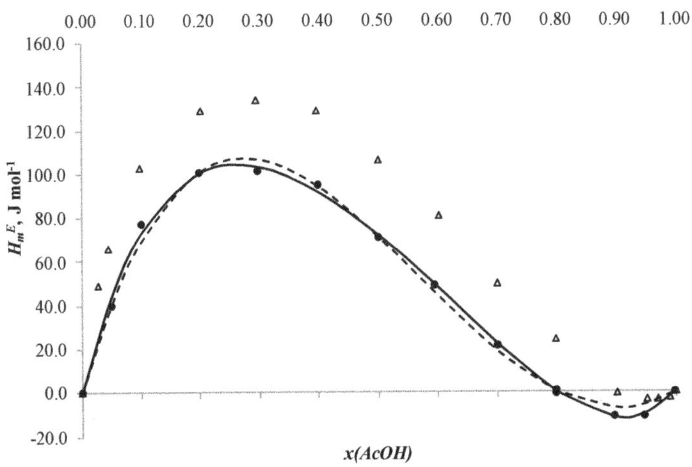

Figure 3. Molar excess enthalpies for binary system acetic acid + n-butyl acetate, (J mol^{-1}): experimental solid circles (•) at 313.15 K, redrawn from Ref. [13], open triangles (△) at 313.15 K, calculated by Redlich–Kister equation (—) and NRTL model (- - -), x—mole fraction of acetic acid.

The molar excess enthalpies for binary systems n-butanol + n-butyl acetate and acetic acid + n-butanol (Figures 1 and 2) are positive in all ranges of mole fraction. The maximums are H_m^E = 1556.3 J mol^{-1} at x_{BuOH} = 0.4967 in the n-butanol + n-butyl acetate system and H_m^E = 475.2 J mol^{-1} at x_{AcOH} = 0.6092 in the acetic acid + n-butanol system. The curves change almost symmetrically. This shape of the curves can be explained by the fact that at the beginning of the mixing processes the hydrogen bonds are stronger than at the end of the processes [33]. The dependence of the molar excess enthalpy on composition for the system acetic acid + butyl acetate (Figure 3) has an S-shape with small exothermic effect on the side of acetic acid. The minimum is $H_m^E = -11.2$ J mol^{-1} at x_{AcOH} = 0.9496 and the maximum is H_m^E = 100.7 J mol^{-1} at x_{BuOH} = 0.2994. Such a change in the shape of the curve of dependence of molar excess enthalpy on composition is explained by the breakage of hydrogen bonds in acetic acid and their formation in the final mixture between acid and ether [33,34].

As can be seen from Figure 1, the data for the n-butanol + n-butyl acetate system have some discrepancy. Figure 1 shows that the results at 298.15 K given in [22] (solid rhomb (♦)) coincide with the data obtained in [24] (open rhomb (◊)) at the same temperature only at low and high concentrations of substances. The points from [22] have lower values of heats in the area of equal component ratios. The results reported by the authors in [23] (open triangle (△)) are shifted to the right according to the schedule. The results presented in [25] (plus (+)) at 303.15K are lower than all the presented data obtained at 298.15 K, and these data lie close to the results provided by the authors of [24] (open rectangle (□)) at 313.15 K in the area of high n-butanol concentrations. The thermogram taken from [24] at 313.15 K lies much lower than all other thermograms obtained at this temperature. The data obtained in this work (solid circle (•)) are in good agreement with the data results from [13] (open circle (o)).

The dependence of molar excess enthalpy on the concentration of n-butanol for the acetic acid + n-butanol system is shown in Figure 2. The data obtained for this system have a strong discrepancy among themselves. The values of molar excess enthalpy obtained in [11] (plus (+)) at 298.15 K are much lower than the data obtained in [12] (open circle (○)) at this temperature. The thermograms plotted in [12] (open rectangle (□)) at 318.15 K lie very close to the data obtained in [13] (open triangle (△)) at 313.15 K. The results given in our work (solid circle (•)) at 313.15 K lie below the data obtained in [13] at the same temperature, which indicates data inconsistency.

3. Materials and Methods

3.1. Materials

The purities of acetic acid, n-butanol, n-butyl acetate and water used in the study were determined by a gas chromatography (GC) method with a Shimadzu GC-2010 plus including comparison of the measured refractive indexes, n_D, with the literature values [35]. The refractive indexes were determined with the IRF-454B2M refractometer ("KOMZ", Russia). The chemical specifications are summarized in Table 5.

Table 5. The purities of the chemicals.

CAS Number	Substance	Symbolic Name	Source	Purity, Mole Fraction	Purification Method	Analysis Technique
64-19-7	Acetic acid	AcOH	LenReactive (Russia)	0.998 [b]	None	GC [a]
71-36-3	n-Butanol	BuOH	Vekton (Russia)	0.995 [b]	None	GC [a]
123-86-4	n-Butyl acetate	BuOAc	Vekton (Russia)	0.999 [b]	None	GC [a]
7732-18-5	Water	H_2O	Bidistill	0.999 [b]	Distillation	GC [a]

[a] Gas chromatography. [b] Standard uncertainties of mole fraction $u(x) = 0.005$.

3.2. Molar Excess Enthalpy Measurements

The study of excess enthalpy of mixing in binary subsystems of the quaternary system acetic acid + n-butanol + n-butyl acetate + water was performed with use of the C80 isothermal mixing calorimeter (Setaram). The measurements were carried out at the temperature 313.15 ± 0.05 K with concentration step of 0.1 in the scale of mole fraction. A membrane mixing cell (material—stainless steel) was used for experimental measurements. At the beginning of the experiment, two pure components were separated from each other by a membrane. Upon achievement of calorimeter signal stabilization, the membrane was destroyed by special rod inside the cell, and the resulting heat flow was precisely measured by a Calvet sensor. The calculation of the heat effect of mixing was fulfilled with the use of the coefficient obtained from Joule effect calibration (electrical calibration). The standard system hexane + cyclohexane was used for testing of the apparatus and procedure. The experimental process was described in more detail in our previous works [34,36]. The relative uncertainty U_r for excess enthalpies is $U_r(H^E_m) = 0.03$.

The work Investigates the heats of mixing of the acetic acid + n-butanol reaction system. The esterification reaction proceeds very slowly in the absence of a strong acid as a catalyst, this is discussed in [11,36,37]. In addition, experiments were carried out in [34] for the system n-propanol + acetic acid, which has proven to have an extremely slow chemical reaction in the absence of a catalyst. In this regard, the effect of the reaction enthalpy on calorimetric measurements can be neglected.

3.3. Calculation

3.3.1. Redlich–Kister

Obtained experimental data were correlated with use of the Redlich–Kister equation in order to check their values for thermodynamic correspondence [38]:

$$H^E_{m,ij} = x_i x_j \sum_{k=0}^{N} A_k \left(x_i - x_j\right)^k \quad (1)$$

where x_i, x_j—mole fraction of component i and j, A_k—the adjustable parameters, N—the polynomial degree. The classical formula of the Redlich–Kister equation is used to correlate the symmetric dependencies for the systems acetic acid + n-butanol, acetic acid + n-butyl acetate and n-butanol + n-butyl acetate. The simulation results are shown in Figures 1–3, respectively.

To characterize the best description of the Redlich–Kister equation by a polynomial for a set of experimental points, the standard deviation parameter was used:

$$\sigma\left(H^E\right) = \sqrt{\frac{\sum_{i=1}^{n}\left(H^E_{calc,i} - H^E_{exp,i}\right)^2}{n - N}} \quad (2)$$

where n is number of experimental points, N is number of coefficients of the polynomial. The average calculation error was estimated using the formula

$$ARD(\%) = \frac{100}{n} \sum_{i=1}^{n} \frac{\left|H^E_{calc,i} - H^E_{exp,i}\right|}{\left|H^E_{exp,i}\right|} \quad (3)$$

Parameters of these equations, average relative deviation (ARD) and standard deviation ($\sigma(H^E)$) are presented in Table 6.

Table 6. Fitting parameters A_k, for Equations (1)–(3) for binary mixtures of acetic acid + n-butanol, acetic acid + n-butyl acetate and n-butanol + n-butyl acetate with ARD and standard deviations, $\sigma(H^E_m$, J mol^{-1}).

Coeff.	AcOH (*i*) + BuOH (*j*)	AcOH (*i*) + BuOAc (*i*)	BuOH (*i*) + BuOAc (*i*)
A_0	1867.1	287.6	6199.4
A_1	−416.4	448.1	624.2
A_2	547.9	60.1	952.0
A_3	−160.0	213.5	239.9
A_4		19.8	
ARD (%)	0.4	5	0.8
$\sigma(H^E)$	1.1	1.3	6

3.3.2. NRTL

The NRTL model [39] was used to approximate experimental results on the enthalpies of mixing binary systems:

$$H^E = x_1x_2\left[\frac{G_{21}\Delta g_{21}(x_1+x_2G_{21}-x_1\tau_{21}\alpha_{12})}{(x_1+x_2G_{21})^2}+\frac{G_{12}\Delta g_{12}(x_2+x_1G_{12}-x_2\tau_{12}\alpha_{12})}{(x_1+x_2G_{21})^2}\right], \quad (4)$$

where

$$G_{12} = exp(-\alpha_{12}\tau_{12}),\ G_{21} = exp(-\alpha_{12}\tau_{21}),\ \tau_{12} = \frac{\Delta g_{12}}{RT},\ \tau_{21} = \frac{\Delta g_{21}}{RT},\ G_{12} = exp\frac{\Delta g_{12}}{RT}, \quad (5)$$

where $\Delta g_{12} = g_{12} - g_{22}$ and $\Delta g_{21} = g_{21} - g_{11}$ are adjustable binary parameters, and α_{12} is the non-randomness parameter.

When finding the coefficients of the equation, the objective function, *OF*, was minimized:

$$OF = \sum_{i=1}^{n}\left(\frac{H^E_{calc,\,i} - H^E_{exp,\,i}}{H^E_{exp,\,i}}\right)^2 \quad (6)$$

where the summation is over all *i* data points.

Parameters of the NRTL model and *ARD* values are given in the Table 7 and plotted in Figures 1–3.

Table 7. Binary interaction parameters of the NRTL model.

i	j	α_{ji}	Δg_{ij}	Δg_{ji}	*ARD*/%
Acetic acid	n-Butanol	0.685	2425.3	1397.6	0.5
Acetic acid	n-Butyl acetate	0.024	−11,171.8	14,584.2	24
n-Butanol	n-Butyl acetate	0.213	4777.1	6170.1	1.0

3.3.3. Estimation of Thermodynamic Functions

For the system acetic acid + n-butanol, we made an attempt to perform estimation of other thermodynamic properties using literature data for H^E at different temperatures. It is a matter of common observation that all the correlations between enthalpy and other thermodynamic functions are true for the molar excess functions and molar functions of mixing as well, because of the fact that molar excess functions are the differences between values of a thermodynamic property of a real system and an ideal one:

$$H^E_m = H^{real}_m - H^{id}_m \quad (7)$$

At the same time, the functions of mixing are the differences between a function of a real system and a corresponding linear combination of functions of pure components:

$$\Delta_{mix}H_m = H^{real}_m - x_1H^*_{m,1} - x_2H^*_{m,2} \quad (8)$$

It is known that the excess enthalpy and molar enthalpy of mixing are equal.

$$\Delta_{mix}H_m = H^E_m \quad (9)$$

That is why, if there are H^E_m data for different temperatures, some other thermodynamic properties for systems can be estimated. For example, molar excess heat capacity is defined as following derivation:

$$C^E_{p,m} = \left(\frac{\partial H^E_m}{\partial T}\right)_{p,x} \quad (10)$$

Then, it can be shown with Legendre transformation that:

$$C_{p,m}^E = \left(\frac{\partial H_m^E}{\partial T}\right)_{p,x} = \left(\frac{\partial (G_m^E + TS_m^E)}{\partial T}\right)_{p,x} = \left(\frac{\partial G_m^E}{\partial T}\right)_{p,x} + S_m^E\left(\frac{\partial T}{\partial T}\right)_{p,x} + T\left(\frac{\partial S_m^E}{\partial T}\right)_{p,x} \tag{11}$$

Excess entropy is a derivation of molar excess Gibbs energy:

$$\left(\frac{\partial G_m^E}{\partial T}\right)_{p,x} = -S_m^E \tag{12}$$

Combining (11) and (12), we have:

$$C_{p,m}^E = T\left(\frac{\partial S_m^E}{\partial T}\right)_{p,x} \tag{13}$$

Then, at p,x = const we have the following equation for the molar excess entropy:

$$dS_m^E = \frac{C_{p,m}^E}{T}dT \tag{14}$$

After integration of (14), we have:

$$S_{T_2,m}^E - S_{T_1,m}^E = \int_{T_1}^{T_2} \frac{C_{p,m}^E}{T}dT \tag{15}$$

For the estimation of the excess entropy with Equation (15), the additional data on S_m^E at one temperature are necessary. They can be calculated from the fundamental equation for the molar excess Gibbs energy:

$$G_m^E = H_m^E - TS_m^E \tag{16}$$

$$S_m^E = \frac{H_m^E - G_m^E}{T} \tag{17}$$

H_m^E is experimentally measurable and G_m^E can be calculated from the vapor–liquid equilibrium data.

For the estimation of the molar excess heat capacity, we used our data at 313.15 K and the literature ones. Data for 298.15 were taken from [11] and for 318.15 from [12]. It was observed that, for the concentration range of acetic acid mole fraction between 0.1 and 0.9, the H_m^E temperature dependence tends to be linear. As the $H_m^E - x$ curves have extreme points and values of the H_m^E for the high and low mole fractions of acid are rather low, the influence of the experimental error becomes rather significant if the mole fraction of any component is extremely high. This can be seen in Figure 2 as well. Therefore, we assumed that the dependence of the excess enthalpy remains linear in those regions as well. The $H_m^E - T$ lines were plotted by processing literature H_m^E data with the least square method for every composition. Using linear approximation, we have the following equation for the H_m^E:

$$H_m^E = kT + b \tag{18}$$

According to (10), the molar excess heat capacity for every composition can be calculated as a slope of the $H_m^E - T$ line:

$$C_{p,m}^E = k \tag{19}$$

The uncertainty of the estimated $C^E_{p,m}$ value for each composition was determined as the uncertainty of the slope calculated with least square method:

$$\Delta k = \sqrt{\frac{1}{N-2}\left(\frac{S^2_H}{S^2_T} - k^2\right)} \tag{20}$$

where number of experimental points $N = 3$, $S^2_H = \langle H^{E\,2}_m \rangle - \langle H^E_m \rangle^2$, $S^2_T = \langle T^2 \rangle - \langle T \rangle^2$, $< \ldots >$ denote values averaged over experimental ones. The uncertainties of $C^E_{p,m}$ were estimated for every experimental composition (12 compositions) and then averaged over the composition:

$$\Delta C^E_{p,m} = \frac{\sum_{i=1}^{n} \Delta k_i}{n} \tag{21}$$

where number of experimental compositions $n = 12$. The uncertainty of the estimated $C^E_{p,m}$ was 0.5 JK^{-1} mol^{-1}. Due to the quite low value of $C^E_{p,m}$ itself (average value is 3.4 JK^{-1} mol^{-1}), the relative standard uncertainty is quite high (15%), however, because of the fact that literature data on molar excess enthalpy are rather limited, such a result tends to be reasonable.

The calculated $C^E_{p,m}$ values at 313.15 K are listed in Table 8 and shown in Figure 4. The dotted lines were obtained with polynomial approximation of the corresponding calculated values. The values of $C^E_{p,m}$ are given with accuracy of one tenth according to the estimated uncertainty. According to (9) and (10), it can be shown that $C^E_{p,m}$ and $\Delta_{mix}C_{p,m}$ are equal, therefore the curve for $\Delta_{mix}C_{p,m}$ should be the same. Molar excess functions and molar functions of mixing must tend to zero at the points corresponding to pure components, therefore in such an estimation of molar heat capacity, molar entropy seems to be inaccurate for a mole fraction of acid more than 0.9, and less than 0.1, which can indicate that the assumption of the linear H^E_m temperature shift becomes unreliable in extreme ranges. Consequently, estimation of all other thermodynamic properties remains reliable only within this concentration range.

Table 8. Estimated values of thermodynamic functions for the system acetic acid + n-butanol at 313.15 K [a].

x(AcOH)	$C^E_{p,m}$/JK^{-1} mol^{-1}	S^E_m/JK^{-1} mol^{-1}	$\Delta_{mix}S_m$/JK^{-1} mol^{-1}	G^E_m/J mol^{-1}	$\Delta_{mix}G_m$/J mol^{-1}
0.0520	3.4	2.4	4.2	-6.5×10^2	-11.8×10^2
0.1046	3.4	2.5	5.3	-6.1×10^2	-14.8×10^2
0.2067	3.7	2.6	6.9	-5.3×10^2	-18.6×10^2
0.3109	4.4	2.7	7.9	-4.6×10^2	-20.7×10^2
0.4107	4.9	2.6	8.4	-3.9×10^2	-21.5×10^2
0.5078	5.2	2.5	8.5	-3.3×10^2	-21.3×10^2
0.6092	4.9	2.3	8.1	-2.6×10^2	-20.0×10^2
0.7082	4.2	2.0	7.2	-1.9×10^2	-17.6×10^2
0.8063	3.0	1.6	5.8	-1.2×10^2	-14.0×10^2
0.8563	2.2	1.3	4.8	-0.9×10^2	-11.6×10^2
0.9032	1.3	0.9	3.6	-0.6×10^2	-8.8×10^2
0.9521	0.4	0.5	2.1	-0.3×10^2	-5.3×10^2

[a] Standard uncertainties of mole fraction $u(x) = 0.0001$, molar excess heat capacity is $u(C^E_{p,m}) = 0.5$ JK^{-1} mol^{-1}, molar excess entropy is $u(S^E_m) = 0.6$ JK^{-1} mol^{-1}, molar entropy of mixing is $u(\Delta_{mix}S_m) = 0.6$ JK^{-1} mol^{-1}, molar excess Gibbs energy is $u(G^E_m) = 1.0 \times 10^2$ J mol^{-1}, molar Gibbs energy of mixing is $u(\Delta_{mix}G_m) = 1.0 \times 10^2$ J mol^{-1} (95 % level of confidence).

The linear temperature dependence of the molar excess enthalpy leads to molar excess capacity remaining constant with the temperature shift which is a widespread approximation for a small temperature range. According to (15), the difference between S_m^E for two temperatures may be calculated as:

$$S_{T_2,m}^E - S_{T_1,m}^E = C_{p,m}^E \int_{T_1}^{T_2} \frac{dT}{T} = C_{p,m}^E \ln \frac{T_2}{T_1} \quad (22)$$

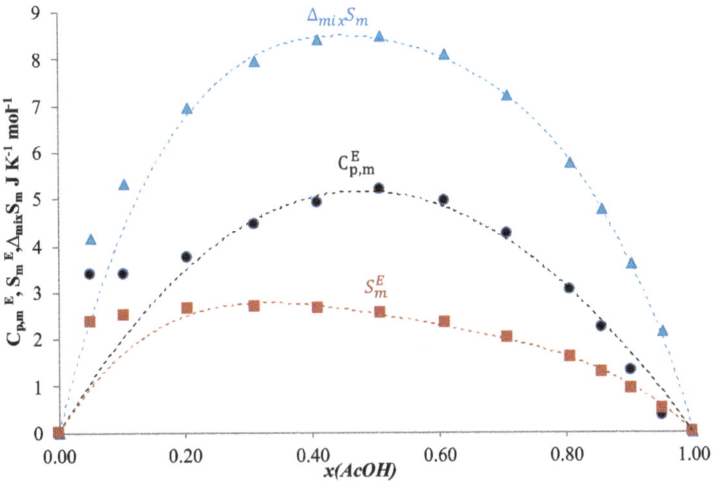

Figure 4. Calculated molar excess heat capacity $C_{p,m}^E$, JK^{-1} mol^{-1} (•), molar excess entropy S_m^E, JK^{-1} mol^{-1} (■) and molar entropy of mixing $\Delta_{mix}S_m$, JK^{-1} mol^{-1} (▲) for the system acetic acid + n-butanol at 313.15K. The dotted lines (- - -) are the approximation polynomials, x—mole fraction of acetic acid.

As the $\ln \frac{T_2}{T_1}$ is a constant for two defined temperatures, the shape of the $S_{T_2,m}^E - S_{T_1,m}^E$ dependence on composition will be similar to the $C_{p,m}^E - x$ curve.

We used literature data on vapor–liquid equilibria for the acetic acid + n-butanol system at 323.15 K [40] to calculate molar excess Gibbs energy at 323.15 K. We calculated the values of G_m^E using the Raoult law assuming that the vapor of n-butanol is an ideal gas, that was proved to be a good approximation in some works [41]. The deviations of the acetic acid vapor from the ideal gas were taken into account by using associated ideal gas model.

Then, we assumed that the discussed temperature range is small enough for extrapolating the linear dependence of H_m^E on T. Thus, we could estimate the value of H_m^E at 323.15 K. The S_m^E at 323.15 K was calculated using Equation (17). Equation (22) was applied to calculate molar excess entropy at 313.15K. Since we have experimental H_m^E values and estimated S_m^E values at 313.15 K, Equation (16) allows estimating molar excess Gibbs energy at the same temperature. Molar excess Gibbs energy of a binary system relates to the molar Gibbs energy of mixing by the following equation:

$$\Delta_{mix}G_m = G_m^E + RTx_1 \ln x_1 + RTx_2 \ln x_2 \quad (23)$$

Molar entropy of mixing was estimated by an equation similar to Equation (17). Calculated values of all the functions at 313.15 K are listed in Table 8. Dependence of estimated S_m^E and $\Delta_{mix}S_m$ on composition of the solution is shown in Figure 4. Figure 5 represents the dependence of estimated G_m^E and $\Delta_{mix}G_m$ on composition of the solution in comparison with the dependence of the H_m^E. The dotted lines are the approximation polynomials.

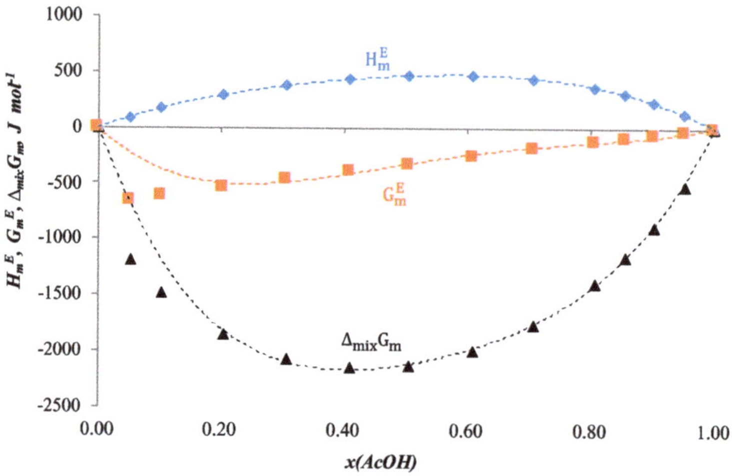

Figure 5. Molar excess enthalpy H_m^E, J mol^{-1} (♦) (our data), estimated molar excess Gibbs energy G_m^E, J mol^{-1} (■) and molar Gibbs energy of mixing $\Delta_{mix}G_m$, J mol^{-1} (▲) for the system acetic acid + n-butanol at 313.15K. The dotted lines (- - -) are the approximation polynomials, x—mole fraction of acetic acid.

For the estimation of uncertainties of calculated thermodynamic properties, we used the formula for uncertainty for a function of many variables:

$$\Delta f(X_1, X_2, \ldots, X_n) = \sqrt{\left(\frac{\partial f}{\partial X_1}\Delta X_1\right)^2 + \left(\frac{\partial f}{\partial X_2}\Delta X_2\right)^2 + \ldots + \left(\frac{\partial f}{\partial X_n}\Delta X_n\right)^2} \quad (24)$$

Using (17) and (20), we estimated the uncertainty of S_m^E at 323.15 K as a function of G_m^E, S_m^E and T. After that, we used calculated uncertainty of S_m^E at 323.15K, calculated by Equation (21) of the uncertainty of $C_{p,m}^E$ and Equations (22) and (24) to estimate the uncertainty of calculated value of S_m^E at 313.15K, which was 0.6 JK^{-1} mol^{-1}. The values of molar excess entropy are even lower than molar excess heat capacity (average value = 2.0 JK^{-1} mol^{-1}), which means that the relative standard uncertainty for calculated entropy will be even more (30%). A reasonable increase in the relative standard uncertainty is observed. Such an increase is caused by the accumulation of the uncertainty with further calculations. However, the main reason is the lack of experimental data presented in the literature.

Substituting (24) in (16), we evaluated the uncertainty of the calculated G_m^E, which was 100 J mol^{-1}. Due to the low contribution of uncertainty of the composition, the uncertainty of the molar Gibbs energy of mixing could be assumed to be equal to the uncertainty of the molar excess Gibbs energy. According to the average values of Gibbs energy (−308 J mol^{-1} for G_m^E and −1550 J mol^{-1} for $\Delta_{mix}G_m$), an increase in the relative standard uncertainty can be observed as well (32% G_m^E and 6% for $\Delta_{mix}G_m$). Such a low uncertainty for the molar Gibbs energy of mixing is caused by a high accuracy of the composition according to (23) and (24).

Although the relative uncertainties of the estimated functions are quite high, in the case of a limited data bank the estimation could be useful for qualitative assessments.

For the other two binary systems, the same estimation was not performed. For the system acetic acid + n-butyl acetate, only data at 313.15 K are available, so there is no possibility to calculate other properties. In the case of the n-butanol + n-butyl acetate binary system, literature data for different temperatures are rather sparse and the calculation seems to be unreliable.

According to Figures 4 and 5, it could be presumed that the estimations of the thermodynamic functions for the binary system acetic acid + n-butanol seem to be reliable at the concentration range $0.2 < x(\text{AcOH}) < 0.8$.

4. Conclusions

According to the results of the study, new experimental data on molar excess enthalpy in binary subsystems of the system with the synthesis of n-butyl acetate were obtained. All experimental values were verified for consistency by the Redlich–Kister equation. The *ARD* for system acetic acid + n-butanol is 0.4%, for system acetic acid + n-butyl acetate it is 5% and for system n-butanol + n-butyl acetate it is 0.8%. The molar excess enthalpies data were also correlated with the NRTL model. The *ARD* between the experimental molar excess enthalpies and the predicted ones for acetic acid + n-butanol is 0.5%, for acetic acid + n-butyl acetate it is 24% and for the system n-butanol + n-butyl acetate it is 1.0%. In addition, a broad comparative analysis of the literature data was carried out, which showed that the results available in the literature on the molar heats of mixing for the acetic acid + n-butanol + n-butyl acetate + water system are not consistent with each other and require additional confirmation. According to our data and literature ones, some thermodynamic properties ($C^E_{p,m}$, S^E_m, $\Delta_{mix}S_m$, G^E_m and $\Delta_{mix}G_m$) of the system acetic acid + n-butanol were estimated using equations of classical thermodynamics. The estimation seems to be reliable for quite a high concentration, however, for extremely high and low mole fractions of the acetic acid the error of the evaluation might be significant due to the lack of data that are necessary for the calculations. The knowledge of the presented excess thermodynamic properties is necessary in quantitative description of the deviations from ideality of the thermodynamic functions of solutions, which arise as a result of interaction between molecules through van der Waals forces, hydrogen bonds and others.

Author Contributions: Physicochemical investigation, analysis of molar excess enthalpies were carried out by A.G., A.S. and Y.A.; methodology, A.G. and M.T.; investigation, A.S. and Y.A.; calculation, G.M. and A.T.; writing—original draft preparation, A.G., G.M. and M.T.; writing—review and editing, A.G., A.T. and I.Z.; supervision, A.T. and I.Z.; funding acquisition, A.G. and A.T. All authors have read and agreed to the published version of the manuscript.

Funding: All experiments were carried out with financial support of the Russian Science Foundation (grant No. 22-73-00159). All calculations were carried out with financial support of the Russian Science Foundation (grant No. 21-13-00038).

Institutional Review Board Statement: Not applicable.

Informed Consent Statement: Not applicable.

Data Availability Statement: Not applicable.

Acknowledgments: The authors acknowledge the Resource Centre of Saint Petersburg State University. The investigations were carried out using the equipment of the Resource Centre of Thermogravimetric and Calorimetric Research. In commemoration of the 300th anniversary of St Petersburg State University's founding.

Conflicts of Interest: The authors declare no conflict of interest.

References

1. He, J.; Xu, B.; Zhang, W.; Zhou, C.; Chen, X. Experimental Study and Process Simulation of N-Butyl Acetate Produced by Transesterification in a Catalytic Distillation Column. *Chem. Eng. Process. Process Intensif.* **2010**, *49*, 132–137. [CrossRef]
2. Mangili, P.V.; Prata, D.M. Improvement of the Butyl Acetate Process through Heat Integration: A Sustainability-Based Assessment. *Chem. Eng. Process.-Process Intensif.* **2019**, *135*, 93–107. [CrossRef]
3. Hanika, J.; Kolena, J.; Smejkal, Q. Butylacetate via Reactive Distillation—Modelling and Experiment. *Chem. Eng. Sci.* **1999**, *54*, 5205–5209. [CrossRef]
4. Ali, S.H.; Al-Rashed, O.; Azeez, F.A.; Merchant, S.Q. Potential Biofuel Additive from Renewable Sources—Kinetic Study of Formation of Butyl Acetate by Heterogeneously Catalyzed Transesterification of Ethyl Acetate with Butanol. *Bioresour. Technol.* **2011**, *102*, 10094–10103. [CrossRef]

5. Dang, W.; Zhao, W.; Schoegl, I.; Menon, S. A Small-Volume, High-Throughput Approach for Surface Tension and Viscosity Measurements of Liquid Fuels. *Meas. Sci. Technol.* **2020**, *31*, 095301. [CrossRef]
6. Smirnov, A.; Samarov, A.; Toikka, M. Liquid–Liquid Equilibrium, Solubility, and Critical States in an Acetic Acid–*n*-Butyl Alcohol–*n*-Butyl Acetate–Water System at 328.15 K and 101.3 KPa: Topology of Phase Diagrams and NRTL Modeling. *J. Chem. Eng. Data* **2021**, *66*, 1466–1474. [CrossRef]
7. Smirnov, A.; Sadaeva, A.; Podryadova, K.; Toikka, M. Quaternary Liquid-Liquid Equilibrium, Solubility and Critical States: Acetic Acid–n-Butanol–n-Butyl Acetate–Water at 318.15 K and Atmospheric Pressure. *Fluid Phase Equilibria* **2019**, *493*, 102–108. [CrossRef]
8. Samarov, A.; Toikka, M.; Toikka, A. Liquid–Liquid Equilibrium and Critical States for the System Acetic Acid+n-Butanol+n-Butyl Acetate+Water at 308.15K. *Fluid Phase Equilibria* **2015**, *385*, 129–133. [CrossRef]
9. Toikka, A.M.; Samarov, A.A.; Toikka, M.A. Phase and Chemical Equilibria in Multicomponent Fluid Systems with a Chemical Reaction. *Russ. Chem. Rev.* **2015**, *84*, 378–392. [CrossRef]
10. Toikka, A.M.; Toikka, M.A.; Pisarenko, Y.A.; Serafimov, L.A. Vapor-Liquid Equilibria in Systems with Esterification Reaction. *Theor. Found. Chem. Eng.* **2009**, *43*, 129–142. [CrossRef]
11. Zhao, J.; Bao, J.; Hu, Y. Excess Molar Enthalpies of (an Alkanol + a Carboxylic Acid) at 298.15 K Measured with a Picker Calorimeter. *J. Chem. Thermodyn.* **1989**, *21*, 811–818. [CrossRef]
12. Haase, R.; Lorenz, R. Enthalpies of Mixing for Binary Liquid Mixtures of Monocarbonic Acids and Alcohols. *Z. Nat. A* **1985**, *40*, 947–951. [CrossRef]
13. Bender, M.; Heintz, A.; Lichtenthaler, R.N. Excess Enthalpy of the System Butyl Acetate + Butan-1-Ol + Acetic Acid + Water at 313 K and Ambient Pressure. *Thermochim. Acta* **1991**, *187*, 79–94. [CrossRef]
14. Haase, R.; Steinmetz, P.; Ducker, K.-H. Notizen:Mischungsenthalpien Beim Flüssigen System Wasser + Essigsäure/Heats of Mixing for the Liquid System Water + Acetic Acid. *Z. Nat. A* **1972**, *27*, 1527–1529. [CrossRef]
15. Nagata, I.; Miyamoto, K. Thermodynamics of Liquid Mixtures of Acids: Phase Equilibria and Excess Molar Enthalpies of Acetic Acid Mixtures with Active Non-Associating Components. *Thermochim. Acta* **1993**, *228*, 319–334. [CrossRef]
16. Nagata, I.; Tanimura, T. Thermodynamics of Associated Solutions. Vapour-Liquid Equilibrium and Excess Enthalpy for Acetic Acid-Polar Unassociated Component Mixtures. *Thermochim. Acta* **1990**, *168*, 241–252. [CrossRef]
17. Giraldo, L.; Rojas, J.L.; Cubillos, G.I.; Valencia, N.; Moreno, J.C. The transfer enthalpy of acetic acid from water to benzene by isoperaibolic calorimetry is determined. *Afinidad* **2009**, *66*, 227–231.
18. Campbell, A.N.; Gieskes, J.M.T.M. Heats of Mixing and Heat Capacities in the System: Acetic Acid–Chloroform–Water, at 25°. *Can. J. Chem.* **1965**, *43*, 1004–1011. [CrossRef]
19. Khurma, J.; Fenby, D. Calorimetric Study of Deuterium Isotope Effects in Water-Acetic Acid Systems. *Aust. J. Chem.* **1980**, *33*, 2349. [CrossRef]
20. Haase, R.; Pehlke, M. Thermodynamic Excess Functions for the Liquid System Water + Acetic Acid from Calorimetric Data. *Z. Nat.* **1977**, *32*, 507–510. [CrossRef]
21. Letyanina, I.; Tsvetov, N.; Toikka, A. Application of the UNIFAC Models for Prediction and Description of Excess Molar Enthalpies for Binary Mixtures of N-Propanol, Acetic Acid, n-Propyl Acetate, and Water. *Fluid Phase Equilibria* **2016**, *427*, 202–208. [CrossRef]
22. Bravo, R.; Paz-Andrade, M.I.; Kehiaian, H.V. Microcalorimetria de Mezclas Liquidas. Estudio Teorico-Experimental de Los Sistemas Ester + Alcohol. *Acta Cient. Compostel.* **1979**, *16*, 141–163.
23. Khanlarzadeh, K.; Iloukhani, H. Measurement and Correlation of Excess Molar Enthalpy of Binary Mixtures Containing Butyl Acetate + 1-Alkanols (C1–C6) at 298.15 K. *J. Solut. Chem.* **2014**, *43*, 1259–1269. [CrossRef]
24. Kehlen, H.; Fuhrmann, E. Die Exzeßenthalpien in Butylacetat-Butanol-Mischungen. *Z. Phys. Chem.* **1978**, *259*, 373–376. [CrossRef]
25. Yan, W.-D.; Lin, R.-S.; Yen, W.-H. Excess Enthalpies of Seven Binary Liquid Systems. *Thermochim. Acta* **1990**, *169*, 171–184. [CrossRef]
26. Suntsov, Y.K.; Kuprienko, G.Y.; Zadorozhnyi, V.P. Liquid-Vapor Equilibrium in n-Butanol-n-Alkyl Acetate Systems. *Russ. J. Phys. Chem. A* **2003**, *77*, 125–128.
27. Peeters, D.; Huyskens, P. Endothermicity or Exothermicity of Water/Alcohol Mixtures. *J. Mol. Struct.* **1993**, *300*, 539–550. [CrossRef]
28. Belousov, V.P.; Panov, M. Teploty Smesheniya Zhidkostej. XI. Entalpii Rastvoreniya H_2O v Spirtah. Heats of Mixing Liquids. XI. Enthalpies of Solution H_2O in Alcohols. *Vestn. Leningr. Univ. Fiz. Khim.* **1976**, *10*, 149–150. (In Russian)
29. Goodwin, S.R.; Newsham, D.M.T. A Flow Calorimeter for the Determination of Enthalpies of Mixing of Liquids: Enthalpies of Mixing of Water-I-n-Propanol and Water + n-Butanol at 30 °C. *J. Chem. Thermodyn.* **1971**, *3*, 325–334. [CrossRef]
30. Belousov, V.P.; Ponner, V. Teploty Smesheniya Zhidkostej. VI. Teploty Smesheniya v Binarnyh Sistemah n. Butilovyj Spirt-Voda, Izobutilovyj Spirt—Voda i Vtor-Butilovyj Spirt-Voda. V. Heats of Mixing Liquids. VI. Heats of Mixing in Binary Systems n-Butyl Alcohol—Water, Isobutyl Alcohol-Water and Sec-Butyl Alcohol–Water. *Vestn. Leningr. Univ. Fiz. Khim.* **1970**, *10*, 111–115. (In Russian)
31. Richon, D.; Viallard, A. Les Systemes Eau/Ester. I. Etude Calorimetrique Des Systemes Eau/Acetate d'alcoyle. *Can. J. Chem.* **1976**, *54*, 2584–2593. [CrossRef]
32. Christensen, J.J.; Hanks, R.W.; Izatt, R.M. *Handbook of Heats of Mixing*; Thermochemical Institute, Brigham Young University: Provo, UT, USA; Wiley: New York, NY, USA, 1982; ISBN 978-0-471-07960-6.
33. Ott, J.B.; Sipowska, J.T. Applications of Calorimetry to Nonelectrolyte Solutions. *J. Chem. Eng. Data* **1996**, *41*, 987–1004. [CrossRef]

34. Letyanina, I.; Tsvetov, N.; Zvereva, I.; Samarov, A.; Toikka, A. Excess Molar Enthalpies for Binary Mixtures of N-Propanol, Acetic Acid, and n-Propyl Acetate at 313.15K and Atmospheric Pressure. *Fluid Phase Equilibria* **2014**, *381*, 77–82. [CrossRef]
35. SciFinder-Chemical Abstracts Service. Available online: http://Scifinder.Cas.Org/ (accessed on 7 February 2023).
36. Golikova, A.; Tsvetov, N.; Anufrikov, Y.; Toikka, M.; Zvereva, I.; Toikka, A. Excess Enthalpies of the Reactive System Ethanol + Acetic Acid + Ethyl Acetate + Water for Chemically Equilibrium States at 313.15 K. *J. Therm. Anal. Calorim.* **2018**, *134*, 835–841. [CrossRef]
37. Nuryoto, N.; Amaliah, A.R.; Puspitasari, A.; Ramadhan, A.D. Study of Esterification Reaction between Ethanol and Acetic Acid Using Homogeneous and Heterogeneous Catalyst. *WCEJ Untirta* **2020**, *4*, 51. [CrossRef]
38. Redlich, O.; Kister, A.T. Algebraic Representation of Thermodynamic Properties and the Classification of Solutions. *Ind. Eng. Chem.* **1948**, *40*, 345–348. [CrossRef]
39. Renon, H.; Prausnitz, J.M. Local Compositions in Thermodynamic Excess Functions for Liquid Mixtures. *AIChE J.* **1968**, *14*, 135–144. [CrossRef]
40. Miyamoto, S.; Nakamura, S.; Iwai, Y.; Arai, Y. Measurement of Isothermal Vapor-Liquid Equilibria for Binary and Ternary Systems Containing Monocarboxylic Acid. *J. Chem. Eng. Data* **2001**, *46*, 1225–1230. [CrossRef]
41. Smirnova, N.A.; Morachevskij, A.G.; Storonkin, A.V. Issledovanie ravnovesiya zhidkost-par i zhidkost-zhidkost-par v sisteme n-propilovyj spirt-n-propilacetat-voda. Investigation of liquid-vapor and liquid-liquid-vapor equilibrium in the n-propyl alcohol-n-propyl acetate-water system. *Vestn. Leningr. Univ. Fiz. Khim.* **1963**, *22*, 97–104. (In Russian)

International Journal of
Molecular Sciences

MDPI

Article

Anomalous Concentration Dependence of Surface Tension and Concentration-Concentration Correlation Functions of Binary Non-Electrolyte Solutions

Carlo Carbone [1], Eduardo Guzmán [1,2] and Ramón G. Rubio [1,*]

[1] Departamento de Química Física, Facultad de Ciencias Químicas, Universidad Complutense de Madrid, Ciudad Universitaria s/n, 28040 Madrid, Spain
[2] Instituto Pluridisciplinar, Universidad Complutense de Madrid, Paseo de Juan XXIII 1, 28040 Madrid, Spain
* Correspondence: rgrubio@quim.ucm.es; Tel.: +34-91-3944123

Abstract: The concentration dependence of the surface tension of several binary mixtures of non-electrolytes has been measured at 298.15 K. The mixtures have been chosen since they presented a so-called "W-shape" concentration dependence of the excess constant pressure heat capacity and high values of the concentration-concentration correlation function. This behavior was interpreted in terms of the existence of anomalously high concentration fluctuations that resemble those existing in the proximities of critical points. However, no liquid-liquid phase separation has been found in any of these mixtures over a wide temperature range. In this work, we have extended these studies to the liquid-air interfacial properties. The results show that the concentration dependence of the surface tension shows a plateau and the mixing surface tension presents a "W-shape" behavior. To the best of our knowledge, this is the first time that this behavior is reported. The weak anomalies of the surface tension near a liquid-liquid critical point suggest that the results obtained cannot be considered far-from-critical effects. The usual approach of substituting the activity by the concentration in the Gibbs equation for the relative surface concentration has been found to lead to large errors and the mixtures to have a fuzzy and thick liquid/vapor interface.

Keywords: binary mixtures; critical point; fluctuations; non-electrolytes; surface tension

1. Introduction

In the last few years, several works showed that the excess constant pressure heat capacity, C_p^E, of some binary mixtures of non-electrolytes presented an anomalous concentration dependence [1–3]. More specifically, this dependence had a more or less pronounced W-shape. The molecules of the studied mixtures did not have any chemical structural correlation. A very extensive description of the fluctuation theory of mixtures and its consequences on the thermodynamic properties of liquid mixtures is provided by Matteoli and Mansoori [4]. Depolarized light scattering experiments performed by Rubio et al. [5] allowed for the determination of the so-called concentration-concentration correlation function, S_{cc}, which is the zero-wave vector limit of the structure factor. The authors pointed out that, for these binary mixtures, S_{cc} took values significantly higher than those corresponding to an ideal mixture, thus indicating that the concentration fluctuations were high. Therefore, at the molecular level, the composition was clearly non-random, thus differing from the overall one. Experimental evidence of the difference between bulk and local concentrations has been provided by different authors using spectroscopic methods [6–8], as well as computer simulations [9–11] and theoretical calculations [12–15]. Notably, S_{cc} is the inverse darken stability [16–19] and is closely related to the osmotic compressibility, $\left(\frac{\partial \mu_i}{\partial x_i}\right)_{T,P}$, with μ_i being the chemical potential. It is well known that $S_{cc}(0)$ values much higher than the one for ideal mixtures (max. $S_{cc}(0) = 0.25$ for $x = 0.5$) are typical of the systems near a critical point. However, the decrease in the temperature of the studied mixtures did not lead to any phase separation; therefore, far-from-critical effects are difficult to be considered for justifying the results found.

It is well known that in binary mixtures the osmotic compressibility gives the excess surface concentration of component i at the vapor/liquid interface. In most cases reported in the literature, an ideal behavior is defined as $\mu_i(T,P) = \mu_i^0(T,P = 1\text{bar}) + RT\ln x_i$, with x_i being the mole fraction of component i. However, Coto et al. [20] have shown that assuming an ideal behavior to calculate the osmotic compressibility may be inadequate for calculating surface relative adsorption, $\Gamma_{2,1}$, which is also observed for dilute water-alcohol mixtures. Using small angle neutron scattering, Almasi et al. [21] have shown that small errors in the excess Gibbs energy, G^E, may lead to large errors in $\left(\frac{\partial \mu_2}{\partial x_2}\right)_T$. Ritacco et al. [22,23] reported that aqueous trisiloxane solutions are an example of this problem from ellipsometry measurements. Similar conclusions were obtained from Llamas et al. [24,25] by combining surface tension and neutron reflectometry measurements.

Concentration fluctuations have been found to have strong consequences on other thermodynamic properties, even relatively far from any critical point, and for mixtures in which strong interactions, e.g., hydrogen bonds, exist. An example is the mixture that shows W-shape C_p^E curves. Similar conclusions have been reached for aqueous solutions [26–34].

The adsorbed amount of component 2 relative to component 1, $\Gamma_{2,1}$, is directly related to the composition dependence of the surface tension and chemical potential, μ_2, through the Gibbs equation:

$$\Gamma_{2,1} = -\left(\frac{\partial \gamma}{\partial \mu_2}\right)_T. \tag{1}$$

The difficulty in precisely calculating the chemical potential from experimental data, has most frequently led researchers to assume an ideal behavior for the mixtures, thus Equation (1) can be rewritten as [35]:

$$\Gamma_{2,1} = -\frac{1}{RT}x_2\left(\frac{\partial \gamma}{\partial x_2}\right)_T. \tag{2}$$

In general, it is clear that Equation (2) cannot be applied to concentrated mixtures, such as those studied in this work. Despite some approximate methods that have been proposed for calculating μ_2 [36], we have preferred to calculate $\Gamma_{2,1}$ using experimental data, thus not relying on any theoretical approximation.

More direct methods for measuring $\Gamma_{2,1}$ have been described in the literature. Ellipsometry [37,38] and neutron or X-ray reflectometry [39–43], as well as radiative decomposition of tritium-doped molecules [44] have provided reliable values of $\Gamma_{2,1}$ for binary mixtures, including dilute solutions of soluble surfactants. However, in some cases, e.g., for ellipsometry or reflectometry, it is necessary to assume the validity of a structural model for the interface, and then to fit the experimental data to the predictions of the model. Moreover, in the case of ellipsometry, the small difference in the refractive indexes of the components in each mixture does not allow one to obtain reliable values of the surface thickness.

Concentration fluctuations, as measured by the concentration-concentration correlation functions (zero wave-vector structure factor), are directly related to the osmotic compressibility by

$$S_{cc}^{-1}(0) = \frac{1}{RTx_1x_2}\left(\frac{\partial \mu_2}{\partial \ln x_2}\right)_T \tag{3}$$

A brief description of the relationship of $S_{cc}(0)$ with the integral theory of fluids is provided in Appendix A.

It is well known that $S_{cc}(0)$ diverges close to the critical points, whereas it takes a simple expression for ideal solutions, $S_{cc}(0) = x_1{\cdot}x_2$. The values of $S_{cc}(0)$ for the studied systems have been obtained from depolarized light scattering experiments, as described in detail in Appendix.

The overall goal of this manuscript is to study whether the systems that presented W-shape C_p^E curves also show an anomalous behavior on the concentration dependence of the surface tension. The reason for this possibility is that, in equilibrium, μ_i (bulk) = μ_i (interface); therefore, strong concentration fluctuations in the bulk might affect the surface properties. Furthermore, we will discuss the excess consequences for the relative surface concentration behavior.

2. Materials and Methods

All solvents used in this work were purchased from Sigma-Aldrich (Saint Louis, MO, USA) and were of the maximum purity available, always exceeding 99.8 wt%. However, in the case of chloronaphtalene, the commercial purity was below 99 wt% in order that it was purified by five melting-thawing cycles until no impurities were detected by gas chromatography. Moreover, constant surface tension and density were observed after two consecutive cycles.

The surface tension was measured using three different tensiometers: Pendant drop, Wilhelmy plate, and de Nouy ring. The first was home made and the ADSA analysis was used for the analysis of the drop profile, as in previous works [24,25]. The Wilhelmy plate measurements were performed using a Nima model 702 Langmuir balance (Biolin, Göteborg, Sweden) and the ring experiments were carried out in a Krüs K-10 tensiometer (Krüs GmbH, Hamburg, Germany). Since the presence of impurities may lead to a time dependent surface tension due to the adsorption of the impurities, we have performed adsorption kinetic measurements for three mole fractions, 0.2, 0.5, and 0.8, for 2-butanone + n-decane, as well as chloronaphtalene + 2,2,4-trimethylpentane. The equilibration time in the plate and ring tensiometers were less than 2 min when each component has been thermostated prior to mixing. This is in accordance with the low viscosity and high diffusion coefficients of the solvents. Our techniques do not allow us to follow a sufficiently detailed kinetic measurement for these short stabilization times. Moving barrier measurements using a Langmuir balance induced a 400% reduction in the initial area. No change in the surface tension was detected within the experimental uncertainty. Therefore, we can conclude that the compounds were sufficiently pure for the surface tension experiments.

The density measurements were carried out using an Anton Paar vibrating tube tensiometer DMA 4200-M (Anton Paar, Graz, Austria) with a precision of 3×10^{-5} g/cm^{-3}.

The light scattering experiments were performed using an ALV/CGS-03 precision photogoniometer that uses a ALV-5000 correlator and a Glass-Thompson polarizer (ALV-Laser Vertriebsgesellschaft GmbH, Langen, Germany). This equipment was used in previous works [45] and includes a Coherent laser beam (Coherent Inc., Santa Clara, CA, USA) working at 532 nm. The refractive index was measured using a Carl-Zeiss refractometer (Jena, Germany) at 546 nm, with a precision of 10^{-5} units. Since it is important to obtain reliable values of $S_{cc}(0)$, independent values of $\left(\frac{\partial n}{\partial x_2}\right)$ were obtained using a BI-DNDC differential refractometer (Brookhaven Instrument Corporation, Holtsville, NY, USA) with a precision of 10^{-3} units.

Milli-Q water was used for cleansing with a resistivity higher than 18 MΩ·cm and an organic material content lower than 6 ppm. The glassware was cleaned with piranha solution (caution: Piranha solutions are dangerous), then thoroughly rinsed with pure water and dried in a vacuum oven.

3. Results and Discussion

Figure 1 shows the mole fraction dependence of the surface tension for the different studied mixtures. It is important to note the existence of a more or less extended pseudoplateau, which to date, has not been described in the literature. Even for mixtures close to their lower critical solution temperature (LCST), the pseudoplateau is not more pronounced [46]. Nevertheless, it must be noted that the experimental γ vs. X data do not show any maximum, minimum, or inflection point where $\left(\frac{\partial \gamma}{\partial x}\right)_{T,P} = 0$, thus leading to $\Gamma_{2,1} = 0$.

This situation is found, e.g., in the case of aneotropic points, and has been discussed in detail in a book by Lyklema on binary mixtures of simpler compounds [47]. For completeness, we have included the results for the system methanol + methy-tertbutyl-ether (MTBE), in which no pseudoplateau is observed. Figure 1d shows that for the mixtures containing 1,2-dibromoethane, the plateau shifts to lower values of x_1 as the chain length of the hydrocarbon decreases. On the other hand, the decrease in T shifts the pseudoplateau toward higher values of x1. Furthermore, as an example, Figure 1e shows that by lowering the temperatures of the mixtures no phase separation was found, as for any of the other mixtures. It has not been possible to describe the behavior observed using any of the available theoretical models within the experimental precision. The shape of the isotherms is similar to the type 2 gas adsorption isotherms (BET model) if γ-γ_2 is plotted. Using this model as an empirical equation no good fit was obtained, even by using a maximum likelihood algorithm that includes the uncertainty in both dependent and independent variables [48]. Therefore, a pure empirical fitting approach has been used, while taking care that no numerical artifacts were introduced in the calculation of the derivative of γ vs. X.

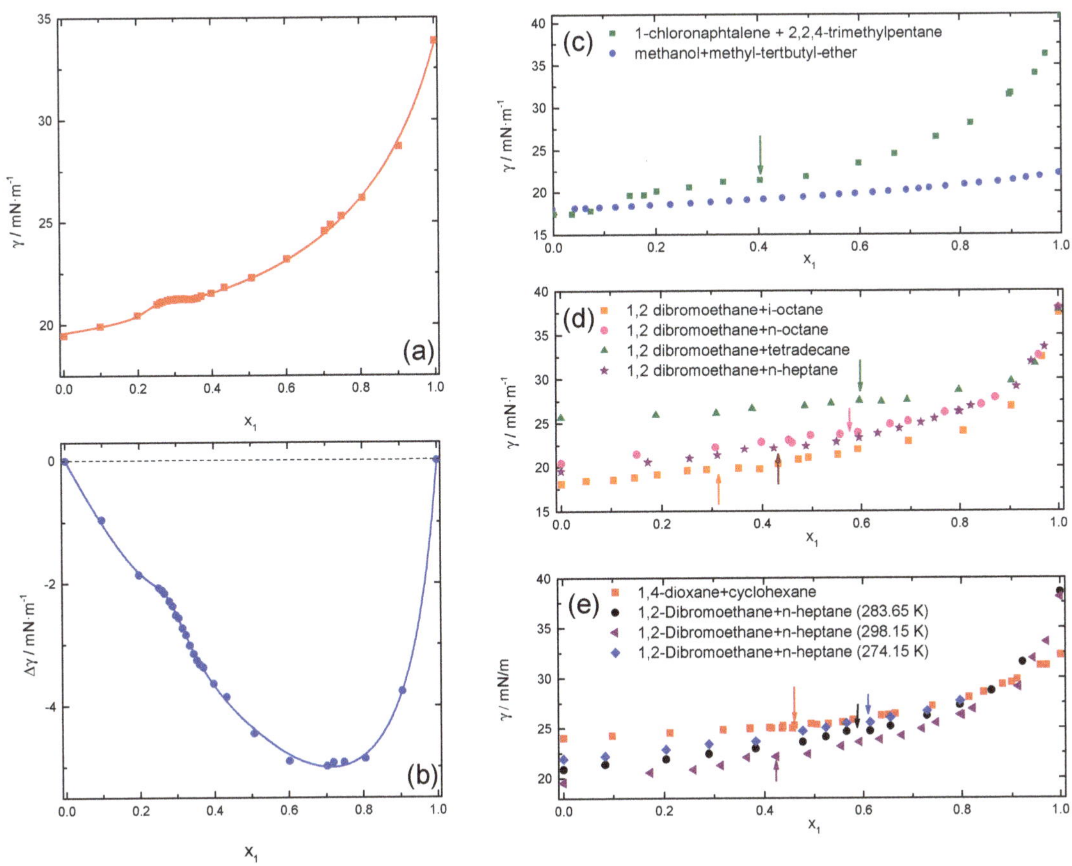

Figure 1. Composition of dependences of the surface tension (**a**) and mixing surface tension, $\Delta\gamma$, (**b**) at 298.15 K for 1,4-dichlorobutane + n-heptane mixtures. Symbols are the experimental results and curves that fit into Equation (4) with n = m = 3. In panels from (**c**–**e**), the composition dependences of the surface tension for the different studied systems are displayed. The insets identify the mixtures and the temperatures of the measurements, 298.15 K, unless the other value is specified. The arrows indicate the positions of the pseudo plateaus.

In order to calculate $\left(\frac{\partial\gamma}{\partial x_2}\right)_{T,P}$ with sufficiently high precision, both the surface tension and the mixing surface tension, $\Delta\gamma = \gamma - X_1{\cdot}\gamma_1 - X_2{\cdot}\gamma_2$, have been fitted to Padé approximants:

$$Y = \Lambda\frac{\sum_{i=0}^{n} A_i x_1^i}{1+\sum_{j=1}^{m} B_j x_1^j},. \tag{4}$$

where A_i and B_i are fitting parameters, $\Lambda = 1$ for the surface tension, γ, and $\Lambda = x_{1\times2}$ for the mixing surface tension, $\Delta\gamma$. Maximum values of 3 for n and m have been used for the sum of Equation (4), while ensuring that the uncertainties of the parameters used were less than 5%. For example, Figure 1a,b shows the experimental results and the corresponding fitted curves for the 1,4-dichlorobutane + n-heptane system, in which the plateau is more visible and has been experimentally mapped in more detail. It is clear that the fits are very good, and lead to randomly distributed residuals (not shown). For completeness, it is convenient to indicate that the experimental uncertainty on $S_{cc}(0)$ is about 3%, although it depends slightly on the difference in the refractive index of the compounds and their polarizability tensors.

As explained in Appendix A:

$$S_{cc}^{-1}(0) = \frac{1}{RTx_1x_2}\left(\frac{\partial\mu_2}{\partial \ln x_2}\right)_T, \tag{5}$$

thus leading to

$$\Gamma_{2,1}(x_2) = -x_1\frac{\left(\frac{\partial\gamma}{\partial x_2}\right)_{T,P}}{S_{cc}(x_2)}. \tag{6}$$

Figure 2 shows the composition dependence of $S_{cc}(0)$ for the different studied mixtures. Some of the results were already published in a previous paper [5]; however, in order to have a more precise value for $\Gamma_{2,1}$, we have performed additional experiments for better interpolation at the concentrations measured for γ. In all cases, it can be observed that the values are noticeably higher than the ideal mixture. Indeed, in some cases, the maxima of the curves are one order of magnitude higher, which as discussed below, has drastic consequences on the excess surface calculations.

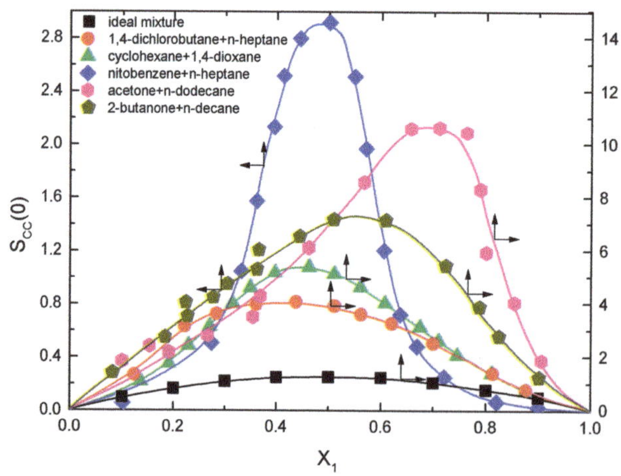

Figure 2. Composition dependence of the concentration-concentration correlation function at 298.15 K. The dashed-dotted line corresponds to an ideal mixture. Notice that some of the curves refer to the left ordinate axis, while the others refer to the right axis. Some of the data shown have been reproduced with permission from our previous publication [5].

Combining the values of $\left(\frac{\partial\gamma}{\partial x_2}\right)_{T,P}$ calculated from the experimental results and Equation (6), the excess surface concentration, $\Gamma_{2,1}$, can be calculated from the experimental results. Notably, since the $\left(\frac{\partial\gamma}{\partial x_2}\right)_{T,P}$ is an experimental result independent of whether the mixture has been considered ideal or real, the ratio $\frac{\Gamma_{2,1}^{real}(x_2)}{\Gamma_{2,1}^{ideal}(x_2)} = \frac{S_{cc}^{ideal}(x_2)}{S_{cc}^{real}(x_2)}$ will give a direct information of how far the ideal mixture assumption is reliable. Since $S_{cc}(0) = x_1 \cdot x_2$ for an ideal mixture, the experimental values shown in Figure 2 clearly point out that for the studied systems in this work, this approximation is not acceptable. An immediate conclusion is that the excess surface adsorption for these mixtures is smaller than the ideal mixtures except for the nitrobenzene + n-heptane in the high x_1 region, as shown in Figure 3. A possible explanation is that the high values of the concentration-concentration correlation function indicate that the concentration which is very close to a molecule of type 2, $X_{2,local}$, is quite different from the average macroscopic value, X_2. However, $X_{2,local}$ around one molecule of type 2 is not exactly the same as around another molecule of the same type at a different position at the same time. Ultimately, of course, the average value of $X_{2,local}$ for all the molecules of type 2 has to be equal to the average value of X_2.

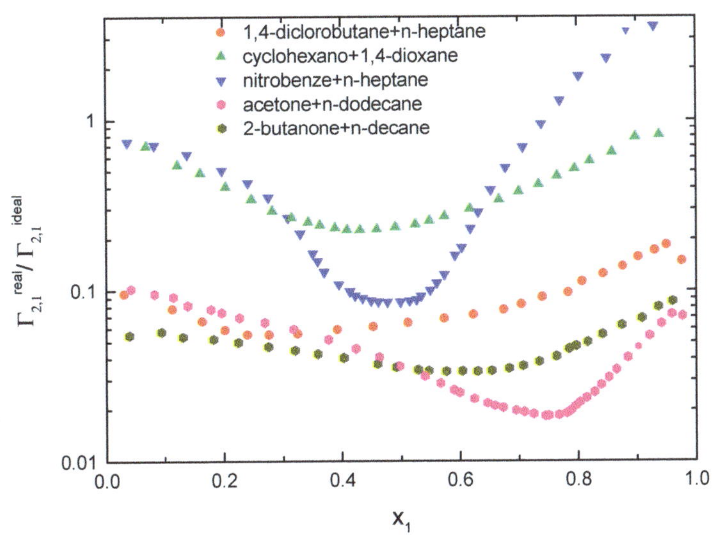

Figure 3. Ratio of the relative surface concentration, $\Gamma_{2,1}$, for real and ideal mixtures.

4. Conclusions

The surface tension and the concentration-concentration has been measured for non-electrolyte binary mixtures that presented a "W-shape" C_p^E-composition dependence. An anomalous concentration dependence has been found both for the surface tension and the mixing surface tension. The first shows a plateau at intermediate compositions, and the mixing surface tension shows a marked shoulder in the plateau region. The concentration fluctuations have been obtained from depolarized light scattering experiments. The values obtained are well above the values corresponding to the ideal mixture. The combination of the two types of results clearly shows that the most usual approximation for calculating the relative excess surface concentration, in which the activities of the two components are substituted by mole fractions, is not acceptable. Indeed, the ratio between the values of the excess concentration calculated using the ideal mixture approximation and those rigorously calculated can be up to two orders of magnitude for the studied mixtures. Considering the high values of the concentration-concentration correlation function for the mixtures, associated with strong concentration fluctuations, these results can be associated with a high thickness of the liquid/vapor interface.

Author Contributions: Conceptualization, R.G.R.; methodology, R.G.R. and C.C.; software, R.G.R. and E.G.; validation, R.G.R.; formal analysis, R.G.R.; investigation, R.G.R., C.C. and E.G.; resources, R.G.R.; data curation, R.G.R.; writing—original draft preparation, R.G.R.; writing—review and editing, R.G.R., C.C. and E.G.; visualization, E.G.; supervision, R.G.R.; project administration, R.G.R. and E.G.; funding acquisition, R.G.R. and E.G. All authors have read and agreed to the published version of the manuscript.

Funding: This work was funded by MICINN (grant number PID2019-106557GB-C21) and E.U. on the framework of the European Innovative Training Network-Marie Sklodowska-Curie Action NanoPaInt (grant agreement 955612).

Institutional Review Board Statement: Not applicable.

Informed Consent Statement: Not applicable.

Data Availability Statement: Data are available upon request.

Acknowledgments: A. Díaz-Pascual is acknowledged for performing some preliminary surface tension measurements.

Conflicts of Interest: The authors declare no conflict of interest. The funders had no role in the design of the study; in the collection, analyses, or interpretation of data; in the writing of the manuscript; or in the decision to publish the results.

Appendix A

The relationships between the osmotic compressibility and the concentration-concentration correlation functions are based on the integral theory of fluids derived by Kirkwood and Buff [49]. In the case of binary mixtures, the structure has been better discussed in terms of the partial structure factors, $a_{ij}\left(\vec{q}\right)$, where $\vec{q}$ is the wave-vector, and i and j refer to the two types of molecules. The partial structure factors are defined as

$$a_{ij}\left(\vec{q}\right) = 1 + \rho \int_0^{\infty} \left[g_{ij}(r) - 1\right] \exp\left(i \cdot \vec{q} \cdot \vec{r}\right) d\vec{r}, \tag{A1}$$

where $g_{ij}(r)$ is the radial distribution function of molecules i and j, $\vec{r}$ is the vector joining the center of the two molecules, and ρ is the density. In the exponential, $i = \sqrt{-1}$.

The long-wave limits of the partial structure factors, q→∞, are directly related to the so-called Kirkwood-Buff integrals [49,50]

$$G_{ij} = \int_0^{\infty} \left[g_{ij}(r) - 1\right] \cdot 4\pi r^2 dr, \tag{A2}$$

which are related to the osmotic compressibility [51] through

$$G_{ij} = RT\kappa_T - \frac{V_i \cdot V_j}{D \cdot V}\ ;\ G_{ii} = G_{ij} + \frac{\left[\frac{V_j}{D} - V\right]}{X_i}, \tag{A3}$$

where V_i and V_j denote the molar volume of component i and j, respectively, V is the molar volume of the mixture, κ_T is the isothermal compressibility, and D is given by

$$D = \frac{X_1 X_2}{S_{cc}(0)} = \frac{X_i}{RT}\left(\frac{\partial \mu_i}{\partial X_i}\right). \tag{A4}$$

Appendix B

$S_{cc}(0)$ is the form factor S(q) in the limit of zero wave vector, q = 0, thus giving long-range information of system fluctuations, which in the case of mixtures, correspond to concentration fluctuations, and thus on the local concentrations. Therefore, from the macroscopic point-of-view, $S_{cc}(0)$ is directly related to the osmotic compressibility of the system. It is known that the long-range fluctuations can be measured from diffraction experiments, typically using synchrotron radiation or depolarized light scattering.

The excess surface concentration at the air liquid/vapor interface is directly related to the osmotic compressibility by

$$\Gamma_{2,1}(x_2) = -\frac{1}{RT}\frac{\left(\frac{\partial \gamma}{\partial x_2}\right)_{T,P}}{\left(\frac{\partial \ln \mu_2}{\partial x_2}\right)_{T,P}} \Rightarrow \Gamma_{2,1}(x_2) = -x_1\frac{\left(\frac{\partial \gamma}{\partial x_2}\right)_{T,P}}{S_{cc}(x_2)}, \tag{A5}$$

where γ is the surface tension, x_2 is the molar fraction of component 2, a_2 is its activity, which is related to its chemical potential by $\mu_2(x_2, T, P) = \mu_2^0 + RT\ln a_2(x_2, T, P)$, and μ_2^0 is the standard chemical potential of component 2. The numerator of Equation (A1) only requires precise measurements of the γ composition dependence, which is not difficult. However, the denominator is more complicated from the experimental point-of-view, as explained below.

The Gibbs-Duhem equation allows one to write

$$\left[\frac{\partial \ln x_2}{\partial\left(\frac{\mu_2}{RT}\right)}\right]_{T,P} = -\frac{x_1}{x_2}\left[\frac{\partial\left(\frac{\mu_1}{RT}\right)}{\partial \ln x_1}\right]_{T,P}. \tag{A6}$$

the left-hand side of Equation (A2) is related to the Rayleigh ratio by

$$\left[\frac{\partial \ln x_2}{\partial\left(\frac{\mu_2}{RT}\right)}\right]_{T,P} = \frac{R_c}{R_{id}}, \tag{A7}$$

with R_c being defined by

$$R_i = R - R_c - R^*, \tag{A8}$$

where R is the Rayleigh ratio for non-polarized incident light, and was obtained from the scattered intensity, I, using toluene as reference according to $R = R_T\frac{I}{I_T}\left(\frac{n}{n_T}\right)^2$, the unsubscribed magnitudes referring to the mixture, and the ones with subscript T to the toluene n are the refractive index, R, which has isotropic and anisotropic contributions, R_i and R_a, respectively. In our case, R_i is the contribution that matters, and can be calculated through $R_i = R\frac{6-7\rho_u}{6+6\rho_u}$, with ρ_u being the depolarization factor, $\rho_u = \frac{I_H}{I_V}$, where I_H and I_V are the scattered intensities of horizontally and vertically polarized laser beams, respectively.

For binary liquid mixtures, R_i contains three contributions $R_i = R_d + R_c + R^*$, where R_d and R_c are due to separate fluctuations of the density and concentration, respectively. R^* gives the correlation between the two types of fluctuations by

$$R_d = \frac{\pi^2}{2\lambda_0^4}k_BT\kappa_T\left[\rho\left(\frac{\partial \varepsilon}{\partial \rho}\right)\right]^2$$

$$R_c = \frac{\pi^2}{2\lambda_0^4}k_BTV\frac{\left(\frac{\partial \varepsilon}{\partial x_2}\right)_T^2}{\left(\frac{\partial \mu_2}{\partial x_2}\right)_T} \tag{A9}$$

$$R^* = \frac{\pi^2}{2\lambda_0^4}k_BT\kappa_T\rho\left(\frac{\partial \varepsilon}{\partial \rho}\right)_{x,T}x_1x_2\left(\frac{\partial \varepsilon}{\partial x_2}\right)_T,$$

where λ_0 denotes the laser wavelength in the vacuum, k_B is the Boltzmann constant, ρ is the density, κT is the isothermal compressibility, V is the molar volume, and ε is the relative dielectric permittivity given, for infinite wavelength, by $\varepsilon = n_2$.

According to Segudovic and Dezelic [52], R^* can be expressed, in a first approximation, as

$$R^* = \frac{\pi^2}{2\lambda_0^4} k_B T \kappa_T \rho \left(\frac{\partial \varepsilon}{\partial \rho}\right)_{x,T} x_1 x_2 \left(\frac{\partial \varepsilon}{\partial x_2}\right)_T . \quad \text{(A10)}$$

Simplifying the previous equations for ideal mixtures, $a_2 = x_2$, the Rayleigh ratio is given by

$$R_{id} = \frac{\pi^2}{2\lambda_0^4} x_1 x_2 \frac{V}{N_A} \left(\frac{\partial \varepsilon}{\partial x_2}\right)^2 , \quad \text{(A11)}$$

where N_A is Avogadro's number. Therefore, for a real mixture, R_c can be expressed as $R_c = R_{id} \left[\frac{\partial(\frac{\mu_2}{RT})}{\partial \ln x_2}\right]^{-1}_{T,P}$, which combined with Equations (A4) and (A7) allow one to obtain $\partial(\frac{\mu_2}{RT})$ and, therefore, $S_{cc}(0)$.

For the density dependence of the relative permittivity one can use the Eykman equation, that assumes $\varepsilon \approx n^2$, which is a high-frequency approximation valid for our experiments, e.g., see Johnson and Smith [53]:

$$\rho \left(\frac{\partial \varepsilon}{\partial \rho}\right)_{x,T} = \frac{2n(n^2-1)(n+0.4)}{n^2+1} + 0.8n. \quad \text{(A12)}$$

References

1. Lainez, A.; Roux-Desgranges, G.; Grolier, J.P.E.; Wilhelm, E. Mixtures of alkanes with polar molecules showing internal rotation: An unusual composition dependence of CpE of 1,2-dichloroethane + an n-alkane. *Fluid Phase Equilibria* **1985**, *20*, 47–56. [CrossRef]
2. Saint Victor, M.-E.; Patterson, D. The W-shaped concentration dependence of CpE and solution non-randomness: Systems approaching the UCST. *Thermochim. Acta* **1990**, *159*, 177–185. [CrossRef]
3. Saint-Victor, M.-E.; Patterson, D. The w-shape concentration dependence of CEp and solution non-randomness: Ketones + normal and branched alkanes. *Fluid Phase Equilibria* **1987**, *35*, 237–252. [CrossRef]
4. Matteoli, E.; Mansoori, G.A. (Eds.) *Fluctuation Theory of Mixtures*; Taylor & Francis: Boca Raton, FL, USA, 1990.
5. Rubio, R.G.; Cáceres, M.; Masegosa, R.M.; Andreolli-Ball, L.; Costas, M.; Patterson, D. Mixtures with "w-Shape" C^{Ep} curves. A light scattering study. *Ber. Bunsenges. Phys. Chem.* **1989**, *93*, 48–56. [CrossRef]
6. Mello, C.; Mello, T.; Sevéri, E.; Coelho, L.; Ribeiro, D.; Marangoni, A.; Poppi, R.J.; Noda, I. Microstructures formation in a seemingly ideal homogeneous mixture of ethanol and methanol: An experimental evidence and two-dimensional correlation spectroscopy approach. *J. Chem. Phys.* **2009**, *131*, 084501. [CrossRef]
7. Phillips, D.J.; Brennecke, J.F. Spectroscopic measurement of local compositions in binary liquid solvents and comparison to the NRTL equation. *Ind. Eng. Chem. Res.* **1993**, *32*, 943–951. [CrossRef]
8. Shulgin, I.L.; Ruckenstein, E. Excess around a central molecule with application to binary mixtures. *Phys. Chem. Chem. Phys.* **2008**, *10*, 1097–1105. [CrossRef]
9. Požar, M.; Perera, A. Evolution of the micro-structure of aqueous alcohol mixtures with cooling: A computer simulation study. *J. Mol. Liquids* **2017**, *248*, 602–609. [CrossRef]
10. Sarkar, S.; Maity, A.; Chakrabarti, R. Microscopic structural features of water in aqueous–reline mixtures of varying compositions. *Phys. Chem. Chem. Phys.* **2021**, *23*, 3779–3793. [CrossRef]
11. Perera, A.; Kežić, B. Fluctuations and micro-heterogeneity in mixtures of complex liquids. *Faraday Discuss.* **2013**, *167*, 145–158. [CrossRef]
12. Chen, H.-F.; Li, J.-T.; Gu, F.; Wang, H.-J. Kirkwood-Buff integrals for hard-core Yukawa fluids. *Eur. Phys. J. E* **2017**, *40*, 93. [CrossRef]
13. Bentenitis, N.; Cox, N.R.; Smith, P.E. A Kirkwood—Buff Derived Force Field for Thiols, Sulfides, and Disulfides. *J. Phys. Chem. B* **2009**, *113*, 12306–12315. [CrossRef]
14. Debenedetti, P.G. The statistical mechanical theory of concentration fluctuations in mixtures. *J. Chem. Phys.* **1987**, *87*, 1256–1260. [CrossRef]
15. Shimizu, S.; Matubayasi, N. Statistical thermodynamic foundation for mesoscale aggregation in ternary mixtures. *Phys. Chem. Chem. Phys.* **2018**, *20*, 13777–13784. [CrossRef]
16. Nishikawa, K.; Hayashi, H.; Iijima, T. Temperature Dependence of the Concentration Fluctuation, the Kirkwood-Buff Parameters, and the Correlation Length of fed-Butyl Alcohol and Water Mixtures Studied by Small-Angle X-ray Scattering. *J. Phys. Chem.* **1989**, *93*, 6559–6565. [CrossRef]
17. Hayashi, H.; Nishikawa, K.; Iijima, T. Small-Angle X-ray Scattering Study of Fluctuations in 1-Propanol-Water and 2-Propanol-Water Systems. *J. Phys. Chem.* **1990**, *90*, 8334–8338. [CrossRef]

18. Hayashi, H.; Morita, T.; Nishikawa, K. Interpretation of correlation length by small-angle X-ray scattering experiments on fluids near critical point. *Chem. Phys. Lett.* **2009**, *471*, 249–252. [CrossRef]
19. Tsuchiya, Y. Elucidation of structural changes and concentration fluctuations in binary mixtures using new thermodynamic relations. *J. Phys. Condens. Matter* **1999**, *11*, 593. [CrossRef]
20. Coto, B.; Mößner, F.; Pando, C.; Rubio, R.G.; Renuncio, J.A.R. Bulk and surface properties for the methanol–1,1-dimethylpropyl methyl ether and methanol–1,1-dimethylethyl methyl ether systems. *J. Chem. Soc. Faraday Trans.* **1996**, *92*, 4435–4440. [CrossRef]
21. Almasi, M.; Khodamoradpoor, M. Study of molecular interactions in binary mixtures by molecular diffusion, thermal diffusion, Soret effect, and separation ratio. *J. Mol. Liquids* **2021**, *335*, 116545. [CrossRef]
22. Ritacco, H.A.; Fainerman, V.B.; Ortega, F.; Rubio, R.G.; Ivanova, N.; Starov, V.M. Equilibrium and dynamic surface properties of trisiloxane aqueous solutions. Part 2. Theory and comparison with experiment. *Colloids Surf. A* **2010**, *365*, 204–209. [CrossRef]
23. Ritacco, H.A.; Ortega, F.; Rubio, R.G.; Ivanova, N.; Starov, V.M. Equilibrium and dynamic surface properties of trisiloxane aqueous solutions: Part 1. Experimental results. *Colloids Surf. A* **2010**, *365*, 199–203. [CrossRef]
24. Llamas, S.; Fernández-Peña, L.; Akanno, A.; Guzmán, E.; Ortega, V.; Ortega, F.; Csaky, A.G.; Campbell, R.A.; Rubio, R.G. Towards understanding the behavior of polyelectrolyte–surfactant mixtures at the water/vapor interface closer to technologically-relevant conditions. *Phys. Chem. Chem. Phys.* **2018**, *20*, 1395–1407. [CrossRef]
25. Llamas, S.; Guzmán, E.; Akanno, A.; Fernández-Peña, L.; Ortega, F.; Campbell, R.A.; Miller, R.; Rubio, R.G. Study of the Liquid/Vapor Interfacial Properties of Concentrated Polyelectrolyte–Surfactant Mixtures Using Surface Tensiometry and Neutron Reflectometry: Equilibrium, Adsorption Kinetics, and Dilational Rheology. *J. Phys. Chem. C* **2018**, *122*, 4419–4427. [CrossRef]
26. Pandey, J.D.; Verma, R. Inversion of the Kirkwood–Buff theory of solutions: Application to binary systems. *Chem. Phys.* **2001**, *270*, 429–438. [CrossRef]
27. Blanco, M.A.; Sahin, E.; Li, Y.; Roberts, C.J. Reexamining protein-protein and protein-solvent interactions from Kirkwood-Buff analysis of light scattering in multi-component solutions. *J. Chem. Phys.* **2011**, *134*, 225103. [CrossRef]
28. Matteoli, E.; Lepori, L. Kirkwood–Buff integrals and preferential solvation in ternary non-electrolyte mixtures. *J. Chem. Soc. Faraday Trans.* **1995**, *91*, 431–436. [CrossRef]
29. Matteoli, E.; Mansoori, G.A. A simple expression for radial distribution functions of pure fluids and mixtures. *J. Chem. Phys.* **1995**, *103*, 4672–4677. [CrossRef]
30. Matteoli, E. A Study on Kirkwood−Buff Integrals and Preferential Solvation in Mixtures with Small Deviations from Ideality and/or with Size Mismatch of Components. Importance of a Proper Reference System. *J. Phys. Chem. B* **1997**, *101*, 9800–9810. [CrossRef]
31. Wilcox, D.S.; Rankin, B.M.; Ben-Amotz, D. Distinguishing aggregation from random mixing in aqueous t-butyl alcohol solutions. *Faraday Discuss.* **2013**, *167*, 177–190. [CrossRef]
32. Chakraborty, S.; Sehanobish, E.; Sarkar, M. A traditional painkiller as a probe for microheterogeneity in 1-propanol–water mixtures. *Chem. Phys. Lett.* **2010**, *501*, 118–122. [CrossRef]
33. Oh, K.-I.; Baiz, C.R. Molecular heterogeneity in aqueous cosolvent systems. *J. Chem. Phys.* **2020**, *152*, 190901. [CrossRef] [PubMed]
34. Marcus, Y. Preferential solvation in mixed solvents. Part 5.—Binary mixtures of water and organic solvents. *J. Chem. Soc. Faraday Trans.* **1990**, *86*, 2215–2224. [CrossRef]
35. Ramírez-Verduzco, L.F.; Romero-Martínez, A.; Trejo, A. Prediction of the surface tension, surface concentration, and the relative Gibbs adsorption isotherm of binary liquid systems. *Fluid Phase Equilibria* **2006**, *246*, 119–130. [CrossRef]
36. Bagheri, A.; Rafati, A.A.; Tajani, A.A.; Borujeni, A.R.A.; Hajian, A. Prediction of the Surface Tension, Surface Concentration and the Relative Gibbs Adsorption Isotherm of Non-ideal Binary Liquid Mixtures. *J. Solut. Chem.* **2013**, *42*, 2071–2086. [CrossRef]
37. Heidel, B.; Findenegg, G.H. Ellipsometric study of the surface of a binary liquid mixture near a critical solution point. *J. Phys. Chem.* **1984**, *88*, 6575–6579. [CrossRef]
38. Privat, M.; Bennes, R.; Tronel-Peyroz, E.; Douillard, J.-M. Ellipsometry and adsorption: The determination of isotherms and the adsorbed layer thickness and fluctuations of the composition in the liquid-vapor interface. *J. Colloid Interface Sci.* **1988**, *121*, 198–207. [CrossRef]
39. Subramanian, D.; Boughter, C.T.; Klauda, J.B.; Hammouda, B.; Anisimov, M.A. Mesoscale inhomogeneities in aqueous solutions of small amphiphilic molecules. *Faraday Discuss.* **2013**, *167*, 217–238. [CrossRef]
40. Shulgin, I.; Ruckenstein, E. Kirkwood−Buff Integrals in Aqueous Alcohol Systems: Aggregation, Correlation Volume, and Local Composition. *J. Phys. Chem. B* **1999**, *103*, 872–877. [CrossRef]
41. Nishikawa, K.; Morita, T. Small-Angle X-ray-Scattering Study of Supercritical Trifluoromethane. *J. Phys. Chem. B* **1997**, *101*, 1413–1418. [CrossRef]
42. Morita, T.; Nishikawa, K. Fluctuations in density and concentration of methanol–water mixtures at 7 MPa and 373, 423 K studied by small-angle X-ray scattering. *Chem. Phys. Lett.* **2004**, *389*, 29–33. [CrossRef]
43. Nishikawa, K.; Kasahara, Y.; Ichioka, T. Inhomogeneity of Mixing in Acetonitrile Aqueous Solution Studied by Small-Angle X-ray Scattering. *J. Phys. Chem. B* **2002**, *106*, 693–700. [CrossRef]
44. Kopf, S.; Bourriquen, F.; Li, W.; Neumann, H.; Junge, K.; Beller, M. Recent Developments for the Deuterium and Tritium Labeling of Organic Molecules. *Chem. Rev.* **2022**, *122*, 6634–6718. [CrossRef] [PubMed]
45. Hernández, M.a.P.; Ortega, F.; Rubio, R.G. Crossover critical phenomena in an aqueous electrolyte solution: Light scattering, density and viscosity of the 3-methylpyridine+water+NaBr system. *J. Chem. Phys.* **2003**, *119*, 4428–4436. [CrossRef]

46. Díez-Pascual, A.; Ortega, F.; Crespo-Colín, A.; Compostizo, A.; Monroy, F.; Rubio, R.G. Concentration Fluctuations and Surface Adsorption in Hydrogen-Bonded Mixtures. *J. Phys. Chem. B* **2004**, *108*, 10019–10024. [CrossRef]
47. Lyklema, J. *Fundamentals of Interface and Colloid Science*; Academic Press: Cambridge, MA, USA, 2000.
48. Rubio, R.G.; Renuncio, J.A.R.; Peña, M.D. Regression of vapor-liquid equilibrium data based on application of the maximum-likelihood principle. *Fluid Phase Equilibria* **1983**, *12*, 217–234. [CrossRef]
49. Kirkwood, J.G.; Buff, F.P. The Statistical Mechanical Theory of Solutions. I. *J. Phys. Chem.* **1951**, *19*, 774–777. [CrossRef]
50. McQuarrie, D.A. *Statistical Mechanics*; Harper & Row: New York, NY, USA, 1973.
51. Gray, C.G.; Gubbins, K.E. *Theory of Molecular Fluids*; Oxford University Press: London, UK, 1984.
52. Segudovic, N.; Dezelic, G. Light Scattering in Binary Liquid Mixtures. I. Isotropic Scattering. *Croat. Chem. Acta* **1973**, *45*, 385–406.
53. Johnson, B.L.; Smith, J. *Light Scattering from Polymer Solutions*; Huglin, M.B., Ed.; Academic Press: New York, NY, USA, 1972.

International Journal of
Molecular Sciences

MDPI

Article

Solubility of Amino Acids in the Eutectic Solvent Constituted by Sodium Acetate Trihydrate and Urea and in Its Mixture with Water

Cristina Gallego, Héctor Rodríguez and Ana Soto *

Cross-Disciplinary Research Center in Environmental Technologies (CRETUS), Department of Chemical Engineering, Universidade de Santiago de Compostela, E-15782 Santiago de Compostela, Spain
* Correspondence: ana.soto@usc.es

Abstract: Industrial fish and aquaculture processing leads to the generation of a huge quantity of by-products, whose accumulation and mismanagement involve serious environmental consequences as well as high economic losses. Taking advantage of these residues as a source of added-value compounds must be a priority in a circular economy. This work is a preliminary study to analyze the possibility of using the eutectic mixture of urea and sodium acetate trihydrate as a solvent for collagen extraction. To that end, the solid–liquid equilibrium of the system was determined in order to define the exact composition and temperature of the eutectic. The solubility in this solvent of the main amino acids that constitute fish collagen was studied at several temperatures and atmospheric pressure. At 308.15 K, solubilities of the major constituents of the target protein, namely L-proline, trans-4-hydroxy-L-proline, and glycine, were 0.19, 0.16, and 0.12 (mass fraction), respectively. These values increased with temperature. Dilution with water (50 wt%) allowed operation at lower temperature and led to an increase in the solubilities. The van 't Hoff model was satisfactorily used to correlate the experimental data and to calculate apparent properties of dissolution. All the dissolution processes studied herein are endothermic, non-spontaneous, and enthalpy-driven. Both the eutectic and its mixture with water are promising solvents for the design of an environmentally benign process for collagen extraction.

Keywords: collagen; amino acids; solubility; eutectic; urea; sodium acetate trihydrate

1. Introduction

In order to meet the growing demand for food, worldwide capture fisheries and aquaculture production have doubled in the last three decades, reaching a value of 180 Mt/year. This expansion in the size of fish-processing industrial capacity is leading concomitantly to an increase in the amount of the low-value by-products generated, which may represent up to 70% of the total mass processed, depending on factors such as the species, the final product, or the type of processing [1]. Such by-products are mainly composed of heads, skin, viscera, bones, or scales, and they have been traditionally diverted to the production of fishmeal, animal feed, or fertilizer, or they have even been simply discarded as waste. In addition to the direct impact on the environment and thus on human health, this strategy also involves tacitly high economic losses, since the aforementioned by-products have the potential to be a valuable source of bioactive compounds [2]. Therefore, in the context of sustainable management of natural resources and rapid growth in public awareness about sustainability, the utilization of fish by-products as a raw material to recover valuable components is emerging as a topic of great interest [3].

One of the most interesting compounds contained in fish by-products is collagen, which is the main structural protein in extracellular connective tissues of the fish body, constituting 15–30% of the total protein content. Although collagen has been classified into at least 28 types based on molecular weight, amino acid composition, or functionality, the most common is the so-called collagen type I, mainly composed of three amino acids: glycine, proline, and hydroxyproline. There is a high demand for collagen type I in biomedicine, pharmaceutical, cosmetic, food, and biomaterial industries due to its biodegradability, biocompatibility, and antigenicity [4–6]. Hydrogen bonds in collagen can be broken upon denaturation through thermal or chemical treatments, significantly affecting its properties and resulting in another widely known product of commercial interest: gelatin.

In an attempt to make extraction processes greener and more effective, eutectics are currently gaining much interest as alternative solvents for recovering bioactive molecules from different food matrices, including fish by-products [7,8]. A eutectic reaction is defined as an isothermal reversible reaction between two (or more) solid phases during the heating of a system, as a result of which a single liquid phase is produced [9]. The eutectic composition has a melting temperature lower than that of any of the parent compounds, enabling its use as a liquid solvent at temperatures at which the individual consideration of the higher-melting parent compounds as solvents would not be possible due to their solid character. Since Abbott's pioneering work [10,11], unfortunately, the meaning of a eutectic solvent has been perverted by the prolific utilization of the term "deep eutectic solvent" to refer to many liquid mixtures without evidence of eutectic behavior supported by the corresponding solid–liquid equilibria. Without intending to neglect the practical interest that composed solvents can have for many varied applications, the work carried out by Coutinho and coworkers aiming to correct some widespread misconceptions in the (ab)use of the label "deep eutectic solvent" must be highlighted [12–14]. In that vein, the adjective "deep" should be restricted to eutectic mixtures for which the melting temperature of the eutectic composition would be lower (how much lower is considered low enough is yet to be shaped by the scientific community through pertinent debate) than one corresponding to the ideal behavior of the liquid mixture. In any case, before referring to a solvent as (deep) eutectic, the solid–liquid equilibrium must be determined in order to verify the eutectic behavior and the corresponding melting temperature and composition of the eutectic point.

Besides enabling high-melting compounds to be used in liquid form as part of the solvent, for example in extraction processes, a further advantage of eutectic solvents is their tunability, since it is possible to select the corresponding parent compounds depending on the desired application. Focusing on the extraction of bioactive compounds, the use of eutectic solvents resulting from the combination of GRAS ("generally regarded as safe") compounds sounds attractive.

A number of eutectics with a melting temperature below or just slightly above room temperature have been investigated in the literature for the recovery of components of interest from bio-residues. Most of these are based on the combination of (2-hydroxyethyl)trimethylammonium chloride ([choline]Cl) with different organic acids, as in the work by Bradić et al. [15], who paired [choline]Cl with lactic, malonic, and citric acid to obtain chitin from shrimp shells. In a similar vein, although in this case for extracting collagen peptides, Bai et al. [16] combined [choline]Cl with lactic, oxalic, and acetic acids, as well as with other organic compounds such as urea, ethylene glycol, or glycerol, to generate eutectic solvents. Bisht et al. [17] tested aqueous solutions of several eutectic solvents to extract collagen from the skin of Atlantic cod, with the system containing urea and lactic acid leading to the best performance, improving on the traditional method with acetic acid. Regarding the eutectic mixture composed of sodium acetate and urea, Nuutinen et al. [18] reported its effectiveness as a solvent for extracting keratin from poultry feathers when diluted with water. The same mixture was successfully used by Rico et al. [19] to valorize melon by-products.

This work aims to be a preliminary study on the use of a eutectic system based on GRAS components for the extraction of collagen from fish residues. Sodium acetate trihydrate was selected as a first component due to the basic character of the acetate anion, which is expected to be able to disrupt the hydrogen-bonding network of the constitutive biopolymers of fish residues. In fact, the acetate anion has proven its capacity to recover bioactive compounds from fish waste as part of ionic liquids [20–22]. Urea was selected as the second component due to its frequent use in the biopharmaceutical industry to solubilize proteins. After ascertaining the eutectic behavior and characteristics, the solubility of representative amino acids of fish collagen and gelatin (see Figure 1) will be determined in this solvent. The possibility of using a ternary mixture comprising the eutectic compounds and water will also be analyzed through the corresponding solubility studies. As far as we know, and despite the huge interest of these combinations, this is the first study about solubilities of amino acids in eutectics.

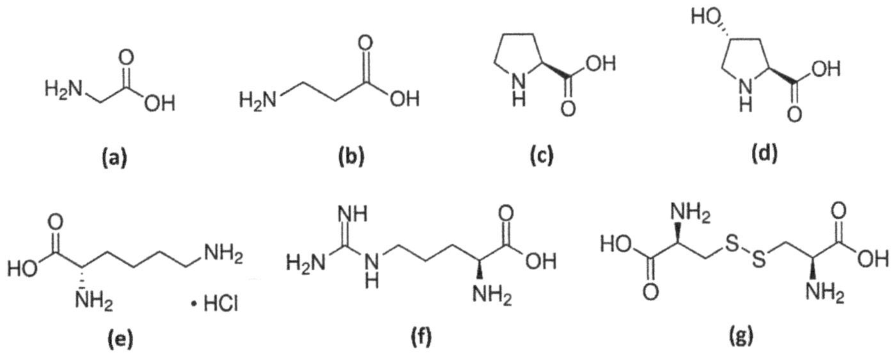

Figure 1. Chemical structures of amino acids: (**a**) glycine (Gly); (**b**) β-alanine (Ala); (**c**) L-proline (Pro); (**d**) trans-4-hydroxy-L-proline (OH-Pro); (**e**) L-lysine monohydrochloride (Lys·HCl); (**f**) L-arginine (Arg); and (**g**) L-cystine (Cys).

2. Results and Discussion

2.1. Determination of the Eutectic

Samples of urea and of sodium acetate trihydrate were independently analyzed by differential scanning calorimetry (DSC) to determine their melting temperatures and enthalpies of fusion. Experimental results are shown in Table 1 along with values reported in the literature. A reasonable agreement can be observed between both sets, with the minor discrepancies being attributable to different experimental conditions during the measurements, to the presence of different types and levels of impurities, or to the moisture of the samples.

Table 1. Melting temperature (T_m) and enthalpy of fusion (ΔH_{fus}) of the eutectic compounds. "Exp." refers to experimentally determined values, and "Lit." refers to literature values.

Compound	T_m (K)		ΔH_{fus} (kJ·mol^{-1})	
	Exp.	Lit.	Exp.	Lit.
Urea	402	406 [23,24]	12.7	14.1 [25], 15.1 [24]
Sodium acetate trihydrate	331	331 [26,27], 332 [24]	37.3	37.0 [27], 36.2 [24]

Both urea and sodium acetate trihydrate melt at temperatures considerably above ambient conditions, thus limiting their practical use as solvents for a number of applications. The existence of eutectic behavior in this system would overcome this issue, expanding the liquid range of the mixture while fully keeping the properties of its constituents. The solid–liquid equilibrium of this system has been previously studied [24,28,29]. However, some discrepancies were found regarding the eutectic composition. Thus, the urea + sodium acetate trihydrate system was investigated via DSC to dispel doubts on its solid–liquid equilibrium and ascertain the characteristics of the eutectic point. The temperature–composition plot is shown in Figure 2, which includes the previously available literature data for comparison purposes. As expected, a eutectic state of solid–solid–liquid equilibrium was found. The urea mole fraction at the eutectic is 0.60 and the eutectic temperature is 304 K. This represents a relevant depression of the melting temperature with respect to those of the individual constituents, thus allowing the mixture to be used as a solvent at extraction temperatures very close to ambient conditions. Comparing our results for the eutectic point with those found in the literature (see Figure 2 and Table 2), the only discrepancy is found with data from Li et al. [24], who obtained a eutectic composition with a much higher urea concentration.

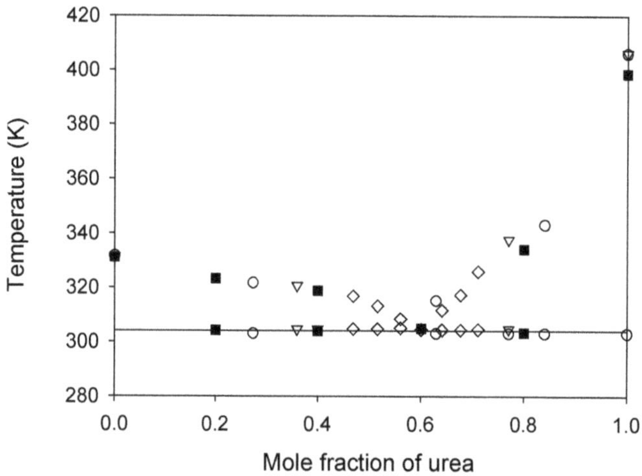

Figure 2. Temperature–composition diagram for the solid–liquid equilibrium of the urea + sodium acetate trihydrate system. Experimental results (■) are compared to literature data: ○, ref. [24]; ▽, ref. [28]; ◇, ref. [29]. The horizontal solid line represents the eutectic temperature found.

Table 2. Urea + sodium acetate trihydrate eutectic composition and melting temperature.

Eutectic Composition (Urea Mole Fraction)	T_m (K)	Reference
0.77	303	[24]
0.60	305	[28]
0.60	305	[29]
0.60	304	This work

2.2. Solubility Studies

The solubility of several amino acids (see Figure 1) was determined in the urea + sodium acetate trihydrate mixture of eutectic composition, as well as in its 50:50 (wt:wt) mixture with water. The addition of water allows the reduction in the viscosity of the solvent and the possibility of working at lower temperatures without risk of solidification of the eutectic-forming compounds. Unfortunately, the solid–liquid equilibrium of the ternary system was only partially studied in the literature [28]. As the study was considered to be beyond the scope of this work, a water content of 50 wt% was estimated as the limit of interest in practical applications and selected instead of the ternary eutectic. Experiments were carried out at four different temperatures, from 308.15 K to 338.15 K when using the eutectic, and from 298.15 K to 328.15 K in the case of the aqueous mixture. All the results, expressed as the mass fraction of amino acid, are shown in Table 3 and represented in Figure 3.

Table 3. Solubilities (mass fraction) of different amino acids in the eutectic solvent and in its mixtures with water, 50:50 (wt:wt), at different temperatures and 0.1 MPa *.

				Eutectic			
T (K)	Glycine	β-Alanine	L-Proline	trans-4-Hydroxy-L-proline	L-Lysine·HCl	L-Arginine	L-Cystine
308.15	0.1201	0.2199	0.1941	0.1629	0.1307	0.1618	0.0019
318.15	0.1329	0.2471	0.2393	0.1739	0.1440	0.1850	0.0047
328.15	0.1439	0.2858	0.2864	0.1939	0.1615	0.2100	0.0113
338.15	0.1553	0.3282	0.3573	0.2166	0.1781	0.2467	0.0257
				Eutectic:Water 50:50 (wt:wt)			
T (K)	Glycine	β-Alanine	L-Proline	trans-4-Hydroxy-L-proline	L-Lysine·HCl	L-Arginine	L-Cystine
298.15	0.1334	0.3202	0.5999	0.1916	0.3633	0.1124	0.0040
308.15	0.1671	0.3448	0.6161	0.2113	0.4000	0.1596	0.0102
318.15	0.2049	0.3801	0.6412	0.2492	0.4371	0.2341	0.0300
328.15	0.2381	0.4204	0.6639	0.2753	0.4760	0.3331	0.0552

* Uncertainties: u(T) = 0.1 K. u(P) = 5 kPa. Eutectic: $u_r(w)$ = 0.004 Gly; $u_r(w)$ = 0.007 Ala; $u_r(w)$ = 0.005 Pro; $u_r(w)$ = 0.006 OH-Pro; $u_r(w)$ = 0.007 Lys; $u_r(w)$ = 0.006 Arg; $u_r(w)$ = 0.009 Cys. Eutectic:water: $u_r(w)$ = 0.003 Gly; $u_r(w)$ = 0.003 Ala; $u_r(w)$ = 0.006 Pro; $u_r(w)$ = 0.004 OH-Pro; $u_r(w)$ = 0.005 Lys; $u_r(w)$ = 0.003 Arg; $u_r(w)$ = 0.008 Cys.

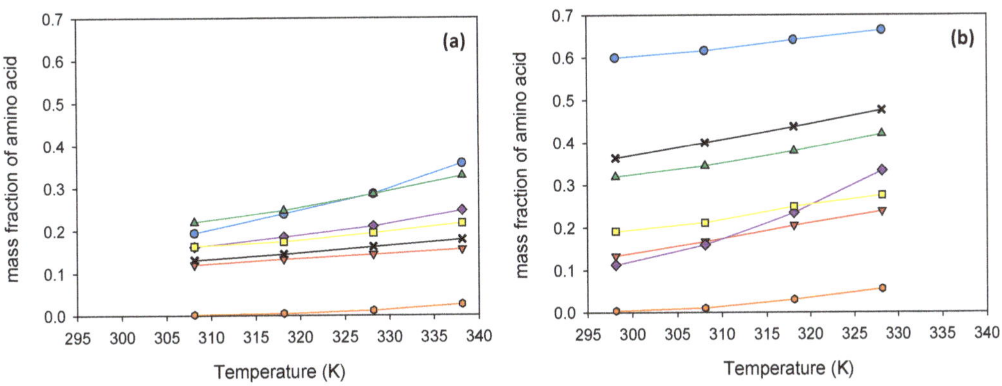

Figure 3. Solubilities of amino acids in the eutectic solvent (**a**) and in its 50:50 (wt:wt) mixture with water (**b**). Results are expressed as mass fraction. Legend: ▼ glycine; ▲ β-alanine; ● L-proline; □ trans-4-hydroxy-L-proline; ✖ L-lysine·HCl; ◆ L-arginine; ⬣ L-cystine.

The urea + sodium acetate trihydrate mixture is characterized by the intensive H-bonding interactions between its components [30]. Thus, the solubilities of amino acids in this mixture can be favored, among other interactions, by the H-bonds between the solute and the solvent components [31]. At temperatures close to ambient conditions, the solubilities of the studied amino acids in the eutectic follow the trend Ala > Pro > Arg ~ OH-Pro > Lys·HCl > Gly > Cys. Cystine is a dimer of two cysteine molecules connected via a disulfide bridge. The high hydrophobicity index, estimated in terms of the solvent-accessible surface area of the disulfide-bonding amino acid [32,33], may explain a much lower solubility than the other amino acids. When water is introduced into the system, the hydrogen bonds among the eutectic constituents are gradually broken, accompanied by the formation of new bonds between water and the eutectic constituents. In the case of a significant quantity of water, the hydrogen bond network of the eutectic is destroyed, and the increase in amino-acid solubilities is due to the presence of water [30]. For this reason, it is not strange that the solubilities of the studied biomolecules in the ternary mixture (urea + sodium acetate trihydrate + water) follow the same trend as in water [34], decreasing according to Pro > Lys·HCl > Ala > OH-Pro > Gly > Arg > Cys. All the solubilities, in the eutectic and its mixture with water, increase with temperature, with this effect being particularly noticeable in the case of the solubility of L-arginine in the aqueous mixture.

Some similarities were found for the solubilities in the eutectic and in the eutectic–water mixture. In both cases, the solubility of glycine (with no hydrophobic side chain) is lower than that of β-alanine; the solubility of trans-4-hydroxy-L-proline is lower than that of L-proline, so the presence of an OH group in the proline ring results in a decrease in solubility; and the existence of a disulfide group in the amino acid (L-cystine) leads to the lowest solubilities among all the amino acids studied. However, some differences must also be highlighted. The polar portions of the L-proline molecule are sufficiently hydrated to ensure the highest solubility in the aqueous solution. However, in the case of the eutectic and at the lowest temperatures, β-alanine is more soluble than L-proline. The second most soluble amino acid in the aqueous mixture is the monohydrochloride form of L-lysine. Nonetheless, its solubility in the eutectic is rather limited.

Focusing on collagen extraction from fish residues as the application, the use of the eutectic mixture is required to make the dissolution of fish biopolymers possible through the disruption of hydrogen bonding, thus facilitating the extraction of components of interest. However, the ternary mixture with water would improve the mass transfer process due to the significant reduction in the viscosity of the solvent and, according to the solubilities here obtained, could favor an increase in the solubility of the protein in the solvent. New solubility studies involving mixtures of amino acids in the compositions found in collagens, and also the protein itself, would be the next stage of fundamental research prior to the optimization of the application.

The mathematical expression of the van 't Hoff equation was used for data correlation:

$$\ln x = \mathrm{A} + \frac{\mathrm{B}}{T(\mathrm{K})} \tag{1}$$

where x is the solubility of the amino acid in the eutectic or its 50:50 (wt/wt) mixture with water, expressed as mole fraction; A and B are the correlation parameters; and T is the absolute temperature.

Table S1 in the Supplementary Materials presents the numerical values of the solubilities expressed in mole fractions. Table 4 shows the correlation parameters and the absolute relative deviation percent (%ARD) of the linear fits of the natural logarithm of x versus the inverse of T. Figure 4 shows these linear fits together with the corresponding experimental solubility data points. It can be seen that, in all cases, this simple and classical model correlates satisfactorily to the experimental data.

Table 4. Regression parameters of the van 't Hoff model obtained from the correlation of the natural logarithm of the experimental solubility data with the inverse of the absolute temperature.

	Eutectic			Eutectic:Water 50:50 (wt:wt)		
Amino Acid	**A**	**B (K)**	**% ARD**	**A**	**B (K)**	**% ARD**
Glycine	−863.5	0.848	0.34	−2133	4.319	1.12
β-Alanine	−1395	3.014	0.94	−1181	1.957	1.45
L-Proline	−2219	5.355	1.12	−629.7	0.835	0.66
trans-4-Hydroxy-L-proline	−1064	1.300	1.55	−1492	2.023	1.63
L-Lysine·HCl	−1154	1.275	0.47	−1317	2.162	0.35
L-Arginine	−1604	2.798	1.35	−4286	10.50	3.07
L-Cystine	−9160	22.46	0.31	−8959	22.47	7.32

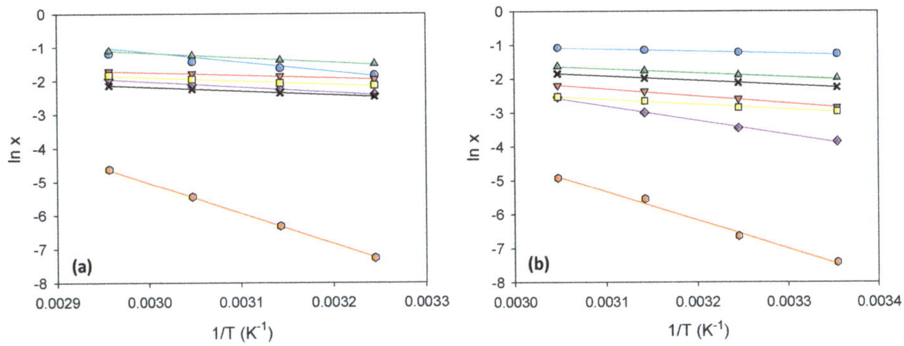

Figure 4. Experimental solubility data correlated by the van 't Hoff equation for different amino acids in the eutectic solvent (**a**) and in its 50:50 (wt:wt) mixture with water (**b**). Legend: ▼ glycine; ▲ β-alanine; ● L-proline; □ trans-4-hydroxy-L-proline; ✖ L-lysine·HCl; ◆ L-arginine; ● L-cystine.

Apparent Properties of Dissolution

The apparent standard enthalpy change for the dissolution of the amino acids in the eutectic solvent or in its 50:50 (wt/wt) mixture with water could be determined from the slope of the modified van 't Hoff equation [35–37]:

$$\frac{\partial \ln x}{\partial\left(\frac{1}{T}-\frac{1}{T_{hm}}\right)_p} = -\frac{\Delta H_d^0}{R} \tag{2}$$

where R is the ideal gas constant, ΔH_d^0 is the apparent standard enthalpy change of dissolution, subscript p denotes constant pressure, and T_{hm} is the harmonic mean temperature calculated as:

$$T_{hm} = \frac{n}{\sum_{i=1}^{n} 1/T_i} \tag{3}$$

with n corresponding to the total number of experimental temperatures investigated. In this study, T_{hm} = 322.76 K for the studies with the eutectic, and T_{hm} = 312.75 K for the studies with the aqueous formulation of the eutectic.

The apparent standard Gibbs energy (ΔG_d^o) and entropy changes (ΔS_d^o) of dissolution can be calculated with the following equations:

$$\Delta G_d^o = -RT_{hm} Intercept \tag{4}$$

$$\Delta S_d^o = \frac{\Delta H_d^o - \Delta G_d^o}{T_{hm}} \tag{5}$$

where *Intercept* refers to the intercept derived from Equation (2). Table 5 shows the apparent properties of dissolution and the contributions of enthalpy and entropy to the Gibbs energy in the dissolution process calculated as:

$$\xi_H^d = \frac{|\Delta H_d^o|}{|\Delta H_d^o| + T_{hm}|\Delta S_d^o|} \tag{6}$$

$$\xi_{TS}^d = \frac{T_{hm}|\Delta S_d^o|}{|\Delta H_d^o| + T_{hm}|\Delta S_d^o|} \tag{7}$$

Table 5. Apparent properties of dissolution.

Amino Acid	ΔH_d^o (kJ·mol^{-1})	$T_{hm}\cdot\Delta S_d^o$ (kJ·mol^{-1})	ΔG_d^o (kJ·mol^{-1})	ξ_H^d	ξ_{TS}^d
		Eutectic			
Glycine	7.18	2.28	4.90	0.76	0.24
β-Alanine	11.60	8.09	3.51	0.59	0.41
L-Proline	18.45	14.37	4.08	0.56	0.44
trans-4-Hydroxy-L-proline	8.85	3.49	5.36	0.72	0.28
L-Lysine·HCl	9.60	3.42	6.18	0.74	0.26
L-Arginine	13.34	7.51	5.83	0.64	0.36
L-Cystine	76.16	60.28	15.87	0.56	0.44
		Eutectic:Water 50:50 (wt:wt)			
Glycine	17.74	11.23	6.51	0.61	0.39
β-Alanine	9.82	5.09	4.73	0.66	0.34
L-Proline	5.24	2.17	3.06	0.71	0.29
trans-4-Hydroxy-L-proline	12.40	5.26	7.15	0.71	0.29
L-Lysine·HCl	10.95	5.62	5.33	0.66	0.34
L-Arginine	35.63	27.30	8.34	0.57	0.43
L-Cystine	74.49	58.42	16.07	0.56	0.44

As clearly shown in Table 5 and Figure 5, ΔH_d^o and ΔG_d^o of the dissolution of the amino acids in both the eutectic and its mixture with water are positive, indicating that the process is endothermic and non-spontaneous, which is in agreement with the increase in solubility with temperature. Moreover, when comparing ΔG_d^o with the solubilities shown in Tabls S1 (in mole fractions), an inverse correlation is observed, as expected. Figure 5 and the contribution parameters presented in Table 5 indicate that the enthalpy is the main contributor to ΔG_d^o. However, in some cases, such as the dissolution of L-proline, L-cystine, and β-alanine in the eutectic or of L-arginine and L-cystine in the ternary mixture, contributions are rather balanced.

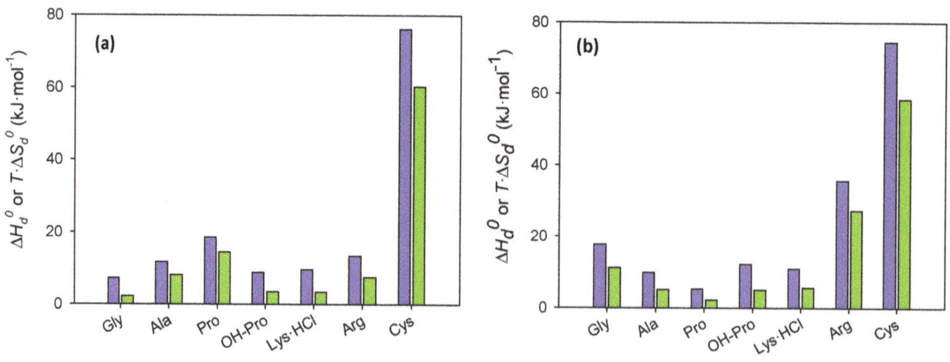

Figure 5. Enthalpic (ΔH_d^o, blue/left bars) and entropic ($T_{hm}\cdot\Delta S_d^o$, green/right bars) contributions to the dissolution process in the eutectic solvent (**a**) and in its 50:50 (wt:wt) mixture with water (**b**).

3. Materials and Methods

3.1. Materials

Sodium acetate trihydrate (>99 wt%) was purchased from Scharlau (Sentmenat, Spain). Urea (>99.5 wt%), glycine (>99 wt%), β-alanine (>99 wt%), trans-4-hydroxy-L-proline (>99 wt%), L-lysine monohydrochloride (>99.5 wt%), L-arginine (>98 wt%), and L-cystine (>98 wt%) were supplied by Sigma-Aldrich (Madrid, Spain). L-Proline, with purity >99 wt%, was supplied by Panreac (Castellar del Vallès, Spain). Table S2 in Supplementary Materials shows sources and purities of the amino acids used in this work. All reactants were used as received, without further purification. Bidistilled water was used throughout the experimental work in preparing the aqueous solutions.

3.2. Methods

3.2.1. Determination of the Eutectic

Samples with a molar composition step of ca. 0.20 were prepared in glass vials, covering the entire composition range of the urea + sodium acetate trihydrate system, and evenly heated to 343 K in an oil bath until total liquefaction and homogenization were achieved. The weighing was carried out on a Mettler Toledo XPE205 analytical balance with a precision of 10^{-4} g. Approximately 5–15 mg of each sample were placed in a hermetically sealed aluminum pan and analyzed by DSC in a Q2000 differential scanning calorimeter (TA Instruments, Cerdanyola del Vallès, Spain) coupled with an RCS 90 cooling system. Each sample was loaded into the measuring chamber with an autosampler, along with an analogous empty capsule used as a reference. A 50 mL/min nitrogen flow was used as sample purge gas. The thermal program consisted in a rapid cooling to 283.15 K followed by three cycles, each of them comprising a heating ramp up to 403.15 K and a cooling ramp down to 203.15 K, at heating/cooling rates of 5 K/min, with 10 min isotherms in between these ramps. After the first heating ramp, all samples became supercooled, and no further identification of endothermic peaks connected with sample melting could be identified in the subsequent heating ramps. For this reason, the thermogram of the first heating ramp was used to determine the melting temperature (with an estimated uncertainty of 1 K) at the onset of the observed endothermic peaks. The analysis of the thermograms was carried out with the software Universal Analysis 2000, version 4.5.0.5, by TA Instruments.

3.2.2. Solubility Studies

The solubility of different amino acids was measured in the eutectic solvent, as well as in its 50:50 (wt/wt) mixture with water. Jacketed glass cells were used, and the temperature was controlled by means of the circulation of water from an Ultratherm-200 P Selecta bath through the cell jackets. For each desired temperature, the solvent was placed together with an excess of the amino acid in the cell and the mixture was magnetically stirred for 24 h and then left to stand for at least 1 h (preliminary tests were carried out to ensure that these times were sufficient to achieve equilibrium conditions). After this, the supernatant was partially taken with a syringe and filtered with nylon filters (pore size 0.45 μm) to ensure removal of any non-dissolved amino acid particles. Both syringe and filters were pre-heated to ensure the absence of amino acid precipitation. Measurement of a physical property was used as the analysis method.

In the case of solubilities in the ternary mixture (eutectic:water, 50:50 wt:wt), the measurement of density was carried out for composition determination. To that end, a vibrating U-tube DMA 5000 density meter (Anton Paar, Madrid, Spain) with automatic viscosity correction and internal temperature control was used. First, several samples with amino acid concentration ranging from zero to a composition close to its solubility were prepared by weight. Then calibration curves were obtained by measuring the density of those samples. Data and correlation curves are presented in the Supplementary Materials (Tables S3–S9). To ensure the validity of this analysis method, new samples were prepared by weight and the composition calculated with the calibration curves, and results were found in good agreement within the uncertainty of the method. Second, to determine solubilities, the equilibrium samples were appropriately diluted (a mass of solvent twice that of the sample was added) for compositional analysis using the density calibration curve. The composition obtained was obviously corrected according to the mass of solvent added.

In the case of the eutectic mixture, and due to the small variation of density with amino acid content, the same analysis method led to high deviations. Therefore, refractive index was selected as the reference property for the calibration curves. An Abbemat 500 Refractometer (Anton Paar, Madrid, Spain) with embedded temperature control was used. The refractive indices of the amino acid solutions in the eutectic solvent and the calibration curves are also presented in the Supplementary Materials (Tables S3–S9). As in the previous case, the dilution of the equilibrium samples was required, and a dilution factor 1:2 with solvent was used.

4. Conclusions

In this work, a study regarding the solubility of different amino acids (namely: glycine, β-alanine, L-proline, trans-4-hydroxy-L-proline, L-lysine·HCl, L-arginine, and L-cystine), being representative constituents of fish collagen, was carried out. The solubilities of these biomolecules were determined in the urea + sodium acetate trihydrate eutectic and its 50:50 (wt/wt) mixtures with water at several temperatures. A discrepancy found in previous studies regarding the composition of the eutectic was solved, identifying a urea mole fraction of 0.60 as the composition that led to the lowest melting temperature of the system (304 K). This value, close to ambient conditions, is promising for the use of the eutectic as solvent in different extraction processes without the need for significant energy input to keep the solvent a liquid.

The solubilities of the most abundant amino acids in fish collagen and gelatin (glycine, proline, and hydroxy-proline) in the aforementioned eutectic were found to be rather high and increased with an increase in temperature. The capacity of the eutectic to establish hydrogen bond interactions along with its capacity to dissolve the target amino acids makes it, in principle, an interesting solvent to disrupt the hydrogen-bonding network in the biopolymers matrix of fish residues, thus potentially facilitating the extraction of the component of interest. When the eutectic was mixed with water, the solubilities of the amino acids in the solvent increased. A favored mass transfer process (viscosity reduction) and higher solubilities also encourage future studies using the ternary mixture as a possible extraction solvent.

All the dissolution processes studied herein were endothermic, non-spontaneous, and enthalpy-driven, although with quite balanced enthalpy–entropy contributions in some cases. Future studies, supported by molecular simulation, are encouraged in order to have a better understanding of how the amino acids are solvated in the eutectic medium.

Supplementary Materials: The supporting information can be downloaded at: https://www.mdpi.com/article/10.3390/ijms24021550/s1.

Author Contributions: Investigation, C.G.; supervision, H.R. and A.S.; writing—original draft, C.G.; writing—review and editing, H.R. and A.S.; funding acquisition: A.S. All authors have read and agreed to the published version of the manuscript.

Funding: Grant PID2021-123622OB-I00 funded by MCIN/AEI/10.13039/501100011033 and by "ERDF A way of making Europe" by the "European Union".

Data Availability Statement: Not applicable.

Conflicts of Interest: The authors declare no conflict of interest.

References

1. FAO. *World Fisheries and Aquaculture*; FAO: Rome, Italy, 2022; ISBN 9789251072257.
2. Furtado, M.; Chen, L.; Chen, Z.; Chen, A.; Cui, W. Development of fish collagen in tissue regeneration and drug delivery. *Eng. Regen.* **2022**, *3*, 217–231. [CrossRef]
3. European Commission. *The EU Blue Economy Report*; Publications Office of the European Union: Luxembourg, 2022.
4. Heidari, M.G.; Rezaei, M. Extracted pepsin of trout waste and ultrasound-promoted method for green recovery of fish collagen. *Sustain. Chem. Pharm.* **2022**, *30*, 100854. [CrossRef]
5. Ahmed, M.; Verma, A.K.; Patel, R. Collagen extraction and recent biological activities of collagen peptides derived from sea-food waste: A review. *Sustain. Chem. Pharm.* **2020**, *18*, 100315. [CrossRef]
6. Chen, Y.; Jin, H.; Yang, F.; Jin, S.; Liu, C.; Zhang, L.; Huang, J.; Wang, S.; Yan, Z.; Cai, X.; et al. Physicochemical, antioxidant properties of giant croaker (Nibea japonica) swim bladders collagen and wound healing evaluation. *Int. J. Biol. Macromol.* **2019**, *138*, 483–491. [CrossRef] [PubMed]
7. Duan, L.; Dou, L.L.; Guo, L.; Li, P.; Liu, E.H. Comprehensive Evaluation of Deep Eutectic Solvents in Extraction of Bioactive Natural Products. *ACS Sustain. Chem. Eng.* **2016**, *4*, 2405–2411. [CrossRef]

8. Della Posta, S.; Gallo, V.; Gentili, A.; Fanali, C. Strategies for the recovery of bioactive molecules from deep eutectic solvents extracts. *TrAC-Trends Anal. Chem.* **2022**, *157*, 116798. [CrossRef]
9. IUPAC. *Compendium of Chemical Terminology*, 2nd ed.; McNaught, A.D., Wilkinson, A., Eds.; the "Gold, Book"; Blackwell Scientific Publications: Oxford, UK, 1997; ISBN 0-9678550-9-8.
10. Abbott, A.P.; Capper, G.; Davies, D.L.; Rasheed, R.K.; Tambyrajah, V. Novel solvent properties of choline chloride/urea mixtures. *Chem. Com.* **2003**, *39*, 70–71. [CrossRef]
11. Smith, E.L.; Abbott, A.P.; Ryder, K.S. Deep Eutectic Solvents (DESs) and Their Applications. *Chem. Rev.* **2014**, *114*, 11060–11082. [CrossRef]
12. Coutinho, J.A.P.; Pinho, S.P. Special Issue on Deep Eutectic Solvents: A foreword. *Fluid Phase Equilibr.* **2017**, *448*, 1. [CrossRef]
13. Martins, M.A.R.; Pinho, S.P.; Coutinho, J.A.P. Insights into the Nature of Eutectic and Deep Eutectic Mixtures. *J. Solut. Chem.* **2019**, *48*, 962–982. [CrossRef]
14. Schaeffer, N.; Silva, L.P.; Coutinho, J.A.P. Comment on "Structural Study of a Eutectic Solvent Reveals Hydrophobic Segregation and Lack of Hydrogen Bonding between the Components". *ACS Sus. Chem. Eng.* **2022**, *10*, 8669–8670. [CrossRef]
15. Bradić, B.; Novak, U.; Likozar, B. Crustacean shell bio-refining to chitin by natural. *Green Process. Synth.* **2020**, *9*, 13–25. [CrossRef]
16. Bai, C.; Wei, Q.; Ren, X. Selective Extraction of Collagen Peptides with High Purity from Cod Skins by Deep Eutectic Solvents. *ACS Sustain. Chem. Eng.* **2017**, *5*, 7220–7227. [CrossRef]
17. Bisht, M.; Martins, M.; Dias, A.C.R.V.; Ventura, S.P.M.; Coutinho, J.A.P. Uncovering the potential of aqueous solutions of deep eutectic solvents on the extraction and purification of collagen type I from Atlantic codfish (Gadus morhua). *Green Chem.* **2021**, *23*, 8940–8948. [CrossRef]
18. Nuutinen, E.M.; Willberg-Keyriläinen, P.; Virtanen, T.; Mija, A.; Kuutti, L.; Lantto, R.; Jääskeläinen, A.S. Green process to regenerate keratin from feathers with an aqueous deep eutectic solvent. *RSC Adv.* **2019**, *9*, 19720–19728. [CrossRef]
19. Rico, X.; Nuutinen, E.M.; Gullón, B.; Pihlajaniemi, V.; Yáñez, R. Application of an eco-friendly sodium acetate/urea deep eutectic solvent in the valorization of melon by-products. *Food Bioprod. Process.* **2021**, *130*, 216–228. [CrossRef]
20. Muhammad, N.; Gonfa, G.; Rahim, A.; Ahmad, P.; Iqbal, F.; Sharif, F.; Khan, A.S.; Khan, F.U.; Khan, Z.U.H.; Rehman, F.; et al. Investigation of ionic liquids as a pretreatment solvent for extraction of collagen biopolymer from waste fish scales using COSMO-RS and experiment. *J. Mol. Liq.* **2017**, *232*, 258–264. [CrossRef]
21. Muhammad, N.; Gao, Y.; Iqbal, F.; Ahmad, P.; Ge, R.; Nishan, U.; Rahim, A.; Gonfa, G.; Ullah, Z. Extraction of biocompatible hydroxyapatite from fish scales using novel approach of ionic liquid pretreatment. *Sep. Purif. Technol.* **2016**, *161*, 129–135. [CrossRef]
22. Shamshina, J.L.; Barber, P.S.; Gurau, G.; Griggs, C.S.; Rogers, R.D. Pulping of Crustacean Waste Using Ionic Liquids: To Extract or Not to Extract. *ACS Sustain. Chem. Eng.* **2016**, *4*, 6072–6081. [CrossRef]
23. Lundström, A.; Andersson, B.; Olsson, L. Urea thermolysis studied under flow reactor conditions using DSC and FT-IR. *Chem. Eng. J.* **2009**, *150*, 544–550. [CrossRef]
24. Li, J.H.; Zhang, G.-E.; Wang, J.Y. Investigation of a eutectic mixture of sodium acetate trihydrate and urea as latent heat storage. *Sol. Energy* **1991**, *47*, 443–445. [CrossRef]
25. Stradella, L.; Argentero, M. A study of the thermal decomposition of urea, of related compounds and thiourea using DSC and TG-EGA. *Thermochim. Acta* **1993**, *219*, 315–323. [CrossRef]
26. Yang, H.; Bao, X.; Cui, H.; Lo, T.Y.; Chen, X. Optimization of supercooling, thermal conductivity, photothermal conversion, and phase change temperature of sodium acetate trihydrate for thermal energy storage applications. *Energy* **2022**, *254*, 124280. [CrossRef]
27. Guion, J.; Physique, L.D.C.; De Nice, U.; Antipolis, S.; Sauzade, J.D.; Laugt, M.; Antipolis, S. Critical examination and experimental determination of melting enthalpies and entropies of salt hydrates. *Thermochim. Acta* **1983**, *67*, 167–179. [CrossRef]
28. Wada, T.; Kimura, F.; Yamamoto, R. Studies on salt hydrates for latent heat storage. II. Eutectic mixture of pseudo-binary system. *Bull. Chem. Soc. Jpn.* **1983**, *56*, 1223–1226. [CrossRef]
29. Li, X.; Fu, Z.; Qiao, Y.; Zhang, Z.; Zhou, Y.; Hai, C.; Shen, Y.; Sun, Y.; Zeng, J.; Ren, X. Preparation, Characterization, and Modification of Sodium Acetate Trihydrate-Urea Binary Eutectic Mixtures as Phase Change Material. *Energy Fuels* **2020**, *34*, 6439–6447. [CrossRef]
30. Cui, Y.; Zhu, Y.; Dai, R.; Shan, Z.; Yi, J.; Chen, H. The solubility and interactions of gelatin in "water-in-sodium acetate trihydrate/urea DES" system. *Colloids Surf. A Physicochem. Eng. Asp.* **2021**, *625*, 126916. [CrossRef]
31. Scheiner, S.; Kar, T.; Pattanayak, J. Comparison of various types of hydrogen bonds involving aromatic amino acids. *J. Am. Chem. Soc.* **2002**, *124*, 13257–13264. [CrossRef]
32. Rose, G.D.; Geselowitz, A.R.; Lesser, G.J.; Lee, R.H.; Zehfus, M.H. Hydrophobicity of amino acid residues in globular proteins. *Science* **1985**, *229*, 834–838. [CrossRef]
33. Nagano, N.; Ota, M.; Nishikawa, K. Strong hydrophobic nature of cysteine residues in proteins. *FEBS Lett.* **1999**, *458*, 69–71. [CrossRef]
34. Cohn, E.J.; Edsall, J.T. *Proteins, Amino Acids and Peptides as Ions and Dipolar Ions*; Renhold Publ. Corp.: New York, NY, USA, 1943.
35. Krug, R.R.; Hunter, W.G.; Grieger, R.A. Enthalpy-Entropy Compensation. 2. Separation of the Chemical from the Statistical Effect. *J. Phys. Chem.* **1976**, *80*, 2341–2351. [CrossRef]

36. Blanco-Márquez, J.H.; Caviedes Rubio, D.I.; Ortiz, C.P.; Cerquera, N.E.; Martínez, F.; Delgado, D.R. Thermodynamic analysis and preferential solvation of sulfamethazine in acetonitrile + water cosolvent mixtures. *Fluid Phase Equilib.* **2020**, *505*, 112361. [CrossRef]
37. Tinjacá, D.A.; Martínez, F.; Almanza, O.A.; Jouyban, A.; Acree, W.E. Effect of N-Methyl-pyrrolidone (NMP) on the Equilibrium Solubility of Meloxicam in Aqueous Media: Correlation, Dissolution Thermodynamics, and Preferential Solvation. *ACS Omega* **2022**, *7*, 37988–38002. [CrossRef] [PubMed]

International Journal of
Molecular Sciences

MDPI

Article

BioMThermDB 1.0: Thermophysical Database of Proteins in Solutions

Mina Nikolić [1,†], Sandi Brudar [1], Evangelos Coutsias [2], Ken A. Dill [2], Miha Lukšič [1], Carlos Simmerling [2] and Barbara Hribar-Lee [1,*]

1 Faculty of Chemistry and Chemical Technology, University of Ljubljana, Večna Pot 113, SI-1000 Ljubljana, Slovenia
2 Laufer Center for Physical and Quantitative Biology, Stony Brook University, Stony Brook, NY 11794-5252, USA
* Correspondence: barbara.hribar@fkkt.uni-lj.si
† Current address: Department of Chemistry, Atomic & Mass Spectrometry—A&MS Research Unit, Campus Sterre, Ghent University, Krijgslaan 281-S12, 9000 Ghent, Belgium.

Abstract: We present here a freely available web-based database, called BioMThermDB 1.0, of thermophysical and dynamic properties of various proteins and their aqueous solutions. It contains the hydrodynamic radius, electrophoretic mobility, zeta potential, self-diffusion coefficient, solution viscosity, and cloud-point temperature, as well as the conditions for those determinations and details of the experimental method. It can facilitate the meta-analysis and visualization of data, can enable comparisons, and may be useful for comparing theoretical model predictions with experiments.

Keywords: database; proteins; solutions; thermodynamics; hydrodynamic radius; electrophoretic mobility; zeta potential; diffusion coefficient; viscosity; cloud-point temperature

1. Introduction

Proteins are the most abundant macromolecules in living cells and represent the building blocks of life. They govern almost all biological processes that define living organisms the way they are. Knowledge of the physical properties of protein solutions can have practical importance for formulating biological agents and drugs [1,2]. In order to maintain their beneficial functions, proteins must remain stable in environments in which they are immersed, usually in different aqueous solutions. Consequently, a wide and diverse set of information on the thermophysical and thermodynamic properties of proteins in aqueous solutions is of critical importance for obtaining a better understanding of the protein structure and its relationship with factors that influence its stability, which is vital for preparing safe pharmaceutical formulations. With the development of biophysical methods for protein characterization and web-based applications, bio-macromolecular studies have become extremely data-rich; thus, the need for data storage, its organization, and interconnection is increasing rapidly these days. Even though a great amount of useful information is available in existing databases, such as ProThermDB [3], MPTherm [4], PROXiMATE [5], and PINT [6], there remains an unmet need for specific data to answer everyday questions that arise in the preparation and modeling of protein solutions; e.g., there are questions such as what will be the phase stability and approximate viscosity of a given protein solution, what type of interactions can be expected in a particular protein solution, and how do its properties change with modifying conditions, such as protein concentration, pH, and temperature. In this study, we developed a database for protein and antibody solutions, called BioMThermDB 1.0, which consists of a broad set of thermophysical and dynamic properties that can help provide adequate answers for such puzzles. The obtained database was created by gathering data from both comprehensive and often scattered scientific literature, as well as from our own experimental results. The database is web-based and enables its users to obtain frequently elusive numerical values of thermophysical quantities. The database is freely available at https://phys-biol-modeling.fkkt.uni-lj.si/biomthermdb.html (current version of 5 October 2022).

2. Results and Discussion

BioMThermDB 1.0 provides thermophysical and thermodynamic data predominantly for globular proteins (e.g., various serum albumins, lysozyme, hemoglobin, etc.) and antibodies but information about other proteins is also available (see Figure 1 and Table 1). Each entry is given as a specific protein solution that contains information about the overall composition of the solution; this includes details about the dissolved protein, such as its concentration and possible PDB-structure code [7]. In addition, information is given on the chemical identity of the buffer, its pH value, the ionic strength of the solution, the temperature, and the possible presence of different excipients (co-solutes) is also taken into account. In its current version, BioMThermDB 1.0 provides several protein-solution properties that are important for determining their stability, such as the hydrodynamic radius, electrophoretic mobility, zeta potential, and the so-called cloud-point temperature, which is the point at which protein solutions separate into two co-existing phases [8–10]. A piece of indispensable information, especially for modeling protein solutions, is also the viscosity of the solvent and protein solution itself [11]. In addition, each entry contains the details of the experimental technique used to obtain the thermophysical data of a certain protein solution, as well as the DOI of the corresponding original article in which the results were first published.

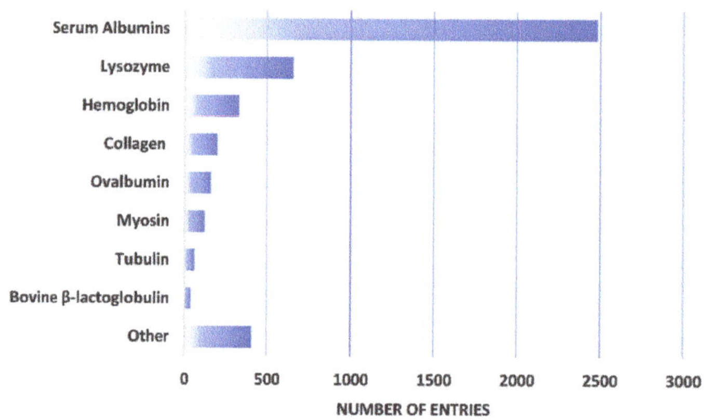

Figure 1. Distribution of thermophysical data entries based on their protein family.

BioMThermDB 1.0 currently is comprised of 5889 specific entries, of which 77.4% belong to globular and other proteins, and the remaining 22.6% are represented by anti-bodies. More than half of all listed protein solutions have their viscosity measured (Figure 2), which makes them very useful in terms of designing protein formulations and verifying calculated results. Figure 2 also reveals that the database already contains at least 500 entries in almost every physical-feature category (the exception being the self-diffusion coefficient) and it will continue to grow further.

Figure 3A demonstrates that the concentration ranges of our database span over all areas, from almost completely diluted to extremely concentrated protein solutions. However, most entries are found in two ranges, namely between 0 and 10, and 101 and 500 mg mL^{-1}, as they together represent 75.15% of all the data. This is due to the well-known fact that interparticle interactions are usually studied in dilute systems; on the other hand, experiments for observing, e.g., liquid–liquid phase separation, are mostly performed at concentrations above 90 mg mL^{-1} [12–14].

Table 1. List of protein entries for group Other in Figure 1.

Protein	Number of Entries
Soy-protein isolate	107
Rice-flour proteins	42
Erythrocytes	30
Horse globulins	26
CP12C75S (C-terminal disulfide bridge mutant)	26
Globulin	24
Conalbumin	19
CP12C31S (N-terminal disulfide bridge mutant)	18
Microtubule-associated Proteins (MAPs)	14
4S α2–β1–glycoprotein	13
Recombinant p53 (1–93)	13
Neurofilaments	11
Serum orosomucoid	10
Fibrinogen	10
Lipoprotein ([4-14C] cholesterol-labeled) in dog's blood serum	10
Wild-type CP12 protein	7
Ovomucoid O	7
Nuclease	6
βL-crystallin	1

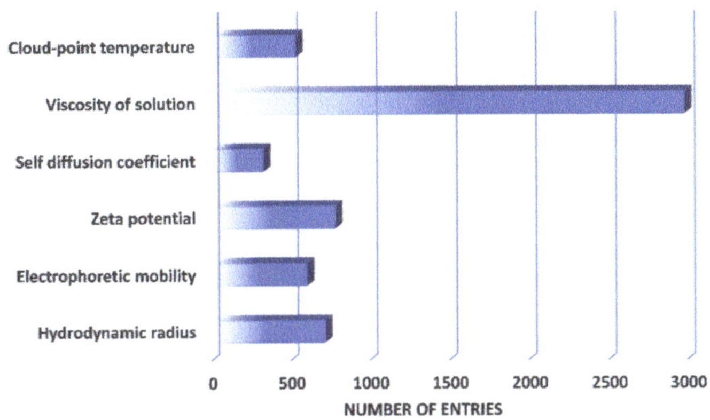

Figure 2. Distribution of thermophysical data entries based on measured properties.

Similar to concentration regimes, BioMThermDB 1.0 covers protein thermophysical data throughout the whole pH range, with experiments carried out even in the harshest known conditions (pH below 2 and above 10), as displayed in Figure 3B. Of course, the majority of entries are in the vicinity of physiological conditions (i.e., 40.61% of entries are between pH = 6 and 8) since they are most important for studying various properties of protein solutions, with an emphasis on the formulations of biological drugs. In terms of pH and buffers, this database is of additional value considering that, given the pH values in combination with buffer identity and the ionic strength of solution, one can more easily shed light on often underestimated buffer-specific effects that could be incorporated into results. Buffers can govern many aspects of protein stability, e.g., conformational, colloidal, and interfacial stability, and, as such, are a non-negligible part of protein solutions [13,15].

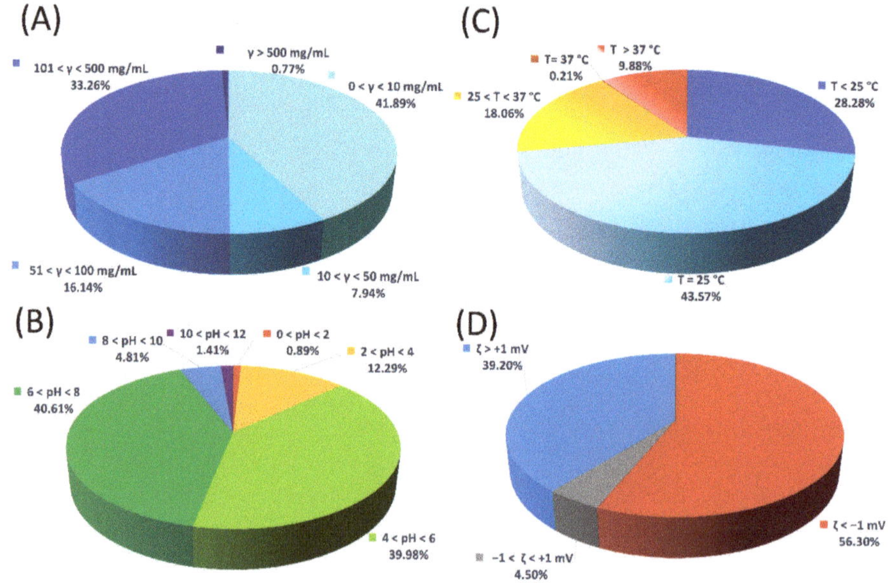

Figure 3. Different distributions of protein thermophysical and thermodynamic data based on (**A**) concentration of protein solutions, (**B**) pH value of protein solutions, (**C**) temperature of protein solutions, and (**D**) zeta potential of protein solutions.

Regarding temperature, one can find most entries (43.57%) at room temperature (Figure 3C) considering that working with those temperatures usually presents the lowest probability of early-protein aggregation onset. However, one is sometimes in search of quite the opposite, namely in the case of when one seeks out the occurrence of protein self-assembly. The self-association of proteins can be achieved by both the cooling and heating of protein solutions. Many such experiments are represented by the database entries whose properties are determined below 25 (28.28%) and above 37 °C (9.88%). The cooling of protein solutions is usually involved in cloud-point measurements, while heating is often necessary for in vitro onset of protein fibrillization [16,17].

Another important physical property that can be used to shorten the time needed to produce trial-protein formulations and help one to both optimize and assess their long-term stability is the zeta potential. The distribution of this indicator of the protein surface charge, as depicted in Figure 3D, shows that only 4.5% of all entries have a zeta potential in the range between −1 and +1 mV, which marks the least stable and most aggregation-prone protein solutions. Protein formulations dominated by repulsive interactions are more stable and, among the entries in BioMThermDB 1.0, these are represented with a pronounced negative (56.30%) or positive (39.20%) zeta potential.

3. Materials and Methods

BioMThermDB 1.0 is currently developed using the HTML programming language and is freely accessible at https://phys-biol-modeling.fkkt.uni-lj.si/biomthermdb.html (current version of 5 October 2022). The editing and ordering of BioMThermDB 1.0 entries, their statistical analysis (calculation of percentages and counting of data), and the final visualization were performed in Microsoft Excel. At the moment, BioMThermDB 1.0 is in tabular form but the database will continue to grow and be upgraded. The development of an efficient data browser is also planned for it in the near future; it will be maintained on a regular basis and both all novelties and upgrades will be published on the BioMThermDB 1.0 homepage.

4. Conclusions

To understand and model the stability of protein solutions, a wide and diverse set of information on the thermophysical and thermodynamic properties of proteins in aqueous solutions is of critical importance. Despite the fact that the thermophysical properties of protein solutions are widely studied, the data are scattered in literature and often not consistent due to different protein batches and different experimental techniques. The BioMThermDB 1.0 database presents an overview of the existing published data and some of our own unpublished thermodynamic and thermophysical data on protein solutions that should help scientists in the theoretical treatment of these systems, as well as help experimentalists in planning new experiments.

Author Contributions: Conceptualization, B.H.-L. and M.L.; methodology, M.N. and B.H.-L.; software, B.H.-L. and M.L.; validation, S.B., M.L. and B.H.-L.; formal analysis, M.N., S.B., M.L. and B.H.-L.; investigation, M.N., S.B. and B.H.-L.; resources, B.H.-L.; data curation, B.H.-L. and M.L.; writing—original draft preparation, S.B.; writing—review and editing, S.B., M.L., B.H.-L., E.C., K.A.D. and C.S.; visualization, S.B.; supervision, B.H.-L.; project administration, B.H.-L.; funding acquisition, B.H.-L. All authors have read and agreed to the published version of the manuscript.

Funding: This research was funded by the National Institutes of Health (NIH) award "Solvation modeling for next-gen biomolecule simulations" (grant no. RM1-GM135136).

Institutional Review Board Statement: Not applicable.

Informed Consent Statement: Not applicable.

Data Availability Statement: Data is available within the article and on the database web-page.

Acknowledgments: M.L. and B.H.-L. acknowledge the support of the Slovenian Research Agency (ARRS) under the core funding nos. P1-0201, BI-US/22-24-125, and BI-US/22-24-063.

Conflicts of Interest: The authors declare no conflict of interest. The funders had no role in the design of the study; in the collection, analyses, or interpretation of data; in the writing of the manuscript; or in the decision to publish the results.

References

1. Zidar, M.; Rozman, P.; Belko-Perkel, K.; Ravnik, M. Control of viscosity in biopharmaceutical protein formulations. *J. Colloid Interface Sci.* **2020**, *580*, 308–317. [CrossRef] [PubMed]
2. Wang, W. Protein aggregation and its inhibition in biopharmaceutics. *Int. J. Pharm.* **2005**, *289*, 1–30. [CrossRef] [PubMed]
3. Nikam, R.; Kulandaisamy, A.; Harini, K.; Sharma, D.; Gromiha, M.M. ProThermDB: Thermodynamic database for proteins and mutants revisited after 15 years. *Nucleic Acids Res.* **2021**, *49*, D420–D424. [CrossRef] [PubMed]
4. Kulandaisamy, A.; Sakthivel, R.; Gromiha, M.M. MPTherm: Database for membrane protein thermodynamics for understanding folding and stability. *Brief. Bioinform.* **2021**, *22*, 2119–2125. [CrossRef] [PubMed]
5. Jemimah, S.; Yugandhar, K.; Gromiha, M.M. PROXiMATE: A database of mutant protein–protein complex thermodynamics and kinetics. *Bioinformatics* **2017**, *33*, 2787–2788. [CrossRef]
6. Kumar, M.D.S.; Gromiha, M.M. PINT: Protein–protein Interactions Thermodynamic Database. *Nucleic Acids Res.* **2006**, *34*, D195–D198. [CrossRef]
7. Berman, H.M.; Westbrook, J.; Feng, Z.; Gilliland, G.; Bhat, T.N.; Weissig, H.; Shindyalov, I.N.; Bourne, P.E. The Protein Data Bank. *Nucleic Acids Res.* **2000**, *28*, 235–242. [CrossRef] [PubMed]
8. McManus, J.J.; Charbonneau, P.; Zaccarelli, E.; Asherie, N. The physics of protein self-assembly. *Curr. Opin. Colloid Interface Sci.* **2016**, *22*, 73–79. [CrossRef]
9. Mason, B.D.; van Enk, J.Z.; Zhang, L.; Remmele, R.L., Jr.; Zhang, J. Liquid-liquid phase separation of a monoclonal antibody and nonmonotonic influence of hofmeister anions. *Biophys. J.* **2010**, *99*, 3792–3800. [CrossRef]
10. Kastelic, M.; Kalyuzhnyi, Y.V.; Hribar-Lee, B.; Dill, K.A.; Vlachy, V. Protein aggregation in salt solutions. *Proc. Natl. Aacd. Sci. USA* **2015**, *112*, 6766–6770. [CrossRef] [PubMed]
11. Kastelic, M.; Dill, K.A.; Kalyuzhnyi, Y.V.; Vlachy, V. Controlling the viscosities of antibody solutions through control of their binding sites. *J. Mol. Liq.* **2018**, *270*, 234–242. [CrossRef] [PubMed]
12. Janc, T.; Kastelic, M.; Bončina, M.; Vlachy, V. Salt-specific effects in lysozyme solutions. *Condens. Matter Phys.* **2016**, *19*, 1–12. [CrossRef]
13. Brudar, S.; Hribar-Lee, B. Effect of Buffer on Protein Stability in Aqueous Solutions: A Simple Protein Aggregation Model. *J. Phys. Chem. B* **2021**, *125*, 2504–2512. [CrossRef] [PubMed]

14. Yadav, S.; Scherer, T.M.; Shire, S.J.; Kalonia, D.S. Use of dynamic light scattering to determine second virial coefficient in a semidilute concentration regime. *Anal. Biochem.* **2011**, *411*, 292–296. [CrossRef] [PubMed]
15. Salis, A.; Monduzzi, M. Not only pH. Specific buffer effects in biological systems. *Curr. Opin. Colloid Interface Sci.* **2016**, *23*, 1–9. [CrossRef]
16. Jaklin, M.; Hritz, J.; Hribar-Lee, B. A new fibrillization mechanism of β-lactoglobulin in glycine solutions. *Int. J. Biol. Macromol.* **2022**, *216*, 414–425. [CrossRef] [PubMed]
17. Brudar, S.; Hribar-Lee, B. The Role of Buffers in Wild-Type HEWL Amyloid Fibril Formation Mechanism. *Biomolecules* **2019**, *9*, 65. [CrossRef] [PubMed]

International Journal of
Molecular Sciences

MDPI

Article

Reaction-Diffusion Systems: Self-Balancing Diffusion and the Use of the Extent of Reaction as a Descriptor of Reaction Kinetics

Miloslav Pekař

Faculty of Chemistry, Brno University of Technology, Purkyňova 118, 612 00 Brno, Czech Republic; pekar@fch.vut.cz

Abstract: Self-balancing diffusion is a theoretical concept that restricts the introduction of extents of reactions. This concept is analyzed in detail for general mass- and molar-based balances of reaction-diffusion mixtures, in relation to non-self-balancing cases, and with respect to its practical consequences. Self-balancing is a mathematical restriction on the divergences of diffusion fluxes. Fulfilling this condition enables the proper introduction of the extents of (independent) reactions that reduce the number of independent variables in thermodynamic descriptions. A note on a recent generalization of the concept of reaction and diffusion extents is also included. Even in the case of self-balancing diffusion, such extents do not directly replace reaction rates. Concentration changes caused by reactions (not by diffusion) are properly described by rates of independent reactions, which are instantaneous descriptors. If an overall descriptor is needed, the traditional extents of reactions can be used, bearing in mind that they include diffusion-caused changes. On the other hand, rates of independent reactions integrated with respect to time provide another overall, but reaction-only-related descriptor.

Keywords: diffusion; extent of reaction; independent reactions; reaction rate

1. Introduction

The simultaneous occurrence of chemical reactions and diffusion is a common phenomenon encountered both in engineered and natural (biochemical) processes and covered by numerous studies and books (for example, [1–8]). The proper (thermodynamic) modelling of reaction-diffusion systems is thus important in the design of technological processes as well as in the study of processes of natural origin. Reactions and diffusion cause local changes in amounts of species (concentrations); in the former case, these changes are constrained by the reaction stoichiometry. The stoichiometric links to diffusion are rarely studied. Both reactions and diffusion are of molecular origin; their theoretical description can be based either on microscopic (statistical, kinetic theory) or macroscopic (phenomenological, continuum) approaches. In this text we are concerned with the latter type. We only mention a recent work on the statistical theory of kinetics altered by diffusion, which operates with stoichiometry [9] but not in the sense of constraints as described in this work. Several reviews have recently been published overviewing the basics of a thorough mathematical and thermodynamic description of reaction-diffusion systems at the macroscopic level. We start with a brief report on them focusing particularly on the stoichiometric impacts on reaction-diffusion systems.

Datta and Vilekar [10] based their detailed treatment of diffusion driving forces on the balance of (linear) momentum, from which reaction rates were eliminated using the equation of the continuity of mixture components, i.e., their mass balances. Thus, they do not discuss the links between the reaction rate and diffusion and the impacts of stoichiometry on these linked variables. Their approach uses barycentric velocity as the reference for component diffusion velocities but is not restricted to this reference velocity.

They placed a variety of diffusion driving forces into a remarkably unified framework of continuum theory.

Whitaker [11] analyzed the balance equations of continuum mechanics and thermodynamics to discuss the relationship between diffusion forces and fluxes. This work is based on the Stokesian fluid mixture model for the partial (species) stress tensor and on the caloric equation of state, which supposes that entropy and species densities are independent variables of the total specific energy of a mixture. Though the mixture is supposed to be a reacting mixture, no other links between chemical reactions and diffusion are developed. The author concludes that the gradient of the chemical potential is not justified as a (single or leading) driving force for the diffusive flux.

Morro [12] overviewed the balance equation for reacting fluid mixtures and presents the equations for diffusion fluxes. The author came to the conclusion that the driving term for diffusion is the gradient of chemical potential rescaled by temperature. The author follows a similar line in a later paper [13]. Both papers involve considerations on the constitutive equations for diffusion fluxes and an analysis of the consequences of entropic inequality (the second law of thermodynamics). No account of the impact of stoichiometry is given in either one.

Yet another contribution by the same author [14] is concerned with the same general balances and equations, but in the context of a reacting mixture of thermoelastic solid constituents. The possibility to express the rate of reaction in terms of the extent of the reaction is briefly touched upon. However, this note is restricted only to cases in which the diffusion is negligible, and no stoichiometric consequences are described.

Already several decades ago, Bowen published a detailed mathematical analysis of the effects of the permanence of atoms (mass conservation) on reaction rates—essentially an analysis of the linear algebra consequences of stoichiometry [15]. He not only showed that rates of individual independent reactions are a consequence of this linear algebra, but also derived a limitation on the use of reaction extent as a descriptor of reaction rates. His results limit this use to non-diffusing reacting mixtures only. Later, Truesdell in his comprehensive treatment on the rational thermodynamics [16] noted briefly that this restriction is too strong and showed that reaction extent can also be used for mixtures with diffusion, providing that the diffusion is self-balancing. These interesting and important results seem not to have been analyzed or applied in chemical reaction engineering or in the thermodynamic modelling of reaction-diffusion systems. This is probably because both authors present only general results, though Bowen gives an example of a mixture of different water phases.

The aim of this work is to look at self-balancing diffusion and the applicability of the reaction extent in more detail. The self-balancing diffusion is the principal object of this study; this work shows what are practical manifestations of the general definition of the self-balancing diffusion (Truesdell gave only the definition), how it could be detected in reality, reformulated into molar description (in contrast to the original, mass-based approach), and what are the consequences of the diffusion not being self-balancing. Another aim is to note the links to a recent generalization of extents [17].

2. Results and Discussion

2.1. Self-Balancing Diffusion and the Extent of Reaction

The charm of self-balancing diffusion is that it enables the extent of reaction to be introduced properly as a descriptor of reaction kinetics, instead of the reaction rate J^α in (S14), or r_α in (40). Then, we should consider only $n-h$ extents instead of n rates. Let us select some reference point, usually the starting point of a reaction, in which the vector $\boldsymbol{\omega}$ has a (constant) value $\boldsymbol{\omega}_0$, and define a new vector:

$$\boldsymbol{\xi} = \boldsymbol{\omega} - \boldsymbol{\omega}_0. \quad (1)$$

If the diffusion is self-balancing then, according to (42), the vector $\dot{\boldsymbol{\xi}}$ is in the reaction space and thus also the vector $\boldsymbol{\xi}$ lies there [16]. The latter vector can therefore be expressed in the basis $\mathbf{g}^p$ of the $(n-h)$-dimensional reaction space [15]; [18] (p. 153):

$$\boldsymbol{\xi} = \sum_{p=1}^{n-h} \xi_p \mathbf{g}^p. \tag{2}$$

This basis is defined as follows [18] (p. 153):

$$\mathbf{g}^p = \sum_{\alpha=1}^{n} P^{p\alpha} \mathbf{e}_\alpha, \ p = 1, \ldots, n-h \tag{3}$$

where, $P^{p\alpha}$ is the stoichiometric coefficient of component α in (independent) reaction p. Combining (S16), (1)–(3) we obtain

$$\sum_{p=1}^{n-h} P^{p\alpha} \xi_p = \frac{w_\alpha - w_\alpha^0}{M_\alpha}, \quad \alpha = 1, \ldots, n. \tag{4}$$

Vector $\boldsymbol{\xi}$ is called the vector of extents of reactions because its components (coordinates) ξ_p in the reaction space are the extents of (independent) reactions p. The condition of self-balancing diffusion (45) can be thus understood as a condition required for the introduction of reaction extents in a reaction-diffusion system [16]. This condition seems to be unknown in the area of chemical reaction engineering and thus has never been tested. Definition (1) shows that

$$\dot{\boldsymbol{\xi}} = \dot{\boldsymbol{\omega}} \tag{5}$$

and n components (in the mixture space) of the vector $\dot{\boldsymbol{\omega}}$ can be replaced by $n-h$ components (in the reaction space) of the vector $\dot{\boldsymbol{\xi}}$. In this way, the number of quantities necessary for the mathematical description of thermodynamics of a reaction-diffusion system is reduced. Further, if the diffusion is self-balancing, we can express the mass fractions in arbitrary reaction time from (4) as a function of extents only:

$$w_\alpha = w_\alpha^0 + M_\alpha \sum_{p=1}^{n-h} P^{p\alpha} \xi_p. \tag{6}$$

The components (coordinates) ξ_p are equivalent to the extents of reactions used traditionally in chemical (engineering) kinetics. The existence of extents enables the number of independent variables to be reduced. Thus, according to (4) or (6), n mass fractions w_α can be replaced only by $n-h$ reaction extents ξ_p. Remember that the vector $\boldsymbol{\xi}$ lies in the n-dimensional mixture space and, in the case of the self-balancing diffusion, also in its $(n-h)$-dimensional subspace (the reaction space) at the same time. The question of its (and similarly located vectors) dimension is then pointless. From a practical point of view, extents of reactions express concentration changes (caused by reaction or by diffusion) in a different manner; whereas reaction rates in (S20) express actual concentration changes per unit of time, reaction extents, as defined by (1), express these changes relatively to some fixed point in time, i.e., they are not of the 'per time' dimension. In other words, reaction rates are a sort of differential quantity, whereas extents are a sort of integral quantity.

Generally, the time derivative of the vector of extents in (5) still embraces diffusion rates, not only reaction rates, as (42), (S19) and (5) show: $\dot{\boldsymbol{\xi}} = \boldsymbol{\sigma} + \boldsymbol{\omega}^+$. When there is no diffusion ($\boldsymbol{\sigma} = 0$), reaction extents are equivalent to reactions rates: $\dot{\boldsymbol{\xi}} = \boldsymbol{\omega}^+$. Equation (5) does not imply identity or one-to-one proportionality between reaction rates and reaction extents. This lack of identity or one-to-one proportionality is seen if the derivatives of extents, which for self-balancing diffusion are also located in the reaction space, are expressed on the basis of this space and combined with $\dot{\boldsymbol{\xi}} = \boldsymbol{\sigma} + \boldsymbol{\omega}^{+}$, (3) and (S16). Accordingly,

$$M_\alpha J^\alpha - \mathrm{div}\rho_\alpha \mathbf{u}_\alpha^w = \rho M_\alpha \sum_{p=1}^{n-h} P^{p\alpha} \dot{\xi}_p. \tag{7}$$

The one-to-one proportionality is achieved only in non-diffusing mixtures ($\mathbf{u}_\alpha^w = 0$) or, at least theoretically, in "divergence-less" diffusion ($\mathrm{div}\rho_\alpha \mathbf{u}_\alpha^w = 0$). The latter case can be illustrated by a simple one-dimensional diffusion along the x-axis, when the y- and z-components in the diffusion velocity vector are zero. Accordingly,

$$\frac{\partial(\rho_\alpha u_{\alpha x}^w)}{\partial x} = 0 \Rightarrow \rho_\alpha u_{\alpha x}^w = const. \Rightarrow u_{\alpha x}^w = \frac{const.}{\rho_\alpha}, \tag{8}$$

where *const.* can be a function of the other two space coordinates (and time).

Note that the whole analysis was based only on two very general principles—the mass conservation and the permanence of atoms, and thus is valid for any specific reaction-diffusion system or model. Further, it shows general impacts of the reaction stoichiometry, which is closely related to the permanence of atoms, on diffusion.

2.2. Self-Balancing Diffusion in Practice

What does self-balancing diffusion mean and when is diffusion self-balancing? The answer is very simple, when the rank of the matrix of atomic composition (for details see [18] (p. 151)) is equal to one. In this case, there is only one basis vector $\mathbf{f}_1$ with components $S_{1\alpha}$. The atomic mass of the only atomic element (or pseudoatomic substance—see the example of NO_2 in dimerization below), E^1, can be expressed using the molar mass of any component—without any loss of generality, let us choose M_1. Then $E^1 = M_1/S_{11}$ and the general condition (44) is

$$M_1\sigma_1 + \frac{M_1}{S_{11}} S_{12}\sigma_2 + \cdots + \frac{M_1}{S_{11}} S_{1n}\sigma_n = 0. \tag{9}$$

Multiplying (9) by S_{11}/M_1 we obtain

$$S_{11}\sigma_1 + S_{12}\sigma_2 + \cdots + S_{1n}\sigma_n \equiv \boldsymbol{\sigma}\cdot\mathbf{f}_1 = 0 \tag{10}$$

which is the self-balancing diffusion condition (45) for the case $h = 1$ and in this case every diffusion is self-balancing. As an example, let us use the reacting mixture of NO_2 and N_2O_4, which describes the dimerization of nitrogen dioxide. If NO_2 is selected as the pseudoatomic substance (numbered as 1), the matrix of atomic composition $\|S_{\sigma\alpha}\|$ is $\begin{bmatrix} 1 & 2 \end{bmatrix}$. The self-balancing condition is then

$$\sigma_{NO_2} + 2\sigma_{N_2O_4} = 0. \tag{11}$$

Thus, the divergences of the diffusion fluxes of the individual components are combined in this condition according to the representation of the pseudoatomic substance in these components. The general condition (44) is really very close to the self-balancing condition in this simple case: $M_{NO_2}\sigma_{NO_2} + M_{N_2O_4}\sigma_{N_2O_4} = 0$. Note that in this example there is only one independent reaction because $n - h = 1$.

As an example of a mixture where $h = 2$, let us select the mixture for ammonia synthesis: N_2 (1), H_2 (2), and NH_3 (3). There are two atomic substances: N (1) and H (2). The compositional matrix is

$$\begin{bmatrix} 2 & 0 & 1 \\ 0 & 2 & 3 \end{bmatrix}.$$

The two basis vectors are: $\mathbf{f}_1 = (2;0;1)$ and $\mathbf{f}_2 = (0;2;3)$. There are two self-balancing diffusion conditions:

$$2\sigma_{N_2} + \sigma_{NH_3} = 0, \tag{12a}$$

$$2\sigma_{H_2} + 3\sigma_{NH_3} = 0. \tag{12b}$$

The first condition restricts the diffusion fluxes according to the representation of nitrogen atoms in all components, whereas the second restricts the diffusion fluxes according to hydrogen atoms. Self-balancing diffusion means that the divergences of diffusion fluxes are balanced with respect to the atomic composition of corresponding components:

$$\frac{2}{M_{N_2}}\mathrm{div}\mathbf{j}^w_{N_2} + \frac{1}{M_{NH_3}}\mathrm{div}\mathbf{j}^w_{NH_3} = 0, \tag{13a}$$

$$\frac{2}{M_{H_2}}\mathrm{div}\mathbf{j}^w_{H_2} + \frac{3}{M_{NH_3}}\mathrm{div}\mathbf{j}^w_{NH_3} = 0. \tag{13b}$$

The first Equation (13a) refers to the balance with respect to nitrogen, whereas the second, (13b), refers to the balance with respect to hydrogen. Both equations contain the diffusion fluxes of reactant and product.

The self-balancing condition can be combined with the general condition (44). First, let us modify the general condition:

$$\mathbf{M}\cdot\boldsymbol{\sigma} = -\sum_{\alpha=1}^{n} M_\alpha \frac{1}{\rho M_\alpha}\mathrm{div}\ \mathbf{j}^w_\alpha = -\frac{1}{\rho}\sum_{\alpha=1}^{n}\mathrm{div}\ \mathbf{j}^w_\alpha = 0. \tag{14}$$

Consequently,

$$\sum_{\alpha=1}^{n}\mathrm{div}\mathbf{j}^w_\alpha = 0. \tag{15}$$

The divergence of the ammonia diffusion flux can be eliminated from (13), giving

$$\frac{3}{M_{N_2}}\mathrm{div}\mathbf{j}^w_{N_2} = \frac{1}{M_{H_2}}\mathrm{div}\mathbf{j}^w_{H_2}. \tag{16}$$

This equation expresses the consequences of the self-balancing condition for diffusion in terms of the diffusion fluxes of reactants. Equation (15) can then be written by eliminating, for example, the hydrogen flux divergence using (16); the result is Equation (13a). This example thus shows that conditions (14) and (13) are consistent, but not equivalent—the specific condition (13a) does not follow from the general condition (14); the former is stronger than the latter.

Equation (16) can be transformed into molar diffusion fluxes $\mathbf{j}'^w_\alpha$, which have units mol m^{-2} s^{-1}:

$$\frac{\mathbf{j}^w_\alpha}{M_\alpha} = \frac{\rho_\alpha \mathbf{u}^w_\alpha}{M_\alpha} = c_\alpha \mathbf{u}^w_\alpha = \mathbf{j}'^w_\alpha. \tag{17}$$

The result is:

$$3\mathrm{div}\mathbf{j}'^w_{N_2} = \mathrm{div}\mathbf{j}'^w_{H_2}. \tag{18}$$

Thus, self-balancing diffusion means the balancing of reactant molar diffusion fluxes (their divergences) as stated by (18) in this example.

Generally, the number of self-balancing conditions is equal to the number of atoms (or pseudoatomic substances) present in the reacting mixture. Each condition balances the (divergences of the) molar diffusion fluxes of all components containing the given atom γ with respect to the number of the atom in each component. Thus,

$$\sigma_1 f_{\gamma 1} + \sigma_2 f_{\gamma 2} + \cdots + \sigma_n f_{\gamma n} = 0 \tag{19}$$

where $f_{\gamma\alpha}$ represents the number of atom γ (or pseudoatomic substance) in component α. It would be desirable to look at published experimental data if such diffusion was observed and in which circumstances. Equation (18) suggests that such diffusion could occur when the initial (input) reaction mixture contains reactants in stoichiometric ratio.

2.3. Molar-Based View

The above analysis was based on mass balances formulated in terms of densities (mass concentrations or mass fractions). In chemistry or chemical engineering, molar amounts and molar concentrations (molar fractions) are more common. However, molar balances do not enable similarly simple and clear equations to be formulated. This is because of the fact that—in contrast to mass—molar amounts are not conserved in chemical reactions. Because $\rho_\alpha = c_\alpha M_\alpha$, balance (36) is transformed as:

$$\frac{\partial c_\alpha}{\partial t} + \mathrm{div} c_\alpha \mathbf{v}_\alpha = \frac{r_\alpha}{M_\alpha} = J^\alpha. \tag{20}$$

The divergence term can be modified using the barycentric velocity:

$$\mathrm{div} c_\alpha \mathbf{v}_\alpha = \mathrm{div} c_\alpha \mathbf{u}_\alpha^w + c_\alpha \mathrm{div}\, \mathbf{v}^w + \mathbf{v}^w.\mathrm{grad}\, c_\alpha. \tag{21}$$

Thus, instead of (40), we have

$$\dot{c}_\alpha = -\mathrm{div} c_\alpha \mathbf{u}_\alpha^w - c_\alpha \mathrm{div}\, \mathbf{v}^w + J^\alpha \tag{22}$$

which, in contrast to (40), also contains the (divergence of) the barycentric velocity and is not of the form of (42). This form results only when the barycentric velocity is zero (or divergence-less):

$$\dot{c}_\alpha = -\mathrm{div} c_\alpha \mathbf{u}_\alpha^w + J^\alpha. \tag{23}$$

Note that $c_\alpha \mathbf{u}_\alpha^w = \mathbf{j}_\alpha'^w$, the molar diffusion flux in the barycentric reference.

Another way to transform mass balance into molar balance is to apply the material derivative with respect to the corresponding component α:

$$\grave{c}_\alpha^\alpha = \frac{\partial c_\alpha}{\partial t} + \mathbf{v}_\alpha.\mathrm{grad}\, c_\alpha, \quad \alpha = 1, \ldots, n. \tag{24}$$

The result is

$$\grave{c}_\alpha^\alpha = -c_\alpha \mathrm{div}\, \mathbf{v}_\alpha + J^\alpha \tag{25}$$

but does not contain diffusion velocity. It can be introduced generally with respect to an arbitrary referential velocity: $\mathbf{u}_\alpha^{\mathrm{ref}} = \mathbf{v}_\alpha - \mathbf{v}^{\mathrm{ref}}$. Then

$$\grave{c}_\alpha^\alpha = -\mathrm{div} c_\alpha \mathbf{u}_\alpha^{\mathrm{ref}} - c_\alpha \mathrm{div}\, \mathbf{v}^{\mathrm{ref}} + J^\alpha \tag{26}$$

which, however, is of the form of (22) and not (42), unless the referential velocity is zero (or divergence-less). Finally, we need not strive to have the material derivative in molar balances; thus, we can write (20) as

$$\frac{\partial c_\alpha}{\partial t} + \mathrm{div} c_\alpha \mathbf{u}_\alpha^{\mathrm{ref}} + \mathrm{div} c_\alpha \mathbf{v}^{\mathrm{ref}} = J^\alpha. \tag{27}$$

This again is of the form of (22), and the form of (42) can be achieved for zero referential velocity or for the divergence-less "referential molar diffusion flow" $c_\alpha \mathbf{v}^{\mathrm{ref}}$.

The concept of self-balancing can be transferred to various balances using a proper definition of vectors in the general balance form of (42). Several examples are given in Table 1.

Table 1. Definitions of vectors to be used in addressing self-balanced diffusion in various balance frameworks.

Balance	$\dot{\boldsymbol{\omega}}$	$\boldsymbol{\omega}^{+}$	$\boldsymbol{\sigma}$
(40)	$\sum_{\alpha=1}^{n} \frac{\dot{w}_\alpha}{M_\alpha} \mathbf{e}_\alpha$	$\sum_{\alpha=1}^{n} \frac{J^\alpha}{\rho} \mathbf{e}_\alpha$	$-\sum_{\alpha=1}^{n} \frac{1}{\rho M_\alpha} (\mathrm{div}\rho_\alpha \mathbf{u}_\alpha^w) \mathbf{e}_\alpha$
(23)	$\sum_{\alpha=1}^{n} \dot{c}_\alpha \mathbf{e}_\alpha$	$\sum_{\alpha=1}^{n} J^\alpha \mathbf{e}_\alpha$	$-\sum_{\alpha=1}^{n} (\mathrm{div} c_\alpha \mathbf{u}_\alpha^w) \mathbf{e}_\alpha$
(22)	"	"	$-\sum_{\alpha=1}^{n} (\mathrm{div} c_\alpha \mathbf{u}_\alpha^w + c_\alpha \mathrm{div}\, \mathbf{v}^w) \mathbf{e}_\alpha$
(27)	$\sum_{\alpha=1}^{n} \frac{\partial c_\alpha}{\partial t} \mathbf{e}_\alpha$	"	$-\sum_{\alpha=1}^{n} (\mathrm{div} c_\alpha \mathbf{u}_\alpha^{\mathrm{ref}} + \mathrm{div} c_\alpha \mathbf{v}^{\mathrm{ref}}) \mathbf{e}_\alpha$

2.4. When Diffusion Is Not Self-Balancing

Of course, definition (1) can be used generally, but if diffusion is not self-balancing, it yields nothing special. Combining (1) with (42), and after integration, we obtain only

$$\boldsymbol{\omega} - \boldsymbol{\omega}_0 == \int_0^t \left(\boldsymbol{\sigma} + \boldsymbol{\omega}^{+}\right) dt, \tag{28}$$

i.e., the extent is just another denomination of the integral comprising diffusion and (component) reaction rates. Note that in chemistry a reaction network with specified stoichiometric coefficients is designed first, and then the extents of individual reactions in the network are defined by relationships similar to (4):

$$\xi_i = \frac{w_\alpha - w_\alpha^0}{P^{i\alpha} M_\alpha}. \tag{29}$$

However, in (29), w_α should refer only to the concentration of component α reacting in reaction i, which is practically indeterminable. In contrast, the technique described in this work first naturally derives a set of acceptable and independent reactions satisfying the permanence of atoms (and related linear algebra) together with their stoichiometric coefficients and only then introduces the extents by (4); this is the mathematically correct procedure. In the general case, there is probably no need to introduce the extents of reactions, their role could be played by rates of independent reactions (independent in the sense of linear algebra [15]) derived by the reported technique. Introducing these rates into balances (S14), we obtain:

$$\frac{\dot{w}_\alpha}{M_\alpha} = -\frac{1}{\rho M_\alpha} \mathrm{div}\rho_\alpha \mathbf{u}_\alpha^w + \frac{1}{\rho} \sum_{p=1}^{n-h} P^{p\alpha} J_p, \quad \alpha = 1, \ldots, n \tag{30}$$

where J_p is the rate of (independent) reaction p [18] (p. 153). Equation (30) illustrates that changes in the concentration of each component are caused by diffusion (the first term on the right hand side) and the reactions in which it takes place (the second term)—only $n-h$ independent reactions can be considered. In contrast, balance (S14) does not directly show individual reactions and their rates.

2.5. Note on Generalized Extents; Summarizing Notes

Rodrigues et al. [17] proposed a generalization of the concept of the reaction extent (and other extents in general) to distributed reaction systems (i.e., space-distributed systems with diffusion). A more detailed comparison can be found in Supplementary Material.

Here, we only note that the approach of Rodrigues et al. [17] can be combined with the methodology presented in this paper by a (formal) splitting of the vector $\boldsymbol{\omega}$ into reaction and diffusion contributions:

$$\boldsymbol{\omega} = \boldsymbol{\omega}_{\mathrm{r}} + \boldsymbol{\omega}_{\mathrm{d}}. \tag{31}$$

Defining

$$\dot{\boldsymbol{\omega}}_{\rm r} = \boldsymbol{\omega}^{+},\ \dot{\boldsymbol{\omega}}_{\rm d} = \boldsymbol{\sigma}, \tag{32}$$

using (42) we obtain

$$\dot{\boldsymbol{\omega}}_{\rm r} + \dot{\boldsymbol{\omega}}_{\rm d} = \boldsymbol{\omega}^{+} + \boldsymbol{\sigma} = \dot{\boldsymbol{\omega}}. \tag{33}$$

The equations under (32) are analogs of (S25) and (S26) taken over from [17]. They seem to bring nothing new to the methodology of this paper. The vector $\boldsymbol{\omega}^{+}$ is always in the reaction space and therefore so is vector $\dot{\boldsymbol{\omega}}_{\rm r}$. If the vector $\boldsymbol{\sigma}$ is not in the reaction space, then neither is $\dot{\boldsymbol{\omega}}_{\rm d}$ and the situation with introducing the extent of reaction with reference to a fixed point in time ($\boldsymbol{\omega}_0$) is the same as that without the splitting described in (31).

On the other hand, Equation (32) can be utilized similarly as in [17] to define generalized extents, i.e., in (S25) and (S26), which is not within the scope of this work. The advantage of introducing generalized reaction ($\mathbf{x}_{\rm r}$) and diffusion ($\mathbf{x}_{\rm d}$) extents should consist of a clear indication of which concentration changes (of a component) are caused by reactions and which by diffusion. Yet, reaction-caused changes are already given by the rates of independent reactions, cf. Equation (S20) and text around Equation (30):

$$\dot{\boldsymbol{\omega}}_{\rm r} \equiv \boldsymbol{\omega}^{+} = \sum_{\alpha=1}^{n} \frac{J^{\alpha}}{\rho} \mathbf{e}_{\alpha} = \sum_{\alpha=1}^{n} \left(\frac{1}{\rho} \sum_{p=1}^{n-h} P^{p\alpha} J_p \right) \mathbf{e}_{\alpha} \equiv \mathbf{J}. \tag{34}$$

If the reaction rates are expressed explicitly as functions of concentrations, e.g., by the mass-action law common in kinetics, a model for the diffusion flux is employed (e.g., Fick's law), the differential equations of mass balance can be solved, concentrations at any time (and space point) can be obtained, and from them their time derivatives and the rates of independent reactions at any time can be calculated, giving the (rate of) change caused by reactions; the remaining change is the result of diffusion, as Equations (S14) or (42) also indicate: $\boldsymbol{\sigma} = \dot{\boldsymbol{\omega}} - \boldsymbol{\omega}^{+}$. The only difference is that the reaction rates (or diffusion fluxes) show actual rates of concentration changes, whereas the extents show changes relative to some reference. Thus, units of extent do not contain time units, as already noted below (6). Concretely, the units of ξ based on (S16) are mol g^{-1}, the units of $\mathbf{x}_{\rm r}$ are mol m^{-3}, and the units of J_p are mol m^{-3} s^{-1}. Extents are thus useful when one is not satisfied with actual or instantaneous characteristics (rates) but, for a particular reason, prefers overall or integral quantities (extents). On the other hand, reaction rates can also be integrated using (34), giving an overall descriptor. Integration in (28) provides a combined (reaction plus diffusion) overall descriptor.

Note that Equation (34) enables the compact balance form (42) to be expressed in terms of reaction rates:

$$\boldsymbol{\sigma} = \dot{\boldsymbol{\omega}} - \boldsymbol{\omega}^{+} = \dot{\boldsymbol{\omega}} - \mathbf{J} = \dot{\boldsymbol{\xi}} - \mathbf{J} \tag{35}$$

(Equation (5) was used in the last equality). Equation (35) shows explicitly that (only) when there is no diffusion, (independent) reaction rates are directly equal to the time derivatives of the extents of these reactions.

3. Methods and Materials

The full theoretical overview is presented in Supplementary Material, here only the basic facts are reproduced (for details see [15,16,18]). A list of symbols is attached at the end; here, we only stress that the composition is expressed using the mass density ρ_{α}, which is the mass of constituent α in unit volume of the whole mixture; in chemistry it is usually called the mass or weight concentration.

The local mass balance for constituent α is:

$$\frac{\partial \rho_{\alpha}}{\partial t} + \mathrm{div}\rho_{\alpha}\mathbf{v}_{\alpha} = r_{\alpha} \tag{36}$$

and for the whole mixture

$$\frac{\partial \rho}{\partial t} + \sum_{\alpha=1}^{n} \mathrm{div}\rho_\alpha \mathbf{v}_\alpha = 0. \tag{37}$$

using the barycentric velocity $\mathbf{v}^w$, balance (37) is rewritten (the material derivative relative to the barycentric velocity is symbolized by a dot):

$$\dot{\rho} + \rho \ \mathrm{div}\ \mathbf{v}^w = 0. \tag{38}$$

The diffusion velocity $\mathbf{u}_\alpha^w$, defined relative to the barycentric velocity:

$$\mathbf{u}_\alpha^w = \mathbf{v}_\alpha - \mathbf{v}^w, \tag{39}$$

is used to rewrite the component balance (36) in the following form:

$$\rho \dot{w}_\alpha = -\mathrm{div}\rho_\alpha \mathbf{u}_\alpha^w + r_\alpha. \tag{40}$$

Defining an abstract n-dimensional vector space, called the mixture space, with a basis $\mathbf{e}^\alpha$ [15] (see also [18] (pp. 151–152)), and

$$\mathbf{M} = \sum_{\alpha=1}^{n} M_\alpha \mathbf{e}^\alpha, \tag{41}$$

the balances (40) can be written in the compact form

$$\dot{\boldsymbol{\omega}} = \boldsymbol{\sigma} + \boldsymbol{\omega}^+, \tag{42}$$

where the vectors $\boldsymbol{\sigma}$ and $\boldsymbol{\omega}^+$ with units mole kg^{-1} s^{-1} are defined in Supplementary Material.

Due to the linear algebra of the permanence of atoms, the mixture space is divided into two orthogonal subspaces—one (of the dimension $h < n$) which has no special name and is denoted as W, the other (of the dimension $n - h$) which is called the reaction space [15]. This division is an outcome of linear algebra of stoichiometry. The reaction rates and, thus, also the vector $\boldsymbol{\omega}^+$ lie in the reaction space. In contrast, the vector of molar masses is located in the subspace W [15] and from the orthogonality of the two subspaces it follows that

$$\mathbf{M}\cdot\boldsymbol{\omega}^+ = 0. \tag{43}$$

From (S17), it follows that $\mathbf{M}\cdot\dot{\boldsymbol{\omega}} = 0$, and combining with (42) and (43) we obtain

$$\mathbf{M}\cdot\boldsymbol{\sigma} = 0. \tag{44}$$

This is a general condition of diffusion in chemically reacting mixtures, which restricts the (divergence of) diffusion fluxes (hidden in $\boldsymbol{\sigma}$).

Diffusion is called self-balancing by Truesdell [16] if the vector $\boldsymbol{\sigma}$ also lies in the reaction space. Then, the vector $\dot{\boldsymbol{\omega}}$ is also located in the reaction space [16]. Any vector from the reaction space is perpendicular to the basis of the subspace W, which we denote $\mathbf{f}_\sigma$ $(\sigma = 1, \ldots, h)$. Consequently, we can formulate the following condition for diffusion to be self-balancing:

$$\boldsymbol{\sigma} \cdot \mathbf{f}_\sigma = 0. \tag{45}$$

4. Conclusions

Self-balancing diffusion is a theoretical condition that enables the proper introduction of extents of (independent) reactions as an overall reference to some fixed point (the composition at this point) conforming to the linear algebra of the permanence of atoms and stoichiometry. Unfortunately, this condition seems to have been ignored in chemistry and chemical engineering. Such extents reduce the number of independent variables in the thermodynamic description of reacting mixtures by replacing concentration variables (mass fractions, e.g.,), because the number of (independent) reactions is lower than the number of components (and component reaction rates). Theoretical descriptions of reaction-diffusion systems, including mathematical modeling of their thermodynamics, can and even should use this reduced number of variables approach. This approach is mathematically sufficient to account for (concentration) changes caused by reactions and diffusion processes. On the other hand, further theoretical work is necessary to analyze the relation of this approach to the traditional one, which ignores mathematical dependences arising from stoichiometry.

The number of conditions of self-balancing diffusion is equal to the number of atoms or pseudoatomic substances present in a reacting mixture, i.e., forming the components of the reacting mixture. Each condition balances the divergences of component diffusion fluxes with respect to the numbers of the corresponding atom (or pseudoatomic substance) in each component.

Even in the case of self-balancing diffusion, there is no direct one-to-one proportionality between extent and reaction rates; generally, extents (their time derivatives) also encompass diffusion rates. Direct proportionality is achieved only for "divergence-less" diffusion and, of course, when there is no diffusion. Regardless of self-balancing, the linear algebra of the permanence of atoms (stoichiometry) puts a general restriction on the divergences of diffusion fluxes (Equation (44)).

The contribution of chemical reactions to concentration changes is expressed by the rates of independent reactions, which are also derived by applying linear algebra to the permanence of atoms and stoichiometry. In this sense, there is no need to introduce extents. However, extents can be useful when there is a need for some overall (integral) characteristic of concentration changes and not only for an instantaneous one. Note that the self-balancing diffusion is not any specific model of diffusion or diffusion flux. It is a general result of two general principles—mass balance and permanence of atoms.

It is hoped that this theoretical work will stimulate experiments aimed at finding and comparing real cases in which diffusion is or is not self-balancing, i.e., experimental works inspecting the divergence of diffusion fluxes. The first step could be to analyze some existing data on a reaction-diffusion system employing and ignoring the self-balancing condition and to compare results. Another way could be to take experimentally determined parameters—rate constants and diffusion coefficients—of a reaction-diffusion system, to perform computer modelling and look at diffusion fluxes and at fulfilling the self-balancing condition.

Supplementary Materials: The following are available online at https://www.mdpi.com/article/10.3390/ijms231810511/s1, details on theoretical background (Section 3), details on relationships to the concept of generalized extents (Section 2.5).

Funding: This research received no external funding.

Institutional Review Board Statement: Not applicable.

Informed Consent Statement: Not applicable.

Data Availability Statement: Not applicable.

Conflicts of Interest: The author declares no conflict of interest.

Nomenclature

c_α	molar concentration of component α
$\mathbf{c}_\mathrm{b}$	vector of formal concentrations defined by Equation (S27)
da	surface element
dv	volume element
e^α, e_α	base vectors of the n-dimensional mixture space, $\alpha = 1, \ldots, n$
$\mathbf{f}_\sigma$	base vectors of the h-dimensional subspace W, $\sigma = 1, \ldots, h$
$\mathbf{g}^p$	base vectors of the $(n-h)$-dimensional reaction space, $p = 1, \ldots, n-h$
h	rank of matrix of atomic composition

$\mathbf{j}_\alpha^w$	mass diffusion flux of component α in the barycentric reference
$\mathbf{j}_\alpha^{\prime w}$	molar diffusion flux of component α in the barycentric reference
$\mathbf{J}$	vector of $n-h$ independent reaction rates
J_p	rate of independent reaction p
J^α	moles of component α produced (or consumed) by chemical reactions in unit volume per unit of time (component molar reaction rate)
$\mathbf{M}$	vector of molar masses
M_α	molar mass of component α
n	number of components
$\mathbf{n}$	normal vector on surface
p	number of independent reactions
$P^{p\alpha}$	stoichiometric coefficient of component α in (independent) reaction p
r_α	mass of component α produced (or consumed) by chemical reactions in unit volume per unit of time (component mass reaction rate)
$S_{\sigma\alpha}$	number of (pseudo)atoms σ in component α
t	time
$\mathbf{u}_\alpha^{\text{ref}}$	diffusion velocity of component α in reference 'ref'
$\mathbf{u}_\alpha^w$	diffusion velocity of component α in barycentric reference
$\mathbf{v}_\alpha$	velocity of component α
$\mathbf{v}^w$	barycentric velocity
w_α	mass fraction of component α
$\mathbf{x}_\text{d}$	vector of generalized extents of diffusion
$\mathbf{x}_\text{r}$	vector of generalized extents of reactions
ξ	vector of extents of reactions
ξ_p	extent of the independent reaction p
ρ	density of (reacting) mixture
ρ_α	mass density (mass or weight concentration) of component α
$\boldsymbol{\sigma}$	vector defined by Equation (S19)
$\boldsymbol{\omega}$	vector defined by Equation (S16)
$\boldsymbol{\omega}^+$	vector defined by Equation (S20)

References

1. Grzybowski, B.A. *Chemistry in Motion: Reaction-Diffusion Systems for Micro- and Nanotechnology*; Wiley: Chichester, UK, 2009.
2. Volpert, V. *Elliptic Partial Differential Equations: Volume 2: Reaction-Diffusion Equations*; Springer: Basel, Switzerland, 2014.
3. Ultman, J.S.; Baskaran, H.; Saidel, G.M. *Biomedical Mass Transport and Chemical Reaction: Physicochemical Principles and Mathematical Modeling*; Wiley: Hoboken, NJ, USA, 2016.
4. Berestycki, H.; Hamel, F. *Reaction-Diffusion Equations and Propagation Phenomena*; Springer: New York, NY, USA, 2017.
5. Scott Fogler, H. *Essentials of Chemical Reaction Engineering*, 2nd ed.; Prentice Hall: Hoboken, NJ, USA, 2018.
6. Demirel, Y.; Gerbaud, V. *Nonequilibrium Thermodynamics: Transport and Rate Processes in Physical, Chemical and Biological Systems*; Elsevier: Amsterdam, The Netherlands, 2019.
7. Mauro, J.C. *Materials Kinetics: Transport and Rate Phenomena*; Elsevier: Amsterdam, The Netherlands, 2021.
8. Mincheva, M.; Siegel, D. Nonnegativity and positiveness of solutions to mass action reaction–diffusion systems. *J. Math. Chem.* **2007**, *42*, 1135–1145. [CrossRef]
9. Gopich, I.V.; Szabo, A. Theory of Diffusion-Influenced Reaction Networks. *J. Phys. Chem. B* **2018**, *122*, 11338–11354. [CrossRef] [PubMed]
10. Datta, R.; Vilekar, S.A. The continuum mechanical theory of multicomponent diffusion in fluid mixtures. *Chem. Eng. Sci.* **2010**, *65*, 5976–5989. [CrossRef]
11. Whitaker, S. Mechanics and thermodynamics of diffusion. *Chem. Eng. Sci.* **2012**, *68*, 362–375. [CrossRef]
12. Morro, A. Governing equations in non-isothermal diffusion. *Int. J. Non-Lin. Mech.* **2013**, *55*, 90–97. [CrossRef]
13. Morro, A. Balance and constitutive equations for diffusion in mixtures of fluids. *Meccanica* **2014**, *49*, 2109–2123. [CrossRef]
14. Morro, A. Diffusion in Mixtures of Reacting Thermoelastic Solids. *J. Elast.* **2016**, *123*, 59–84. [CrossRef]
15. Bowen, R.M. On the Stoichiometry of Chemically Reacting Systems. *Arch. Ration. Mech. Anal.* **1968**, *29*, 114–124. [CrossRef]
16. Truesdell, C. *Rational Thermodynamics*, 2nd ed.; Springer: New York, NY, USA, 1984; pp. 356–357.
17. Rodrigues, D.; Billeter, J.; Bonvin, D. Generalization of the concept of extents to distributed reaction systems. *Chem. Eng. Sci.* **2017**, *171*, 558–575. [CrossRef]
18. Pekař, M.; Samohýl, I. *The Thermodynamics of Linear Fluids and Fluid Mixtures*; Springer: Cham, Switzerland, 2014.

International Journal of
Molecular Sciences

MDPI

Article

Various Approaches to Studying the Phase Transition in an Octamethylcyclotetrasiloxane Crystal: From X-ray Structural Analysis to Metadynamics

Alexander D. Volodin [1,2,*], Alexander F. Smol'yakov [1,3] and Alexander A. Korlyukov [1,2]

1 A. N. Nesmeyanov Institute of Organoelement Compounds, Russian Academy of Science, Vavilova Street 28, 119991 Moscow, Russia
2 Highest Chemical College of RAS, D. Mendeleev University of Chemical Technology of Russia, Miusskaya Square 9, 125047 Moscow, Russia
3 Basic Department of Chemistry of Innovative Materials and Technologies, Plekhanov Russian University of Economics, Stremyannyy Pereulok 36, 117997 Moscow, Russia
* Correspondence: alex.d.volodin@gmail.com

Abstract: The structure, thermodynamic parameters, and the character of thermal motion in octamethylcyclotetrasiloxane (D4) were investigated using the combination of experimental (single-crystal X-ray diffraction, thermochemistry) and theoretical (density functional theory calculations, ab initio molecular dynamics and metadynamics) methods. Single crystals of D4 were grown in a glass capillary in situ and the structures of high- (238–270 K) and low-temperature (100–230 K) phases were studied in detail. In the temperature interval 230–238 K, a phase transition with rather low enthalpy (−1.04(7) kcal/mol) was detected. It was found that phase transition is accompanied by change of conformation of cyclosiloxane moiety from boat-saddle (cradle) to chair. According to PBE0/6-311G(d,p) calculation of isolated D4, such conformation changes are characterized by a low barrier (0.07 kcal/mol). The character of molecular thermal motion and the path of phase transition were established with combination of periodic DFT calculations, including molecular dynamics and metadynamics. The effect of crystal field led to an increase in the calculated phase transition barrier (4.27 kcal/mol from low- to high-temperature phase and 3.20 kcal/mol in opposite direction).

Keywords: in situ cryo-crystallization; X-ray structure determination; periodic DFT calculations; quantum chemical calculations; Born–Oppenheimer molecular dynamics; metadynamics; phase transition path; free energy of phase transition

1. Introduction

Cyclic oligosiloxanes are of great experimental and theoretical interest due to their wide practical use [1–6]. Siloxanes are used in various fields of medical, catalytic, and synthetic chemistry. The research results are used in various fields of industry and medicine. Siloxane monomers usually are viscous colorless liquids. The physical macroscopic properties (viscosity, thermal expansion, heat capacity, temperatures of phase transitions, etc.) of these compounds are well studied, whereas their microscopic ones (liquid microstructure and crystal structures, Debye–Waller factors, etc.) are studied to a lesser extent. Detailed study of microscopic properties is difficult and highly desirable to be carried out together with the use of atomistic modeling in combination with precise experimental measurements by diffraction and thermochemical techniques.

The octamethylcyclotetrasiloxane (D4, $[Me_2SiO]_4$, Scheme 1) attracted our interest due to its molecular flexibility and great importance in the industry of various commercial products, including polymeric materials and lubricants. Hence, the structure and properties of D4 have been discussed in several publications. Indeed, in 1953, the dependence of the dielectric constant on temperature was studied by J. Hoffman [7], who found a phase transition at −16.3 °C. Two years later, in 1955, H. Steinfink et al. carried out a series of X-ray diffraction experiments and established the crystalline structure of the low-temperature phase at −50 °C [8]. Their attempt to determine the structure of the high-temperature phase of D4 was unsuccessful; they were only able to determine the spacegroup ($I4_1/n$) and unit cell parameters. The authors suggested that molecules in a crystal with $I4_1/n$ symmetry should be located on the $\overline{4}$ axes and should probably be disordered. Later, Shklover et al. showed that the structure of the high-temperature phase can be ordered, and the boat-saddle conformation (S_4 symmetry) is the most possible for an octa-membered cyclosiloxane ring in contrast to the chair conformation (C_{2h} symmetry) in the low-temperature phase [9].

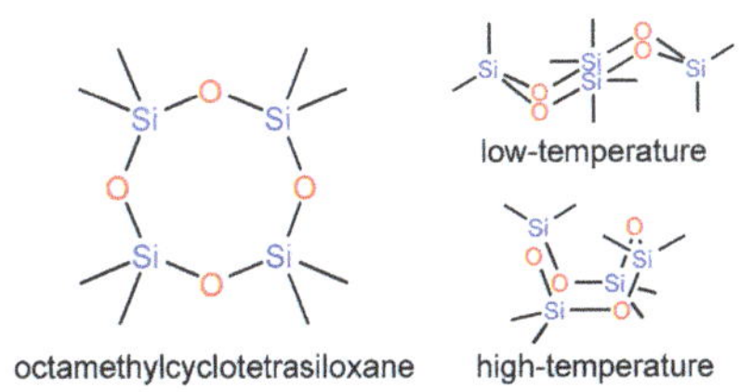

Scheme 1. Octamethylcyclotetrasiloxane and it's conformations in different crystal phases.

One of the most reliable approaches to obtain the molecular structure of a low-melting compound consists of growing a crystal in situ on an X-ray diffractometer and performing subsequent diffraction experiments [10–12]. Inspired by the works of R. Boese, W. Lipscomb, and M. Antipin, we decided to design our in-house heating device suitable for in situ crystallization using nichrome wire as a heater [13–19].

The reliable approaches to study the process pathway are known [20–28]. To determine the crystal structure of the high-temperature phase and gain an insight into the mechanism of phase transition, we carried out the X-ray and differential scanning calorimetry (DSC) study of D4. We performed X-ray experimental series of D4 on a monocrystal diffractometer with in situ crystallized samples. The combination of X-ray diffraction analysis and computational methods allow us to establish energy parameters and the pathway of phase transition.

2. Results and Discussion

2.1. The Crystal Structure of the Low- and High-Temperature Phases of Octamethylcyclotetrasiloxane

At room temperature, octamethylcyclotetrasiloxane is a transparent liquid with fairly high viscosity. Despite the preliminary drying of the D4, the first attempts of crystallization upon gradual cooling of a capillary were unsuccessful. Indeed, after using the zone melting procedure, D4 formed an amorphous phase or a polycrystalline sample consisting of numerous domains. This fact can be explained by overcooling of liquid D4 and ensuing rapid crystallization or solidification process far from equilibrium (up to 40 °C). To decrease the effect of overcooling and then bring crystallization closer to equilibrium, an external nucleation initiator (a tiny NaCl crystal) was added. The addition of an external nucleating agent resulted in accelerated crystallization and a significant decrease in the number of crystals. As a result, a single crystal suitable for X-ray diffraction was grown at 270 K. The crystal structure of D4 was studied at 270, 250, 240, 237, 230, 227, 224, 221, 210, 200, 150, and 100 K.

The spacegroup and unit cell parameters in the high-temperature phase we studied at 270 K completely coincided with Steinfink's observations (s.g. $I4_1/a$, $a = b = 16.250(7)$ Å, $c = 6.943(4)$ Å). However, in contrast to Steinfink's assumption, we found the molecular structure of D4 in the high-temperature phase to be ordered. The asymmetrical part of the unit cell contains a quarter of the D4 molecule (atoms Si1, O1, C1, and C2, Figure 1) since the molecule occupies a fourth-order position (on the $\overline{4}$ axis).

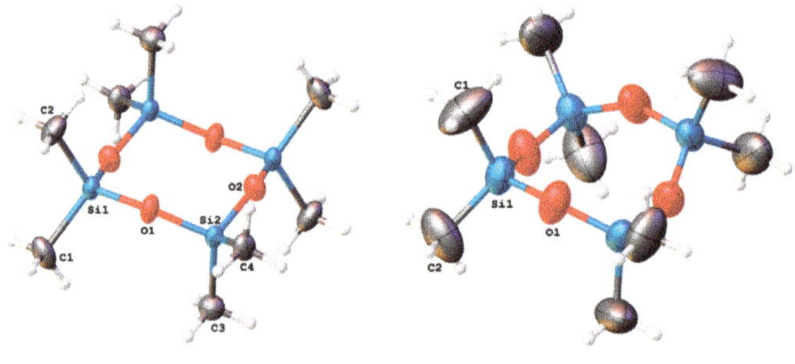

Figure 1. Conformations in low- (**a**) and high-temperature (**b**) phases. Ellipsoids are set at 50% probability.

The highest value of the equivalent atom displacement parameter (U_{eq}) for carbon atoms at 270 K is 0.13 Å^2. At the same time, the positions of carbon atoms could not be separated without applying strong restraints (EADP and DFIX) due to the absence of reflections at high 2θ values (>50°). The U_{eq} values of oxygen and silicon atoms do not exceed 0.1 Å^2. The shapes of atomic displacement parameter (ADP) ellipsoids are far from spherical; they are considerably stretched in a perpendicular (Si and O atoms) or tangential (C atoms) direction to the mean plane of the siloxane ring. Such behavior of ADP ellipsoids is indicative of the intensive molecular vibrations related to the folding of the siloxane cycle. As a consequence of thermal motion, Si1-O1 and Si1-O1[5/4−y,1/4 + x, 5/4−z] bonds (1.611(2) and 1.623(2) Å at 240 K) are somewhat shorter than those published in 1955 by Steinfink et al. for the low-temperature phase (1.64–1.66 Å). The length of Si1-C1 and Si1-C2 bonds (1.835(3) and 1.822(4) Å) also turned out to be shorter than those published earlier [8] by 0.07–0.12 Å.

The cell volume linearly decreased during cooling up to 238 K (Figure 2). The phase transition was observed at the temperature range between 230 and 238 K. Indeed, the picture of diffraction changed drastically at temperatures near 237–238 K; the number of reflections was considerably increased while those shapes were transformed from almost spherical to elliptical. The diffraction patterns came to resemble a pattern of a polycrystalline sample. Such a change led to an increase of R-values and the appearance of reflections, which should be absent for the $I4_1/a$ spacegroup. At 230 K and below, we found the diffraction pattern to be definitely indexing in the $P4_2/n$ spacegroup. To overcome the problems related to diffraction quality below 230 K, we studied several samples and various rates of cooling.

The most accurate structural data were obtained for a crystal grown at a temperature of about 270 K and cooled at a rate of 60 °C/h. Surprisingly, at a cooling rate slower than 60 °C/h during the phase transition, the crystal is partially or completely destroyed. In this case, the diffraction pattern changes and becomes similar to the diffraction of a powder or even an amorphous body.

The X-ray study allowed us to establish the temperature of the phase transition with sufficient accuracy. According to data obtained, the phase transition occurs at 236–238 K on cooling and 240–242 K on heating of crystal samples of D4. Thus, the temperature of the phase transition in the capillary appears to be lower than the previously established temperature (257 K).

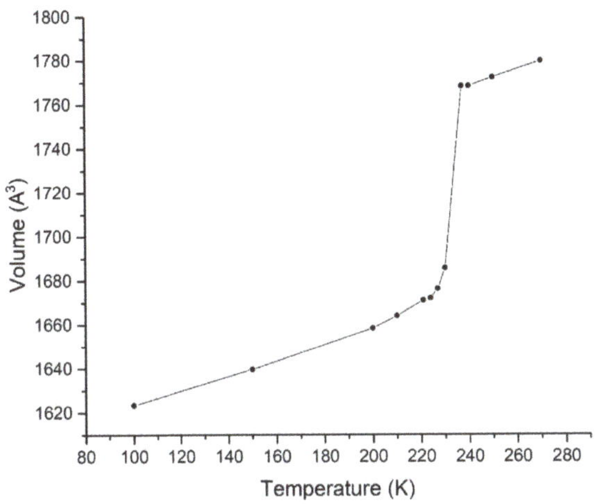

Figure 2. The dependence of unit cell volume on temperature.

The crystal structure of the low-temperature phase was the same as that described by Steinfink; however, we obtained more accurate atomic coordinates and displacement parameters. Cooling the sample to 100 K caused a considerable decrease in thermal motion and subsequent lowering of ADPs. At the same time, the differences between our study and the data published by Steinfink et al. in most cases were insignificant. The only exception was the lengths of Si-C bonds, which differ more strongly (1.850(3)–1.855(3) Å in our study vs 1.90–1.95 Å in [8]). As a result of phase transition, I-centering disappeared, and the fourfold position of the D4 molecule transformed into the center of inversion (Figure 3). Consequently, the change of spacegroup from $I4_1/a$ to $P4_2/n$ can lead to the formation of a disordered structure or a conformation transition. Indeed, the conformation of the eight-membered siloxane cycle in the high-temperature phase of D4 can be described as a boat-saddle, while that in the low-temperature phase presented as a chair conformation.

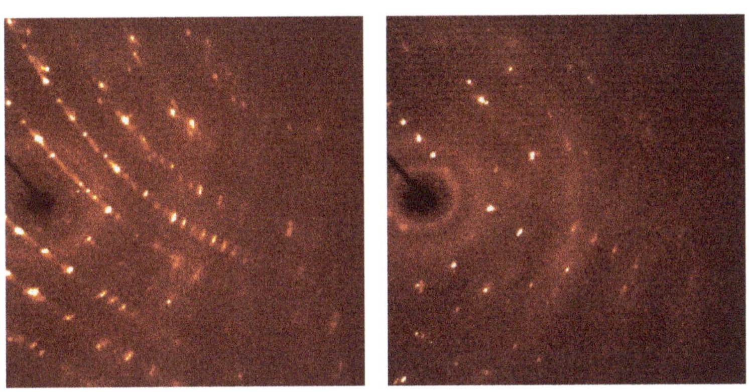

(**a**) (**b**)

Figure 3. Diffraction patterns for low-temperature (**a**) and high-temperature (**b**) phases. Reflections corresponding to I-centering are absent in the high-temperature phase. Upon transition to the low-temperature phase, the reflections broadened.

In the low-temperature phase of D4, the z-coordinates (along c-axis) of centers of two adjacent molecules alternate by half the cell parameter c (3.259 Å). Concurrently, the z-coordinates of centers of two adjacent molecules differ by one-quarter of the cell parameter c (1.721 Å) in the high-temperature phase. Thus, the molecules should move at least 1.53 Å during the phase transition (Figure 4). According to Landau's theory of phase transition types [29], D4 has a shift-type phase transition.

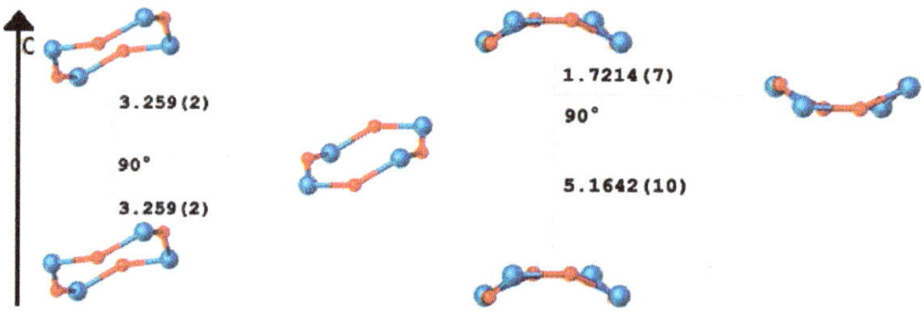

(a) (b)

Figure 4. Mutual position of adjacent molecules in the low-temperature (**a**) and high-temperature (**b**) phases. Image plane equal to the ac plane of the unit cell. Hydrogen and carbon atoms are omitted for clarity. Color code: silicon—blue, oxygen—red.

2.2. Differential Scanning Calorimetry Studies

Two series of differential scanning calorimetry (DSC) experiments were performed for compound D4. The first experiments were carried out with an impure sample (98% purity, Figure 5a). We observed three peaks of differential heat on DSC: at 218–220, 256–258, and 280 K (melting point). The other one was performed with a pure sample (>99.5% purity, Figure 5b). In the second experiment, we only observed two phases and a huge hysteresis loop (>40 °C). At the same time, in an X-ray diffraction experiment, only two phases could be found. Thus, the phase that occurred at a temperature interval from 218 to 258 K is no other than a strange hysteresis loop, which should be a result of a slow asynchronous phase transition. Such a huge difference between DSC data shows how many problems could be caused by sample impurity.

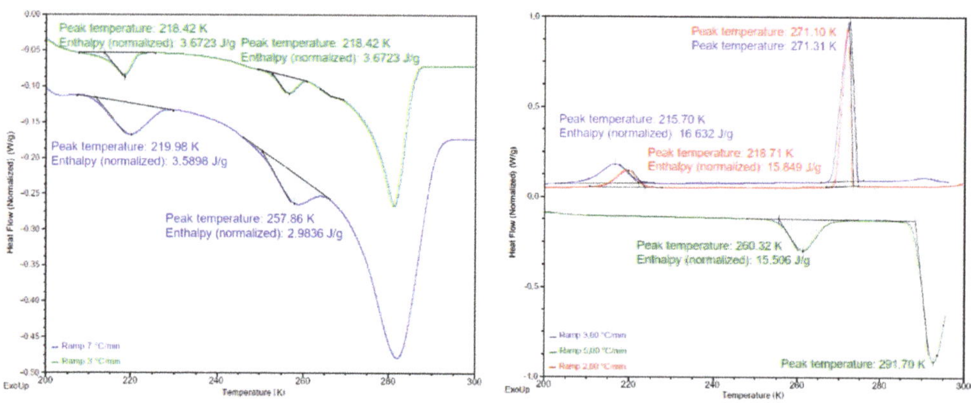

(a) (b)

Figure 5. Differential scanning calorimetry data of impure (**a**) (98%) and pure (**b**) (>99.5%) D4 samples.

The average enthalpy of solid–solid phase transition, calculated from the results of the DSC study, was 1.04(7) kcal/mol, while the calculated difference in total energy between the two solid phases was 3.14 kcal/mol (at 0 K, from VASP optimization).

2.3. The Comparison of Experimental and Calculated ADPs

Comparison of calculated and experimental atom displacement parameters is one of the easiest ways to find anomalies in molecular behavior. Besides, the ADP calculation method can be used to optimize strongly disordered structures and estimate the ability of molecules to move in a crystal. ADPs calculated with molecular dynamics were calculated and compared with experimental data. The average values of experimental and calculated U_{eq} by type of atom (Equation (3)) are listed in Table 1.

Table 1. Average equivalent ADPs by temperature and type of atom.

Temperature, K	Experimental $\langle U_{eq}(t,T)\rangle \times 10^3$ Si	O	C	Calculated $\langle U_{eq}(t,T)\rangle \times 10^3$ Si	O	C
100	16.515	21.937	26.866	17.838	28.643	34.753
150	23.682	30.892	38.581	22.447	32.650	40.655
200	30.027	39.202	49.814	30.779	42.338	57.780
210	31.883	41.93	53.364	37.851	47.740	89.316
221	34.277	44.518	56.858	38.193	53.145	74.892
230	35.827	48.093	60.622	75.543	93.211	142.686
240	54.897	69.820	109.33	65.665	97.464	134.831
250	57.437	72.773	114.994	56.689	68.092	151.815
270	68.383	85.827	131.237	45.922	70.727	107.603

The ADPs obtained from molecular dynamics (MD) calculations $\langle U_{eq}^{calc}\rangle$ at 100–200 K completely coincide with the experimentally measured ADPs $\langle U_{eq}^{exp}\rangle$. The $\langle U_{eq}^{calc}\rangle$ obtained at higher temperature values stand out from the general dependence. This happens because conformations significantly change and whole molecules move during MD calculation. Calculated molecule trajectories at 210 K and higher contain an intensive oscillation with a period near 4 ps (picoseconds), as described earlier. It causes large $\langle U_{eq}^{calc}\rangle$ values and outliers of this data from main dependence. In this case, the equivalent ADP values cannot be calculated correctly for all temperatures where the oscillation is present (210–250 K).

Another problem that happened during MD calculation was that the crystal system always suffers phase transition during MD calculations of the high-temperature phase (240, 250, and 270 K). The conformations and positions of molecules at the end of their MD trajectories almost coincide with those found in the calculated local minimum of the potential energy surface (PES). Thus, the comparison of ADPs at these temperatures with experimental ones is pointless, because the crystal structures correspond to different phases. The ADP values are high due to intense molecular motion. Molecules change conformations all the time during the simulation and oscillate distinctly along the c-axis.

2.4. Phonon Spectra and Physical Properties

Another way to find ADPs is to calculate them from the phonon spectra. This way can only calculate relative ADPs, but the main axes of the most intensive harmonic oscillations should have the same direction as in the experiment. Phonon spectra for low- and high-temperature cells almost match, but several peaks are shifted. These signals are located at ranges from 400 to 550 cm^{-1} and from 950 to 1100 cm^{-1} and respond to rocking oscillations of the siloxane ring (Figure 6). Oscillations with wavenumbers between 400 and 550 cm^{-1} also move molecular centroids in the same direction as an intensive oscillation in the MD calculations. The multipeak at 1050 cm^{-1} corresponds to the rotational oscillation of the siloxane ring. Many thermodynamic properties (S, C_V, F, etc.) of the crystal phases can be calculated from phonon spectra. The dependence of Helmholtz free energy on temperature is shown in Figure 7. According to the calculation, at temperature 258 K, ΔF is equal to 0.08 kcal/mol.

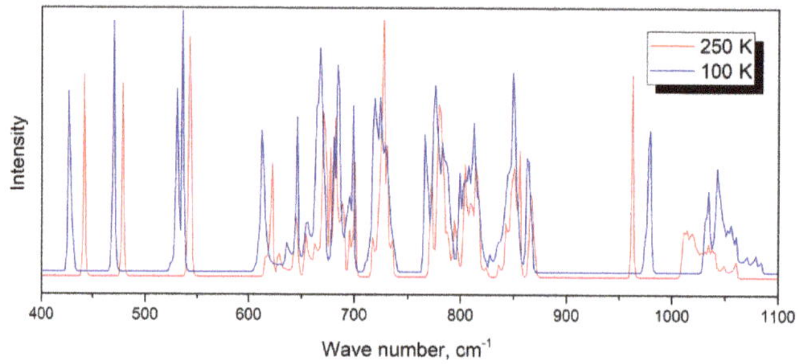

Figure 6. Comparison of the low- (blue) and high-temperature (red) phonon spectra at 100 and 250 K.

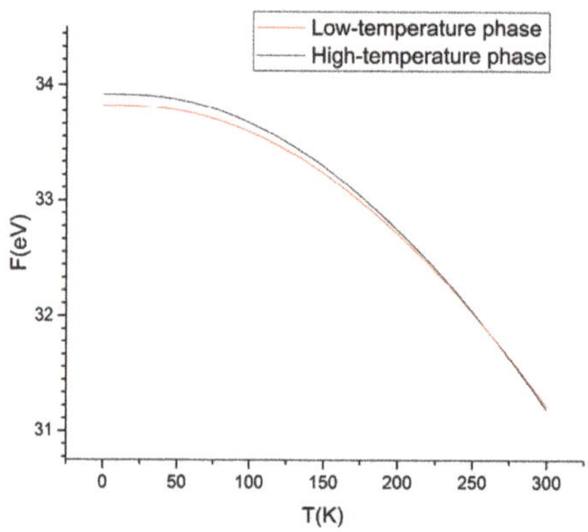

Figure 7. Calculated dependence of Helmholtz free energy on temperature. Curves intersect at 263 K, which is near the phase transition temperature reported in the literature (258.15 K).

2.5. Phase Transition Simulation

Phase transition analysis started with the calculation of transition state conformation. First, we sought to determine the transition state by using the Berny algorithm [24] and an intrinsic reaction coordinate [20] calculation in the Gaussian 09 program. Unfortunately, the potential energy surface (PES) of a set of conformations is very flat, and only approximate conformation of transition state can be calculated. The calculated conformation was boat–boat (D_{2d} symmetry, Figure 8a). The Si1-Si2-Si3-Si4 torsion angle equals 145.35°. Si-O-Si angles are 151.44, 153.84, 153.90, and 158.58°. The tetrahedral environment of silicon atoms persists (O-Si-O angle = 109.45(5)°). Each pair of oxygen atoms was very close to the plane formed by the three nearest silicon atoms (distances to the plane are less than 0.05 Å). The calculated difference between low-temperature conformation energy and calculated transition state energy was 0.07 kcal/mol. This very small value is the activation barrier of the conformation transition of an isolated molecule.

The phase transition energy barrier in a real crystal could be much higher and can only be calculated correctly using a crystal environment and periodic boundary conditions. Thus, the next step to obtain a phase transition pathway was restrained molecular dynamics with periodical boundary conditions. This calculation was carried out with VASP software. Several dihedral angles and nonbonding distances were restrained during molecular dynamics calculation. This time, the conformational pathway found had an unbelievably high energy barrier. This result was obtained due to the abundance of restraints, leading to the collapse of the molecule. The gradual weakening of imposed restrictions solved this problem. When only the internal coordinates of the molecules were restricted, the best result (most equilibrium pathway) was achieved.

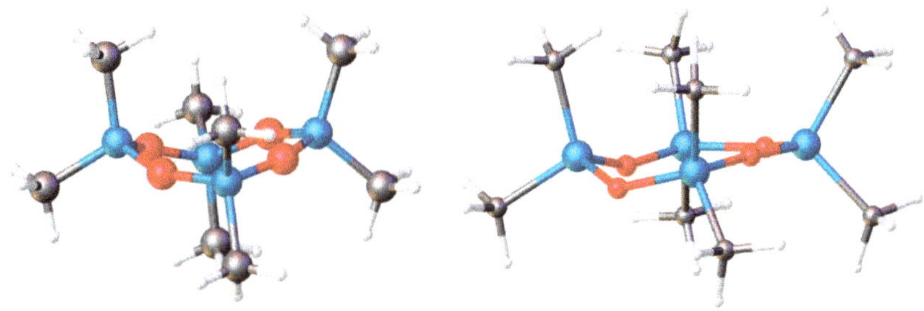

(a) (b)

Figure 8. Transition state conformation of an isolated D4 molecule (**a**) and a conformation found in the local minimum by MD calculation (**b**). Color code: silicon—blue, oxygen—red, carbon—grey, hydrogen—white.

Finally, the best but the most time-consuming way is to the trace the transition path from the molecular dynamics' simulation. The character of molecular movement in a crystal was found and it was reproduced in several separate MD simulations. The oscillation had a period of 4 ps and existed between 210 K and 270 K (inclusive). Several molecules (one or two of four, presented in a unit cell) in a crystal move up to 1.6 Å along c axes, which correlates well with the distance of average shift performed by molecules during phase transition. We can assume that two types of molecular translational movement within a unit cell are possible. The first one implies correlated movement of molecules. The second type is the situation when molecules move intendedly. In this case, we observed that one of four molecules move translationally when others conserve their position despite temperature changes from 210 to 270 K. Such character of molecular movement can be considered asynchronous.

The MD calculations concluded with a notable result: a certain region in the PES was limited by energy borders. All molecules in this "crystal" state had taken different conformations even in a $2 \times 2 \times 1$ supercell. Nevertheless, the conformations of D4 molecules remain similar: at least a half cycle of the molecule is almost flat (Si-O-Si-O-Si chain, Figure 8b). Molecules can change their conformations without leaving a minimum in the PES, and since the ring could become almost flat, they will be able to take the conformation of both high- and low-temperature phases. Thus, the wide flat local minimum of energy cannot be assigned to any of the known crystal phases and could be considered as an amorphous phase.

The addition of bias potential to the molecular dynamics calculations makes it possible to force the system to undergo a phase transition. This allows us to trace one of the transition paths, but, obviously, not the most optimal one. The energy parameters of the phase transition also cannot be determined from the result of molecular dynamics using the bias potential. Nevertheless, the use of the bias potential made it possible to check how well the chosen coordinate corresponds to the process—in this case, the phase transition.

Born–Oppenheimer molecular dynamics calculations (with and without bias potential) have demonstrated the possibility of simulating a phase transition in a crystal by computational methods. However, the phase transition does not have to go through a metastable phase. To establish the energy parameters of the phase transition, as well as to verify the assumption that the metastable phase is a local minimum in the PES, we applied a recent computational approach that has rarely been used to study processes of this type: ab initio metadynamics. Metadynamics made it possible to recognize the dependence of the free energy on the intrinsic coordinate of the process. The intrinsic coordinate of the process is a linear combination of the distances between atoms in the system and was chosen so that the values on the left refer to the system in the high-temperature phase (Figure 9a, point a), and the values on the right relate to the low-temperature phase (Figure 9a, point e). Point c in the diagram (Figure 9a) relates to metastable phase. The calculated energy barrier was only 3.20 kcal/mol during the transition from the high-temperature phase to the low-temperature phase, and 4.27 kcal/mol in the opposite direction at an average temperature of 238 K. The disordered phase found in Born–Oppenheimer molecular dynamics calculations also occurs in metadynamics. In the figure, it is represented by the roughness of the right slope of the high-temperature phase (segment a–b in Figure 9a). The free energy change during such a transition should be about 1.07 kcal/mol. Considering the DSC data (1.04(7) kcal/mol), the entropy of studied phases should be almost equal.

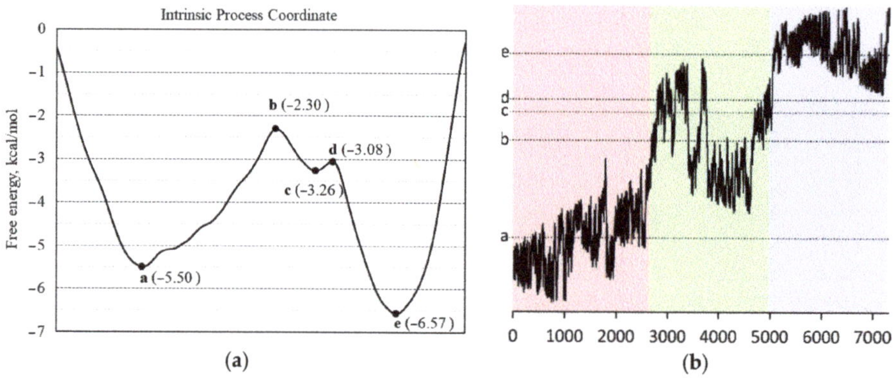

Figure 9. (**a**) The dependence of free energy on intrinsic coordinate during the phase transition in crystal of D4. The high-temperature phase is on the left, and the low-temperature phase is on the right side of plot. (**b**) The path of intrinsic coordinate (vertical) during metadynamics.

The path of intrinsic coordinate during metadynamics can be divided into several areas (Figure 9b). The first area (red, points 0–2700) corresponds to a high-temperature phase and metastable states. The stepwise character of the coordinate change is caused by the existence of several states that are stable under these conditions. In the next area (green, 2700–4600), the crystal system oscillates near point b, passing to the minimum c and returning to the metastable state a–b. After point 5050, the system crosses points b, c, and d and moves into the low-temperature phase. The calculation was terminated when conformations of molecules became unrealistic (the bond lengths (+/− 0.07 Å) and bond angles (+/−30 °) were distorted).

To prove that metadynamics had completed, additional calculations were made using the obtained bias potentials. The calculations were molecular dynamics with tracking of the intrinsic coordinate. The starting point of the calculation was the state of the system at every 1000th step. The system behaved chaotically, in some cases tending to leave the area of interest, bending the molecules into unrealistic conformations. Rarely, the system is stuck near point c, which may indicate an error in the calculated energy. However, since the nature of being stuck was random and rarely encountered, it was decided that the metadynamics could be considered complete.

3. Materials and Methods

3.1. X-ray Experimental Studies

Single crystals of D4 were grown in a sealed glass capillary (0.3 mm diameter) upon slow crystallization using nichrome wire as a heater. Crystallization was performed at 270 K. The movement of the heater along the glass capillary was maintained by a hand-made elevating unit with PC control. We performed 3 experimental series. Cooling rates between X-ray experiments were 60, 30, 20, and 15 °C/h. All equipment for crystal growth was installed into a Bruker APEX II diffractometer with on three-circle Eulean goniometer with a fixed χ axis (Bruker AXS, Karlsruhe, Germany). Cobra (Oxford Cryosystems, Long Hanborough, England) was used for temperature control. To simplify crystal growth on a three-circle goniometer, a special goniometer head was designed and manufactured (Figure 10). With this head, it became possible to turn the capillary into a vertical position from an inclined one (~54.7°) and vice versa. In fact, all crystals grown by this technique consisted of a main domain with several small ones. However, it was possible to distinguish the diffraction patterns for various domains using RLATT and CELL_NOW software (Bruker AXS, Karlsruhe, Germany) [30].

Figure 10. A special goniometer head was designed and manufactured for in situ X-ray studies.

To solve and refine the structures of low- and high-temperature phases of D4 measured at various temperatures, the SHELX-2014 program package (by Sheldrick, Göttingen, Germany [31]) was utilized. All the structures were solved by the intrinsic phasing method (SHELXT program, version 2014/4 [31]) and refined with the least-squares method (SHELXL program, version 2014/6 [31]) in anisotropic approximation for non-hydrogen atoms. The positions of hydrogen atoms were refined using a riding model (AFIX 137 instruction). Crystallographic data and information about the refinement of crystal structures can be found in Supplementary Materials (Table S1). The Cambridge Crystallographic Data Centre (CCDC 2,182,853–2,182,861) contains the supplementary crystallographic data for the D4 crystal at 100, 150, 200, 210, 221, 230, 240, 250, and 270 K. These data can be obtained free of charge from The Cambridge Crystallographic Data Centre via https://www.ccdc.cam.ac.uk/structures (accessed on 2 August 2022).

3.2. Differential Scanning Calorimetry

Differential scanning calorimetry (DSC) analyses were provided on DSC25 (TA Instruments, New Castle, DE, USA) with the cooling system RCS 90. TRIOS Software (TA Instruments, New Castle, USA) was used as a hardware control program. The weighed portions were encapsulated in aluminum pot Tzero (CS Ceramic Co., Ltd., Chashan Town, Liling City, Hunan Province, China). The range of scanned temperatures was from −80 °C to 60 °C. Temperature change rates were 2, 3, 5, and 7 °C/min.

3.3. Calculations

Ab initio calculations of molecular dynamics and metadynamics were performed within the PBE exchange-correlation functional using Vienna Ab initio Simulation Package v. 5.4.1 (VASP, VASP Software GmbH, Vienna, Austria) [32,33]. A combination of pseudopotentials and a set of plane waves was used as a basis set. To improve the description of van der Waals interactions, the D3 correction was applied. The point group symmetry within a unit cell was switched off during all calculations, including molecular dynamics simulations. The atomic cores were described using the "ultrasoft" and "hard" projector augmented wave (PAW) potentials. The basis of plane waves was used to describe valence electrons: the kinetic energy limit was 300 (for ultrasoft pseudopotentials) or 800 eV (for "hard" pseudopotentials). To describe the temperature of the system, calculations were carried out using a Nose–Hoover thermostat [34,35]. The starting point of the calculation corresponds to the experimentally obtained data (elementary cell parameters, atomic positions, and temperature). The time integration step was chosen to be 1 fs (femtosecond).

Born–Oppenheimer molecular dynamics calculations were carried out using "hard" pseudopotentials and a kinetic energy limit of 800 eV. The same applies to molecular dynamics calculations using internal constraints. To study the path of the phase transition, the complex coordinate of the process (weighted sum of interatomic distances) was changed by a small value with each time step. The cell parameters were fixed at the maximum experimental values reached during the phase transition.

The parameters of the anisotropic displacement of atoms were calculated from the trajectories of atoms obtained in molecular dynamics calculations without restrictions. Trajectories of the motion of atoms from 1 to 10,000 steps (the first 10 ps) were not taken into account in this calculation. The analysis of a large amount of data on the movement of atoms was carried out by an application program written personally by the author to solve this problem.

To carry out metadynamics calculations, ultrasoft pseudopotentials and a kinetic energy limit of 300 eV were used. The weighted sum of intra- and intermolecular distances between atoms was used as the process coordinate. The calculation was carried out until the system began to randomly change its state. The total duration of the calculation was 800,000 steps or 800 ps. The dependence of the free energy on the process coordinate was calculated based on 7300 Gaussian functions with a height of 0.005 eV and a width of 0.2 Å.

The molecular dynamics (MD) calculations were performed using the VASP for D4 crystal lattices at 100, 150, 200, 210, 221, 230, 240, 250, and 270 K. The MD simulations started from the corresponding X-ray structures. The periodic boundary conditions with the NVT ensemble (Nose–Hoover thermostat) were used. Trajectories of 15 ps were collected after a 5 ps equilibration period, with 1 fs integration time.

The transition state conformation was determined using the Berny algorithm [24] and an intrinsic reaction coordinate [20] calculation in Gaussian 09 program (rev. A.01, Gaussian, Inc., Wallingford, CT, USA). The exchange-correlation PBE0 functional with the basis 6–311 G(d,p) was used for calculations [36].

Phonon spectrum calculations for 100 K and 250 K cell units were provided with the PHON program (version 1.43, by Dario Alfè, London, UK) [37]. From the phonon frequencies, we can calculate the dependence of the Helmholtz free energy on temperature and atom displacement parameters.

3.4. Mathematical Introduction and Theory

Atomic displacement parameters (ADPs) for each atom can be calculated from MD trajectories as elements of the covariance matrix:

$$U_{ij} = \langle (X_i - \langle X_i \rangle)\,(X_j - \langle X_j \rangle) \rangle\,, \tag{1}$$

where X_i and X_j are the x, y, or z coordinates in a Cartesian coordinate system and angular brackets denote averaging over time, i.e., over all points of the trajectory.

The value of the equivalent displacement in orthogonal coordinate systems equals one-third of a trace of the atomic displacement matrix U_{ij} and represents an average mean square displacement of the atom over three main directions:

$$U_{eq} = \frac{1}{3}(U_{11} + U_{22} + U_{33}), \tag{2}$$

where U_{11}, U_{22}, and U_{33} are the mean square displacements in the x, y, and z directions, respectively.

The average ADPs by atom type were calculated according to (2) and then compared to experimental ones:

$$\langle U_{eq}(t,T)\rangle = \frac{\sum_i U_{eq}(a_i,T)}{N_t}, \quad (3)$$

where a_i is the type of atom (carbon, oxygen, etc.), T is the temperature, and N_t is the number of atoms of type t. Angular brackets denote the average value over all atoms of the same type.

4. Conclusions

The temperature-dependent series of X-ray diffraction experiments with octamethylcyclotetrasiloxane (D4) made it possible to reliably establish the dependence of the cell parameters and atomic positions on temperature. Both indicated phases were determined at several temperatures. The large hysteresis loop on DSC and the slow-cooling crystal destruction phenomenon may be a consequence of the existence of a metastable phase. The Born–Oppenheimer molecular dynamics simulation of phase transition shows the possibility of the presence of an intermediate amorphous phase. Calculating ADPs from atom trajectories could easily detect a presence of anomalous behavior in MD calculations. During all unrestricted molecular dynamics calculations, the system in the high-temperature phase changed into an unknown unsymmetrical phase. The simulated metadynamics of phase transition shows a presence of two different metastable phases. Thus, the phase transition is asynchronous and could pass through unstable amorphous phases. The comparison of easily obtained phonon spectra can help to find changes in the molecular environment and to describe the first steps of a phase transition pathway.

Supplementary Materials: The supporting information can be downloaded at https://www.mdpi.com/article/10.3390/ijms23169073/s1.

Author Contributions: Conceptualization, A.A.K. and A.F.S.; methodology, A.D.V. and A.F.S.; validation, A.D.V. and A.A.K.; formal analysis, A.D.V.; investigation, A.F.S.; resources, A.A.K.; data curation, A.D.V.; writing—original draft preparation, A.D.V.; writing—review and editing, A.F.S. and A.A.K.; visualization, A.D.V.; supervision, A.A.K.; project administration, A.A.K.; funding acquisition, A.A.K. and A.F.S. All authors have read and agreed to the published version of the manuscript.

Funding: The quantum chemical calculations were supported by the Russian Foundation for Basic Research (project 19-33-90196). The X-ray diffraction was performed with the financial support of the Ministry of Science and Higher Education of the Russian Federation (Contract/agreement No. 075-00697-22-00).

Institutional Review Board Statement: Not applicable.

Informed Consent Statement: Not applicable.

Data Availability Statement: The data presented in this study are available on request from the corresponding author.

Acknowledgments: The authors are grateful to Ashot V. Arsumanyan and Anton A. Anisimov (A.N. Nesmeyanov Institute of Organoelement Compounds, Moscow, Russia) for providing samples for research, and to Dmitriy Arkhipov (JSC "Composite", Korolev, Russia) for DSC analysis.

Conflicts of Interest: The authors declare no conflict of interest.

References

1. Shin, M.; Matsuda, K.; Ishii, O.; Terai, H.; Kaazempur-Mofrad, M.; Borenstein, J.; Detmar, M.; Vacanti, J.P. Endothelialized Networks with a Vascular Geometry in Microfabricated Poly(Dimethyl Siloxane). *Biomed. Microdevices* **2004**, *6*, 269–278. [CrossRef] [PubMed]
2. Shams, E.; Yeganeh, H.; Naderi-Manesh, H.; Gharibi, R.; Mohammad Hassan, Z. Polyurethane/Siloxane Membranes Containing Graphene Oxide Nanoplatelets as Antimicrobial Wound Dressings: In Vitro and in Vivo Evaluations. *J. Mater. Sci. Mater. Med.* **2017**, *28*, 75. [CrossRef] [PubMed]

3. Racles, C.; Zaltariov, M.-F.; Iacob, M.; Silion, M.; Avadanei, M.; Bargan, A. Siloxane-Based Metal-Organic Frameworks with Remarkable Catalytic Activity in Mild Environmental Photodegradation of Azo Dyes. *Appl. Catal. B Environ.* **2017**, *205*, 78–92. [CrossRef]
4. Baskaran, T.; Kumaravel, R.; Christopher, J.; Radhakrishnan, S.; Sakthivel, A. Synthesis and Heterogenization of Siloxane Functionalized Cobalt Complex: Potential Catalyst for Oxidation of Alcohols. *Catal. Lett.* **2015**, *145*, 851–859. [CrossRef]
5. Naka, K.; Irie, Y. Synthesis of Single Component Element-Block Materials Based on Siloxane-Based Cage Frameworks: Single Component Element-Block Materials. *Polym. Int.* **2017**, *66*, 187–194. [CrossRef]
6. Saraswathy, M.; Stansbury, J.; Nair, D. Water Dispersible Siloxane Nanogels: A Novel Technique to Control Surface Characteristics and Drug Release Kinetics. *J. Mater. Chem. B* **2016**, *4*, 5299–5307. [CrossRef]
7. Hoffman, J.D. Thermal and Dielectric Study of Octamethylcyclotetrasiloxane. *J. Am. Chem. Soc.* **1953**, *75*, 6313–6314. [CrossRef]
8. Steinfink, H.; Post, B.; Fankuchen, I. The Crystal Structure of Octamethyl Cyclotetrasiloxane. *Acta Crystallogr.* **1955**, *8*, 420–424. [CrossRef]
9. Shklover, V.E.; Kalinin, A.E.; Gusev, A.I.; Bokii, N.G.; Struchkov, Y.T.; Andrianov, K.A.; Petrova, I.M. Crystal Structures of Cyclic Siloxanes: II. 1,1,2,2-Tetramethyl-3,3,4,4-Tetraphenylcyclotetrasiloxane. *J. Struct. Chem.* **1974**, *14*, 639–645. [CrossRef]
10. Brodalla, D.; Mootz, D.; Boese, R.; Osswald, W. Programmed Crystal Growth on a Diffractometer with Focused Heat Radiation. *J. Appl. Crystallogr.* **1985**, *18*, 316–319. [CrossRef]
11. Boese, R. Special Issue on In Situ Crystallization. *Z. für Krist. Cryst. Mater.* **2014**, *229*, 595–601. [CrossRef]
12. Volodin, A.D.; Korlyukov, A.A.; Smol'yakov, A.F. Organoelement Compounds Crystallized In Situ: Weak Intermolecular Interactions and Lattice Energies. *Crystals* **2019**, *10*, 15. [CrossRef]
13. Lipscomb, W.N.; Streib, W.E. Growth, Orientation, and X-Ray Diffraction of Single Crystals near Liquid Helium Temperatures. *Proc. Natl. Acad Sci. USA* **1962**, *48*, 911–913.
14. Smith, H.W.; Lipscomb, W.N. Single-Crystal X-Ray Diffraction Study of β-Diborane. *J. Chem. Phys.* **1965**, *43*, 1060–1064. [CrossRef]
15. Antipin, M.Y. Low-Temperature X-Ray Diffraction Analysis: Possibilities in the Solution of Chemical Problems. *Russ. Chem. Rev.* **1990**, *59*, 607–626. [CrossRef]
16. Boese, R.; Antipin, M.Y.; Nussbaumer, M.; Bläser, D. The Molecular and Crystal Structure of 4-Methoxybenzylidene-4′-*n*-Butylaniline (MBBA) at −163°C. *Liq. Cryst.* **1992**, *12*, 431–440. [CrossRef]
17. Boese, R.; Nussbaumer, M. *In Situ Crystallisation Techniques*; Oxford University Press: Oxford, UK, 1994; Volume 7, p. 20.
18. Bond, A.D.; Davies, J.E. In Situ Crystallization of the Linear Alkynes Cn H2n-2 (N = 7, 8, 9, 10). *Z. Für Krist. Cryst. Mater.* **2014**, *229*, 661–666. [CrossRef]
19. Nath, N.K.; Naumov, P. In Situ Crystallization and Crystal Structure Determination of Chlorobenzene. *Maced. J. Chem. Chem. Eng.* **2015**, *34*, 63–66. [CrossRef]
20. Fukui, K. The Path of Chemical Reactions—The IRC Approach. *Acc. Chem. Res.* **1981**, *14*, 363–368. [CrossRef]
21. Peng, C.; Bernhard Schlegel, H. Combining Synchronous Transit and Quasi-Newton Methods to Find Transition States. *Isr. J. Chem.* **1993**, *33*, 449–454. [CrossRef]
22. Peng, C.; Ayala, P.Y.; Schlegel, H.B.; Frisch, M.J. Using Redundant Internal Coordinates to Optimize Equilibrium Geometries and Transition States. *J. Comput. Chem.* **1996**, *17*, 49–56. [CrossRef]
23. Laio, A.; Rodriguez-Fortea, A.; Gervasio, F.L.; Ceccarelli, M.; Parrinello, M. Assessing the Accuracy of Metadynamics. *J. Phys. Chem. B* **2005**, *109*, 6714–6721. [CrossRef] [PubMed]
24. Li, X.; Frisch, M.J. Energy-Represented Direct Inversion in the Iterative Subspace within a Hybrid Geometry Optimization Method. *J. Chem. Theory Comput.* **2006**, *2*, 835–839. [CrossRef] [PubMed]
25. Barducci, A.; Bonomi, M.; Parrinello, M. Metadynamics. *WIREs Comput. Mol. Sci.* **2011**, *1*, 826–843. [CrossRef]
26. Tiwary, P.; Parrinello, M. From Metadynamics to Dynamics. *Phys. Rev. Lett.* **2013**, *111*, 230602. [CrossRef]
27. Tycko, R.; Dabbagh, G.; Fleming, R.M.; Haddon, R.C.; Makhija, A.V.; Zahurak, S.M. Molecular Dynamics and the Phase Transition in Solid C_{60}. *Phys. Rev. Lett.* **1991**, *67*, 1886–1889. [CrossRef]
28. Schelling, P.K.; Phillpot, S.R.; Wolf, D. Mechanism of the Cubic-to-Tetragonal Phase Transition in Zirconia and Yttria-Stabilized Zirconia by Molecular-Dynamics Simulation. *J. Am. Ceram. Soc.* **2001**, *84*, 1609–1619. [CrossRef]
29. Landau, L.D. On the Theory of Phase Transitions. I. *Zh. Eksp. Teor. Fiz.* **1937**, *11*, 19. [CrossRef]
30. Sheldrick, G.M. *CELL NOW Program for Unit Cell Determination*; Göttingen University: Göttingen, Germany, 2003.
31. Sheldrick, G.M. *SHELXT*–Integrated Space-Group and Crystal-Structure Determination. *Acta Crystallogr. Sect. A Found. Adv.* **2015**, *71*, 3–8. [CrossRef]
32. Kresse, G.; Hafner, J. Norm-Conserving and Ultrasoft Pseudopotentials for First-Row and Transition Elements. *J. Phys. Condens. Matter* **1994**, *6*, 8245–8257. [CrossRef]
33. Kresse, G.; Joubert, D. From Ultrasoft Pseudopotentials to the Projector Augmented-Wave Method. *Phys. Rev. B* **1999**, *59*, 1758–1775. [CrossRef]
34. Nosé, S. A Unified Formulation of the Constant Temperature Molecular Dynamics Methods. *J. Chem. Phys.* **1984**, *81*, 511–519. [CrossRef]
35. Hoover, W.G. Canonical Dynamics: Equilibrium Phase-Space Distributions. *Phys. Rev. A* **1985**, *31*, 1695–1697. [CrossRef] [PubMed]

36. Frisch, M.J.; Trucks, G.W.; Schlegel, H.B.; Scuseria, G.E.; Robb, M.A.; Cheeseman, J.R.; Scalmani, G.; Barone, V.; Mennucci, B.; Petersson, G.A.; et al. *Gaussian 09 Revision, A.01*; Gaussian, Inc.: Wallingford, CT, USA, 2009.
37. Alfè, D. PHON: A Program to Calculate Phonons Using the Small Displacement Method. *Comput. Phys. Commun.* **2009**, *180*, 2622–2633. [CrossRef]

International Journal of
Molecular Sciences

MDPI

Article

Friction in Myocardial Anoxia Leads to Negative Excess Entropy Production, Self-Organization, and Dissipative Structures

Yves Lecarpentier [1,*], Victor Claes [2], Jean-Louis Hébert [3], Xénophon Krokidis [1], Olivier Schussler [4] and Alexandre Vallée [5]

1 Centre de Recherche Clinique, Grand Hôpital de l'Est Francilien, 77100 Meaux, France; xkrokidis@gmail.com
2 Department of Pharmaceutical Sciences, University of Antwerp, 2180 Wilrijk, Belgium; victor.claes@scarlet.be
3 Institut de Cardiologie, Hôpital de la Pitié-Salpêtrière, Assistance Publique-Hôpitaux de Paris, 75013 Paris, France; jean.l.hebert@gmail.com
4 Département de Chirurgie Thoracique, Hôpital Cochin, Hôpitaux Universitaires Paris Centre, Paris-Descartes Université, Assistance Publique-Hôpitaux de Paris, 75014 Paris, France; olivier.schussler@gmail.com
5 Department of Epidemiology-Data-Biostatistics, Delegation of Clinical Research and Innovation, Foch Hospital, 92150 Suresnes, France; alexandre.g.vallee@gmail.com
* Correspondence: yves.c.lecarpentier@gmail.com

Abstract: Contraction of the heart is caused by actin filaments sliding along myosin filaments. This generates a frictional force inducing wear of the contractile apparatus. We postulated that this process could be exacerbated when the heart was submitted to severe anoxia. Anoxia induced dramatic abnormalities in the molecular properties of actin-myosin crossbridges. We applied the formalism of far-from-equilibrium thermodynamics to the left ventricular papillary muscles (LVPMs) of mammalian rat hearts which had been subjected to a prolonged anoxia (3 h). We showed that when subjected to prolonged anoxia, the heart operated far-from-equilibrium as evidenced by the non-linearity between thermodynamic force (F/T: Frictional force/Kelvin temperature) and thermodynamic flow (v0: myofilament sliding velocity). The rate of entropy production (EPR) was the product of (F/T) and v0. The excess entropy production (EEP) was equal to $\frac{\partial \delta^2 S}{\partial t} = \frac{\partial F}{T} \delta vo$; (S: entropy). The tribological system remained stable when EEP was positive and became unstable when EEP became negative, thus characterizing instability of the system and reflecting the occurrence of self-organization and possibly dissipative structures. After 3 h anoxia, re-oxygenation induced significant reversibility. About 20% of the myosin heads did not recover despite re-oxygenation. These results may be of importance in the context of heart transplantation where the delay between the time of sampling from the donor and the time of the graft installation in the recipient should be as short as possible.

Keywords: friction; myocardium; anoxia; far-from-equilibrium thermodynamics; excess entropy production; self-organisation; dissipative structures

1. Introduction

The understanding of the mechanical behavior of striated muscles has been considerably improved after the publication of two pioneering articles which laid the foundations for the theory of sliding filaments [1,2]. In fact, actin filaments slide along myosin filaments, resulting in the Z striations coming together and therefore resulting in shortening of sarcomeres (Figure 1A,B). Myosin heads, or crossbridges (CBs), alternatively hook into and unhook from actin filaments. During heart contraction, the sliding of actin and myosin filaments against each other generates friction. A. Huxley also proposed a theoretical formalism to determine CB molecular properties such as the CB attachment and detachment rate constants, as well as the number of active CBs per volume unit, myofilament sliding velocity, and myosin content [3]. The aim of our study was to assess the consequences of a prolonged cardiac anoxia from a thermodynamic point of view by applying the physical concepts established in tribology, the science of interacting surfaces in relative motion.

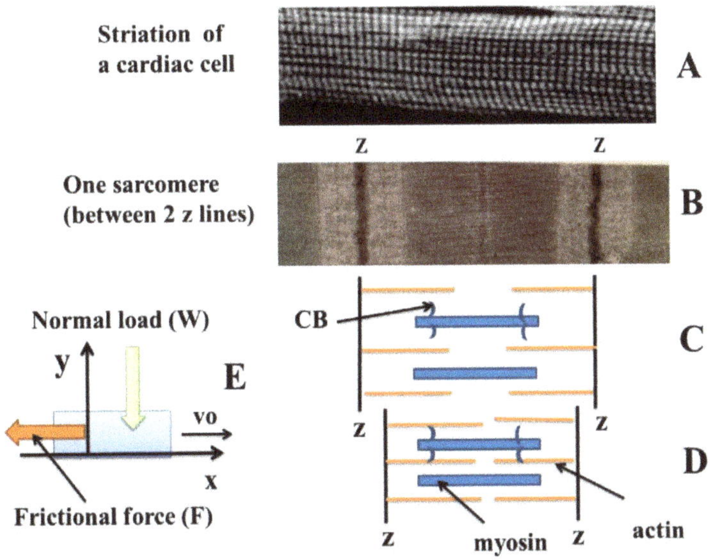

Figure 1. Schematic representations of myofilaments in sarcomeric heart muscle and tribological cardiac system. Cardiomyocytes exhibit a periodic striation characterized by alternating dark A-bands (anisotropic) and light I-bands (isotropic) (**A**,**B**). The central part of the I-band is marked by the Z line. A sarcomere is the smallest functional unit of a striated muscle. It is the repeating unit between two Z-lines (**B**). The I-band is the zone of thin filaments that is not superimposed with thick filaments (myosin). Thin myofilaments are mainly composed of F actin, resulting from the polymerization of many globular actin molecules (G actin). The actin myofilaments are attached by their caudal end to the Z lines. Actin filaments are the major component of the I-band and extend into the A-band. Thick myosin filaments are made up of 200 to 300 molecules of native myosin. Myosin has a long, fibrous tail and a globular head that binds to actin. Myosin filaments are bipolar and extend throughout the A-band. The myosin head also binds to ATP. Muscle contraction results in the sliding of actin filaments along myosin filaments, which is visible only at the level of the I-bands, while the dark A-bands keep a constant length (**C**,**D**) [1,2]. This movement is controlled by the heads of the myosin molecules (CBs), which bind and then detach from the actin molecule. The displacement of myosin on actin is possible through the hydrolysis of ATP molecules. (**E**) refers to the normal (y) and tangential (x) degrees of freedom during dynamic friction. The thermodynamic force was equal to μ W/T, with μ the frictional coefficient, and W the normal load.

Ilya Prigogine and colleagues have developed the concept of self-organization to describe complex thermodynamic processes operating far-from-equilibrium [4–10]. Thus, amplified small fluctuations can spontaneously create patterning, orderliness, self-organization, and possibly dissipative structures. The occurrence of self-organization supposes an open system exchanging energy and matter as well as a non-linear and far-from-equilibrium thermodynamic behavior. Friction and wear are usually considered to generate irreversible processes which lead to energy dissipation and material deterioration. Klamecky introduced the concept of non-equilibrium thermodynamics to describe friction and wear [11]. Barshefsky conducted the first investigations on friction-induced self-organization [12]. Nosonovsky and colleagues developed the thermodynamic principles of irreversible processes used to investigate the formation of spatial and temporal structures induced by friction [13,14]. At the interface, orderliness can increase so that entropy decreases [15,16]. According to Prigogine and colleagues [4,8,10], far-from-equilibrium processes can be a source of order and lead to a new state of matter called dissipative structures. During these processes, the entropy production rate (EPR) can decrease and be followed by the occurrence of a self-organisation process. Due to self-organization, dissipative structures result in reducing EPR, leading to a decreased rate of wear [17,18]. There is no simple criterion to determine if self-organization can occur. However, there is a key condition: self-organization can begin only if the tribosystem has lost its thermodynamic stability, which is characterized by the fact that the excess entropy production (EEP) becomes negative [19]. This is a pre-requisite for self-organization. During friction, self-organization leads to a decrease in wear rate and an increase in the durability of materials.

Huxley's formalism [3] allowed us to calculate the sliding rate of myofilament (vo) (i.e., the thermodynamic flow) and the thermodynamic frictional force (F/T; F: frictional force and T: Kelvin temperature) generated by the heart. We calculated EPR and EEP and sought to determine whether EEP could become negative after a prolonged period of cardiac anoxia. Cardiac ischemia, the most severe form of which is myocardial infarction, is a leading cause of death worldwide. Despite an encouraging decline, coronary heart mortality remains a major health burden, especially among the elderly [20,21]. Moreover, numerous heart transplants are performed each year worldwide [22]. Our study offers valuable insights for furthering our understanding of the thermodynamic aspects of hypoxic heart diseases and could be of interest in the context of heart transplants where the recipient must benefit from the donor's heart in the shortest possible time. The main aim of our study was to use far-from-equilibrium thermodynamics to determine whether a cardiac tribosystem submitted to a prolonged anoxia might result in a beneficial self-organization when submitted to dramatic anoxic conditions.

2. Results

2.1. Mechanical Properties of Left Ventricular Papillary Muscles (LVPMs) and Molecular Myosin CB Characteristics

Some mechanical parameters significantly decreased during prolonged cardiac anoxia. This was the case for the basic mechanical parameters characterizing the whole LVPM, such as maximum shortening velocity (Vmax) (Figure 2A) and total muscle tension (Figure 2C). Other parameters characterizing the myosin CB molecular properties also decreased during anoxia, such as the CB single force (Figure 2B), the CB detachment rate constant g2 (Figure 3A), and the molecular content in myosin per g of tissue (Figure 2D). All these parameters diminished progressively and significantly during anoxia compared to their reference values before anoxia. These parameters are represented in Figure 2A–C and Figure 3A together with the molecular content in myosin per g of tissue (Figure 2D).

2.2. Coefficient of Friction, Frictional Force and Normal Load

Normal load (W) decreased during anoxia (Figure 3B). Other parameters increased during anoxia, such as the friction coefficient (Figure 3C) and frictional force (F) (Figure 3D).

2.3. Thermodynamic Force and Thermodynamic Flow

The myofilament sliding velocity vo, or thermodynamic flow (Figure 4A), increased until reaching a maximum at about 120 min of anoxia and then decreased from 120 to 180 min of anoxia. Figure 4B,C show the absence of relationship between the thermodynamic force (F/T) and the thermodynamic flow (vo). Importantly, F/T varied non-linearly with vo. This means that the cardiac tribosystem operated in a non-linear fashion and thus far-from-equilibrium. In Figure 4B all values are presented, whereas mean values are shown in Figure 4C.

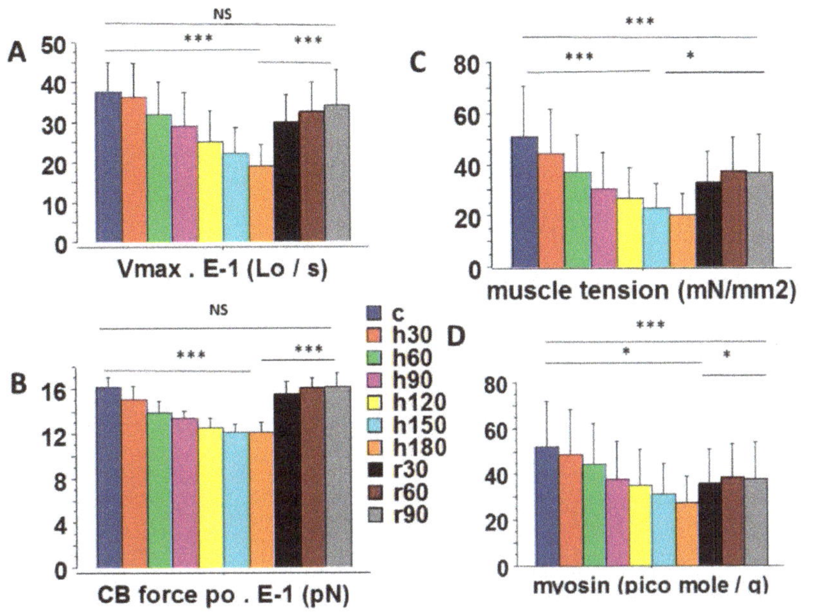

Figure 2. Mechanical parameters of LVPMs and CBs. (**A**): Maximum unloaded shortening velocity (Vmax); (**B**): CB single force (po); (**C**): total muscle tension; (**D**): myosin content. *: $p < 0.05$; ***: $p < 0.001$.

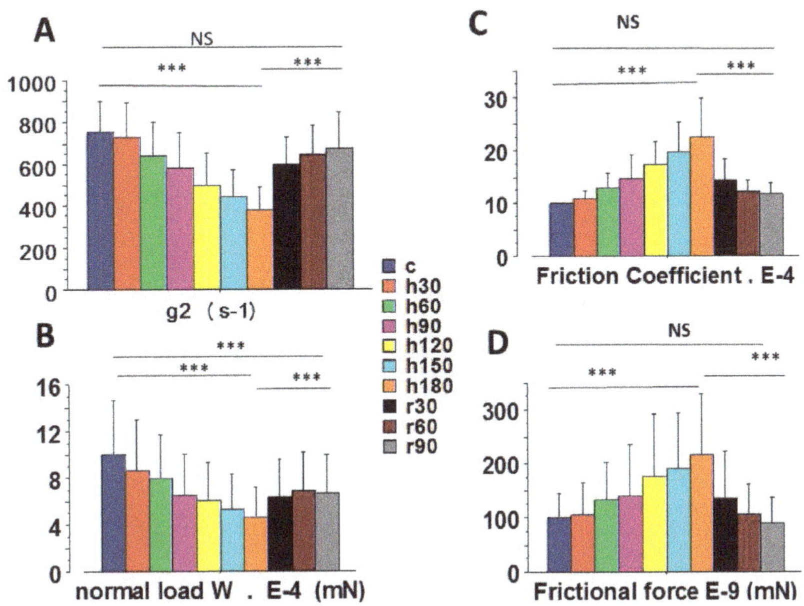

Figure 3. Molecular parameters of CBs. (**A**): CB detachment rate constant (g2); (**B**): normal load (W); (**C**): friction coefficient (μ); (**D**): frictional force (F). ***: $p < 0.001$.

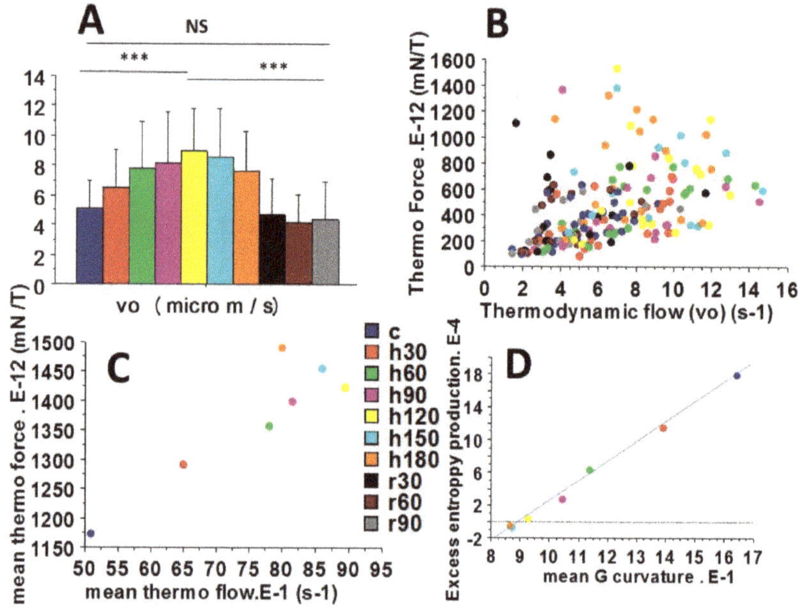

Figure 4. Non-linear thermodynamic force versus thermodynamic flow relationship. The myofilament sliding velocity (vo), or thermodynamic flow, is represented in (**A**); (**B**): Mean thermodynamic force versus mean thermodynamic flow; (**C**): Mean values of thermodynamic force versus thermodynamic flow (vo). ***: $p < 0.001$; (**D**): Linear relationship between EEP and the G curvature of the Hill hyperbola: EEP (10^{-4}) = −22 + 2.4 G (10^{-1}); r = 0.99.

2.4. Partial Time Derivatives of Thermodynamic Force and Thermodynamic Flow

Figure 5 shows the mean values of the thermodynamic flow (vo) (Figure 5A) and thermodynamic force (F/T) (Figure 5B) versus time and their respective partial time derivative (Figure 5C,D). F/T increased over time and its partial time derivative ∂ (F/T) decreased but remained always positive (Figure 5D). Importantly, the partial time derivatives of vo, (∂vo) was positive up until 120 min, and then became negative from 150 to 180 min (Figure 5C).

2.5. Entropy Production Rate (EPR) and Excess Entropy Production (EEP)

Figure 6 shows the mean values of the entropy production rate (δ^2 S) (Figure 6A) and the mean values of the excess entropy production ($\frac{\partial \delta^2 S}{\partial t}$) versus time (Figure 6B). Up to 120 min of anoxia, EPR increased and then decreased from 120 to 180 min. EEP decreased continuously during anoxia but, importantly, remained positive up to 120 min of anoxia, after which it became negative (from 150 to 180 min of anoxia). The fact that EEP became negative after 150 min of anoxia demonstrates the instability of the non-linear cardiac tribosystem. Interestingly, there was a linear relationship between EEP and the CB detachment rate g2 (Figure 4D; EEP = 2.4 g2 − 22; r = 0.99). There was also a linear relationship between EEP and the maximum efficiency (Figure 6C) (EEP = 1.9 max. Efficiency-3 7; r = 0.90).

2.6. Self-Organization and Dissipative Structures

Self-organization can begin only if the system has lost its thermodynamic stability, which is characterized by the fact that EEP becomes negative. This is a pre-requisite for self-organization and this was the case in our study. Dissipative structures occur far-from-equilibrium under the following conditions: (i) the system must be open; (ii) it operates far-from-equilibrium; (iii) under a non-linear regime; (iv) and is submitted to *fluctuations*. Dissipative structures are maintained by thermodynamic processes which take place because of the exchange of energy between the system and its environment. They disappear as soon as that exchange ceases. All these conditions were observed in the studied cardiac system. The heart is an open system. Under continuous anoxia during 3h, it was submitted to slight fluctuations, it operated far-from-equilibrium due to the non-linearity between the thermodynamic force and the thermodynamic flow, and re-oxygenation induced a large reversibility of thermodynamic abnormalities.

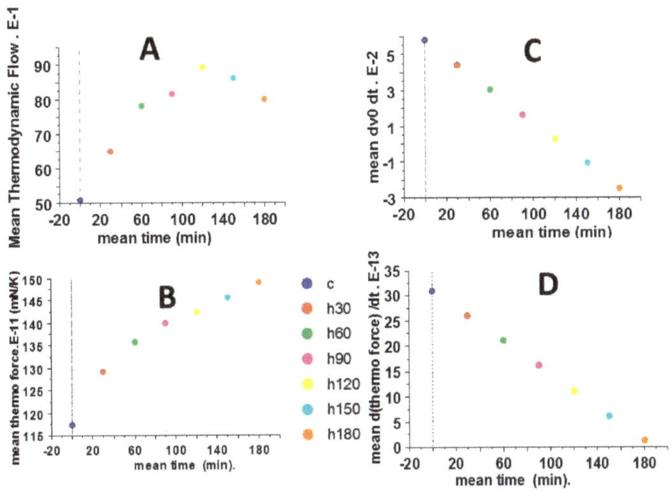

Figure 5. Mean thermodynamic parameters. (**A**): Mean thermodynamic flow versus time; (**B**): mean thermodynamic force versus time; (**C**): mean thermodynamic flow versus partial time derivative; (**D**): mean thermodynamic versus partial time derivative.

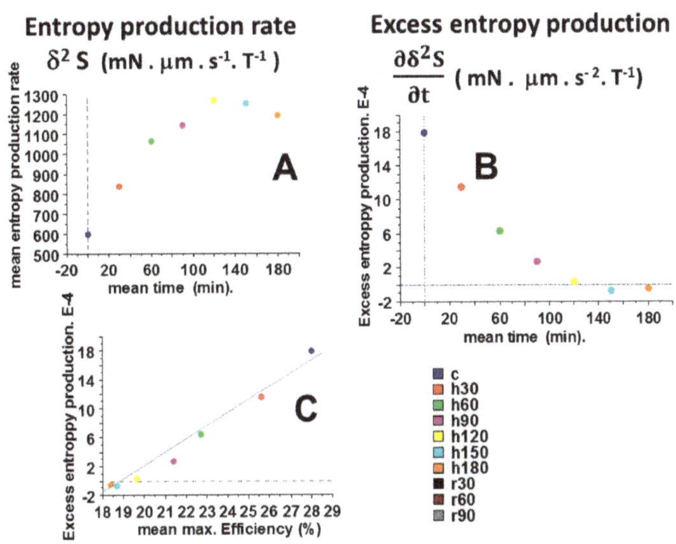

Figure 6. Entropy production rate (EPR) and excess entropy production (EEP). Entropy production rate (EPR) (in mN · µm · s^{-1} · T^{-1}) versus time is represented in (**A**). (**B**) presented excess entropy production (EEP) (in mN · µm · s^{-2} · T^{-1} versus time. Note that EEP became negative from 150 to 180 min of anoxia. (**C**) showed the linear relationship between mean EEP and mean max. Efficiency (EEP = 1.9 max. Efficiency-3 7) (r = 0.90). EEP became negative when max. Efficiency was <19%.

A fundamental point is introduced by the variations of the sliding velocity (i.e., vo, the thermodynamic flow) (Figure 4A); vo increased, reached a maximum and then decreased during cardiac anoxia. Huxley's equations show that vo is inversely proportional to the time stroke (ts) (Equation (7)). This means that, in the face of prolonged anoxia, the myosin head underwent ultrastructural changes, capable of modifying its molecular kinetics. Time stroke is a key step of the actin-myosin cycle. During this state, the myosin head generates a displacement of about 10 nm and a unitary CB force of some picoNewtons (Figure 2B). Both sliding velocity (vo) and time stroke (ts) returned to their control values after re-oxygenation. These changes in molecular kinetics of the myosin head due to changes in friction coefficient corresponded to changes in its molecular ultrastructure. This conferred to the myosin head a new ordered configuration that represents a dissipative structure during the anoxic period.

2.7. Re-Oxygenation

Re-oxygenation induced considerable reversibility of abnormalities of the LVPM mechanical indices due to anoxia. Reversibility was complete for several indices during re-oxygenation. This was the case for the maximum unloaded shortening velocity (Vmax) (Figure 2A), the myofilament sliding velocity (vo) (Figure 4A), the individual CB force (Figure 2B), the detachment rate constant g2 (Figure 3A), the frictional force (Figure 3D), and the friction coefficient (Figure 3C). Other indices, however, exhibited a partial reversibility and thus did not return completely to their control values. This was the case of the normal load (Figure 3B) and the molecular myosin content per volume unit of tissue (Figure 2D). Thus, anoxia induced a definitive CB loss of about 20%. Nevertheless, after re-oxygenation, the molecular characteristics of surviving CBs (i.e., po (Figure 2B) and g2 (Figure 3A) returned to the level of their respective control values.

3. Discussion

Prolonged myocardial anoxia induced dramatic abnormalities in molecular properties of the cardiac myosin. However, a high degree of reversibility of mechanical and thermodynamic defects was observed after re-oxygenation. A. Huxley has proposed the theory of sliding filaments, reflecting the friction of myosin filaments along actin filaments during contraction [1]. Based on this, we were able to apply the laws of tribology to the heart. A. Huxley also established a complex mathematical formalism [3] that made it possible to calculate the molecular mechanical properties of myosin CBs, including the force of a single CB, the maximum rate constant of CB detachment, the myosin content, and the sliding velocity of myofilaments. Myocardial anoxia resulted in a gradual decrease of mechanical indexes including the thermodynamic force (F/T). On the contrary, the myofilament sliding velocity (vo) increased up to 2 h of anoxia and then decreased from 2 h to 3 h of anoxia.

The entropy production rate (EPR) is the product of thermodynamic force (F/T) and thermodynamic flow (vo). We established that F/T and vo evolved in a non-linear fashion (Figure 4B,C). This demonstrated that the cardiac tribosystem operated far-from-equilibrium. Moreover, EEP became negative after 120 min of anoxia (Figure 6B), showing that the system became unstable. These results accounted for the occurrence of self-organization and dissipative structures in the anoxic heart. The stability condition for the tribological system is given by the expression of the second variation of the entropy production rate ($\delta^2S > 0$) (Equation (8)). Otherwise, the tribosystem became unstable and then a transition to a self-organized state could occur when $\delta^2S < 0$. In our study, this was the case after 150 min of anoxia (Figure 6A). Importantly, there was a linear relationship between EEP and max.Efficiency. Instability of the cardiac tribosystem occurred when max.Efficiency became <19% (Figure 6C).

The empirical Amontons–Coulomb law states that the frictional force varies linearly with the normal load ($F = \mu W$) (Figure 1E). Other linear empirical laws of physics, such as Fourier's law of heat conduction, Ohm's law of electrical conductivity, and Fick's law of diffusion, link the thermodynamic flow to the thermodynamic force, the product of which represents the entropy production rate (EPR). The approximate Amontons–Coulomb law remains valid for numerous classes of materials: metals, ceramics, polymers, composites, etc. In tribology, several works on self-organization have been carried out over the past 40 years: [11–13,23–25]. Friction and wear are irreversible processes that lead to energy dissipation and material deterioration. Under certain circumstances, frictional sliding can result in the formation of spatial and temporal patterns, i.e., self-organization. This results from the tendency of energy and matter to achieve a complex disordered state that can lead to self-organization [10,13,24,26,27]. It has been proposed an entropic criterium for friction-induced self-organization [13–16,24].

In our study, the cardiac tribosystem revealed a fine adaptive process that provided a considerable resistance to anoxia. This process delayed cell death of cardiomyocytes and allowed them to remain contractile cells. This behavior was the result of a self-organization process. Several systems with dissipative structures and self-organization have been described in earlier studies, i.e., chemical oscillations, reaction of Belousov–Zhabotinsky, Brusselator, and Oregonator chemical models, Turing structures; turbulent liquid motion, Bénard cells, biomolecular asymmetry, etc. [10]. In tribology, an example of such a system is the beneficial lubricating action of tribofilms during machining [18]. In techniques of material synthesis (for example, physical vapor deposition), coating deposition favors mechanisms that occur far-from-equilibrium, leading to a high density of lattice imperfections [28] and creation of a surface with a highly non-equilibrium state, accelerating beneficial physicochemical reactions through the appearance of dissipative structures [18,19,24]. The occurrence of self-organization can be suggested if the wear rate is reduced [24]. In biological systems and beside friction force, numerous physical forces provide important non-chemical roles particularly during morphogenesis of embryos. Many examples can be mentioned such as gravity [29] and surface tension due to intercellular adhesion [30,31]. A decrease in entropy production rate, i.e., self-organization, is implied in these processes. Thermodynamics in life has been initially raised by Schrödinger [32].

After re-oxygenation, the thermodynamic abnormalities were found to be largely reversible, demonstrating that the deleterious effects induced by anoxia were not totally irreversible. Importantly, re-oxygenation after 3 h of anoxia induced a return towards a thermodynamic status relatively close to that of the initial state. However, two indices did not totally return to their control values, i.e., myosin content (Figure 2D) and normal load (Figure 3B). The occurrence of self-organization is ubiquitous in natural systems, particularly in biological complex open systems [10]. The fact that the myocardium reacted in this way after such prolonged anoxia bears witness to the incredible ability of nature to generate self-organization processes capable of resisting and delaying cardiac cell death. Self-organization probably prevented the death of a significant number of myosin heads, which would have otherwise irreversibly and dramatically impaired the contractile function of the cardiomyocytes after prolonged anoxia.

4. Methods

4.1. Ethical Statement

All experimental procedures conformed to the Guide for Care and Use of Laboratory Animals. The study protocol was approved by the Ethical Committee of the Institut National de la Santé et de la Recherche Médicale (INSERM), Paris, France. The research complied with the commonly accepted '3Rs'.

4.2. General Study Approach

Our study sought to determine EPR generated by the sliding of actin filaments along myosin filaments during the contraction phase (Figure 1A–E). EPR was the product of the thermodynamic frictional force (F/T) and the thermodynamic flow (vo); vo was the sliding rate of myofilament and was calculated from the A. Huxley equations [3]. Once EPR was determined, the next step was to calculate EEP and to see whether it turned negative after a prolonged period of anoxia, characterizing the instability of the cardiac tribosystem and the occurrence of self-organization [10,13,14,26].

4.3. Experimental Procedure

Experiments were conducted on 27 adult male rats from Charles River Laboratories. Rats were anesthetized with pentobarbital sodium (50 mg/kg ip). Left ventricular papillary muscles (LVPMs) were carefully dissected from the heart. Each LVPM was rapidly mounted in a tissue chamber containing a Krebs–Henseleit solution (in mM): 118 NaCl; 24 $NaHCO_3$; 4.7 KCl; 1.2 $MgSO_4$ $7H_2O$; 1.1 KH_2PO_4; 2.5 $CaCl_2$ $6H_2O$; 4.5 glucose. The solution was bubbled with 95% O_2-5% CO_2 and maintained at pH 7.4 and 29 °C. LVPMs were electrically stimulated by means of two platinum electrodes delivering a 5ms stimulus at 0.1 Hz frequency. Experiments were carried out at Lo, i.e., the initial resting length corresponding to the apex of the length-active tension curve. The experimental procedure and the electromagnetic lever system have been previously described [33]. Maximum unloaded shortening velocity (V_{max}, in Lo · s^{-1}) of the LVPM was measured by means of the zero-load clamp technique [34]. Peak isometric tension (T, in mN/mm^2; force per cross-section area) was measured from the fully isometric contraction. The tension-velocity relationship was derived from the peak velocity (V) of 8 to 10 isotonic afterloaded contractions, plotted against the isotonic tension level (iT) and by successive load increments from zero-load up to peak isometric tension (T) [33]. The tension-velocity (iT-V) relationship was fitted according to A.V. Hill's equation (iT + a) (V + b) = [T + a] b where -a and -b are the asymptotes of the hyperbola [35]. For each LVPM, the iT-V relationship was accurately fitted by means of a hyperbola [33,35,36]. The curvature (G) of the hyperbola was T/a = Vmax/b.

4.4. A. Huxley's Formalism

The use of the A. Huxley formalism [3] requires verifying that each LVPM presented a hyperbolic tension-velocity (iT-V) relationship. Values for the asymptotes -a and -b of the iT-V relationship were introduced into the Huxley equations. The rate of total energy release (E_{Hux}) and isotonic tension (iT = F_{Hux}) as a function of the LVPM velocity (V) were obtained by the following equations [3]:

$$E_{Hux} = (N\,e)\,(h/2\,l)\,(f_1/(f_1+g_1))\,\{g_1 + f_1\,(V/\varnothing)\,[(1-\exp(-\varnothing/V)]\} \quad (1)$$

$$F_{Hux} = N\,(w/2l)\,(f_1/(f_1+g_1))\,\{1-(V/\varnothing)\,[(1-\exp(-\varnothing/V))\,(1+(1/2))\,((f_1+g_1)/g_2)^2\,(V/\varnothing)]\} \quad (2)$$

where w was the maximum mechanical work generated by a single CB (w/e = 0.75) and e was the free energy required to split one ATP molecule. In the A. Huxley formalism, only one ATP was split per CB cycle. The standard free energy $\Delta G^{\circ\prime}_{ATP}$ was −60 kJ/mol; e was equal to 10^{-19} J [37]. The myosin CB tilt relative to the actin filament varied from 0 to h; f_1 was the maximum value of the rate constant for CB attachment; g_1 and g_2 were the maximum values of the rate constants for CB detachment; f_1 and g_1 corresponded to a tilt of the CB from 0 to h; g_2 corresponded to a tilt > h; ∅ was equal to $(f_1 + g_1)$ h/2 = b; and N was the number of CBs per cross-sectional area. The molecular step size h corresponded to the distance of translocation of the actin filament after the CB swing. The constant l was the distance between two successive actin sites with which a myosin head could bind. In agreement with the A. Huxley conditions (l >> h), h and l values were h = 10 nm and l = 28.6 nm. The parameters po, G curvature of the iF-V hyperbola, f_1, g_1, and g_2 were calculated using the following equations:

$$G = f_1/g_1 \quad (3)$$

$$g_1 = 2wb/ehG \quad (4)$$

$$g_2 = 2Vmax/h \quad (5)$$

$$po = (w/l) \times [(f_1)/(f_1 + g_1)] \quad (6)$$

where po was the single CB force (in pN). Myosin content was calculated from the number of cycling myosin CB per volume unit of tissue and the Avogadro number. The sliding velocity of myosin filaments (in μm/s) along actin filaments was:

$$vo = h/ts \quad (7)$$

where ts was the time stroke [38]. The rate of mechanical work (WM) is given by the formula: WM = P_{HUX} V. At any given load level, the mechanical efficiency (Eff) of the LVPM is defined as the ratio of WM to E_{HUX}, i.e., Eff = WM/E_{HUX} and maximum efficiency (max. Efficiency) is the peak value of Efficiency.

4.5. Tribology and Heart Muscle

Friction is loosely defined as the resistance that a surface or object encounters when moving over another one. Bio-friction or "bio-tribology" can be defined as a friction applied to biological systems [39]. The principles of friction have been used for centuries in China and Egypt. Scientific studies on friction began with Leonardo da Vinci. Subsequently, Amontons and Coulomb laid the foundation for the current understanding of friction. Kinetic friction is an irreversible dissipative process. Friction represents the general tendency for irreversible energy dissipation in accordance with the Second Law of thermodynamics. The empirical Coulomb law of friction stipulates that the frictional force (F) is proportional to the normal load (W) according to the following formula (Figure 1E):

F = μW where μ the friction coefficient (without dimension) is independent of the normal load (W in mN), the myofilament sliding velocity (vo in μm/s), and the area of contact. The Coulomb frictional force (F in mN) is independent of the myofilament sliding velocity (vo).

The normal load (W) was calculated from Huxley's equation. The myosin content per volume unit was calculated from Huxley's equations [3] by dividing the CB number per volume unit by the Avogadro number. The myosin weight per volume unit was calculated knowing the myosin content and the myosin molecular weight. Cardiac myosin (molecular weight ~ 528 kDa) is a hexameric protein consisting of two myosin heavy chains (MHCs, 223 kDa), two pairs of regulatory light chains (RLCs, ~19 kDa), and two pairs of essential light chains (ELCs, ~22 kDa) [40]. The normal load W was determined by applying Newton's second law, i.e., W = 9.81 × myosin molecular weight.

The coefficient of friction μ has been determined in numerous living tissues [41]. In a myosin filament sliding along an actin filament, μ has been determined to be equal to ~0.001 [42]. The higher the detachment rate constant g2, the easier the sliding of myosin along actin and the lower the friction coefficient μ. The coefficient of friction was equal to 0.001 × (g2 control/g2 anoxia). The thermodynamic flow (vo) was equal to the myofilament sliding velocity and the thermodynamic force was equal to the frictional force F divided by T (T: Kelvin temperature): thermodynamic force = μ W/T.

4.6. Stability Conditions

Thermodynamic variables can be considered as functions of time and position. This is the assumption for assuming the local equilibrium principle. The fluctuations induced by the anoxia were small, and mechanical and thermodynamic changes were very progressive and continuous during 3-h.

In a non-equilibrium system and in the case of friction phenomena, the entropy production rate (EPR) is given by the expression:

$$\delta^2 S = (1/T) \cdot F \cdot v0, \text{ (in mN} \cdot \mu\text{m} \cdot \text{s}^{-1} \cdot \text{T}^{-1}) \quad (8)$$

where S is the entropy of the tribosystem. The transition from a stationary state sliding regime to a regime with self-organization occurred through the destabilization of the stationary state. Equation 8 is considered as a Lyapunov function. The second variation of entropy (δ^2 S) corresponds to the entropy production rate (EPR) which relates to a disturbance of the system due to anoxia. The stability condition for the thermodynamic system is given by the following equation:

$$\frac{\partial \delta^2 S}{\partial t} = \frac{\partial F}{T} \delta vo > 0 \tag{9}$$

$\frac{\partial \delta^2 S}{\partial t}$ is the excess entropy production (EEP).

The system is unstable when the value obtained through Equation (9) is negative.

Here (1/T) δF and δvo are the deviations of the thermodynamic force (F/T) and the thermodynamic flow (vo) from the stationary state, respectively. If EEP remains positive, the system remains stable. When the deviation from the stationary state goes beyond a certain critical value, the system is then able to reach an instability threshold. At a given value, EEP may change its sign, may become negative and thus unstable, and a transition to a self-organized solution with patterns can occur. A criterion based on the loss of thermodynamic stability (EEP < 0) is an indication of the occurrence of self-organization [4,9,10,18,19]. The necessary condition for the occurrence of self-organization is EEP < 0.

4.7. Statistical Analysis

Data were expressed as means ± standard deviations. Comparisons of parameters between groups were performed using Student's unpaired *t*-test. A p value < 0.05 was considered
statistically significant.

5. Conclusions

Friction is a process based on the concepts of non-equilibrium thermodynamics and self-organization. Within a tribosystem, self-organization may lead to dissipative structures that occur in far-from-equilibrium open systems, resulting in a decrease in wear rate [24]. The fundamental concepts of non-equilibrium thermodynamics, entropy, self-organization, dissipative structures, and thermodynamic stability-instability analysis represent powerful tools that have been used to develop innovative materials with substantial benefits. In our study, during prolonged anoxia, the heart was an open non-linear thermodynamic tribosystem operating far-from-equilibrium, as asserted by the non-linearity between the thermodynamic force and the thermodynamic flow. The negative value of the excess entropy production after 2 h of cardiac anoxia enabled self-organization. Anoxia led to dramatic changes in the friction coefficient, inducing modifications to CB kinetics which are largely reversible after re-oxygenation Such an adaptation was related to the occurrence of dissipative structures [24]. This anoxic cardiac tribosystem was thus able to shift from a severe wear mode to a milder one. These results are of particular interest in the context of heart transplants where the time that elapses between heart removal from the donor and its transplantation in the recipient is crucial. Self-organization and dissipative structures appear to prevent, or at least to limit, the appearance of dramatic irreversible damage to the myosin CBs.

Author Contributions: All authors contributed to the study conception and design. Material preparation, data collection and analysis were performed by Y.L., V.C., J.-L.H., X.K., O.S. and A.V. The first draft of the manuscript was written by Y.L. and all authors commented on previous versions of the manuscript. All authors have read and agreed to the published version of the manuscript.

Funding: This research received no external funding.

Institutional Review Board Statement: All experimental procedures conformed to the Guide for Care and Use of Laboratory Animals and the study protocol was approved by the Ethical Committee of the Institut National de la Santé et de la Recherche Médicale, Paris France.

Informed Consent Statement: Not applicable.

Data Availability Statement: Not applicable.

Acknowledgments: We thank Christophe Locher, Director of the Clinical Research Center of the Grand Hôpital de l'Est Francilien (GHEF), Meaux, France, for his valuable support in making the necessary research facilities available for this study. We are grateful to Brian Keogh for his improvement of the final manuscript. Experimental data can be seen at the Clinical Research Center.

Conflicts of Interest: The authors declare no conflict of interest.

References

1. Huxley, A.F.; Niedergerke, R. Structural changes in muscle during contraction; interference microscopy of living muscle fibres. *Nature* **1954**, *173*, 971–973. [CrossRef] [PubMed]
2. Huxley, H.; Hanson, J. Changes in the cross-striations of muscle during contraction and stretch and their structural interpretation. *Nature* **1954**, *173*, 973–976. [CrossRef] [PubMed]
3. Huxley, A.F. Muscle structure and theories of contraction. *Prog. Biophys Biophys Chem.* **1957**, *7*, 255–318. [CrossRef]
4. Nicolis, G.; Prigogine, I. Fluctuations in Nonequilibrium Systems. *Proc. Natl. Acad. Sci. USA* **1971**, *68*, 2102–2107. [CrossRef]
5. Nicolis, G.; Auchmuty, J.F. Dissipative Structures, Catastrophes, and Pattern Formation: A Bifurcation Analysis. *Proc. Natl. Acad. Sci. USA* **1974**, *71*, 2748–2751. [CrossRef]
6. Nicolis, G. Dissipative structures and biological order. *Adv. Biol. Med. Phys.* **1977**, *16*, 99–113.
7. Nicolis, G.; Basios, V.; Nicolis, C. Pattern formation and fluctuation-induced transitions in protein crystallization. *J. Chem. Phys.* **2004**, *120*, 7708–7719. [CrossRef]
8. Prigogine, I.; Nicolis, G.; Babloyantz, A. Nonequilibrium problems in biological phenomena. *Ann. N. Y. Acad. Sci.* **1974**, *231*, 99–105. [CrossRef]
9. Prigogine, I.; Nicolis, G. Biological order, structure and instabilities. *Q. Rev. Biophys* **1971**, *4*, 107–148. [CrossRef]
10. Kondepudi, D.; Prigogine, I. *Modern Thermodynamics from Heat Engines to Dissipative Structures*; Wiley & Sons: New York, NY, USA, 1999; pp. 1–486.
11. Klamecky, B.E. A thermodynamic model of friction. *Wear* **1980**, *63*, 113–120. [CrossRef]
12. Bershadsky, L. On self-organisation and concept of tribosystem self organizing. *J. Fric. Wear* **1992**, *13*, 101–114.
13. Nosonovsky, M. Self-organization at the frictional interface for green tribology. *Philos. Transactions. Ser. A Math. Phys. Eng. Sci.* **2010**, *368*, 4755–4774. [CrossRef] [PubMed]
14. Nosonovsky, M.; Roy, P. Scaling in Colloidal and Biological Networks. *Entropy* **2020**, *22*, 622. [CrossRef] [PubMed]
15. Nosonovsky, M.; Bhushan, B. Green tribology: Principles, research areas and challenges. *Philos. Transactions. Ser. A Math. Phys. Eng. Sci.* **2010**, *368*, 4677–4694. [CrossRef] [PubMed]
16. Nosonovsky, M.; Bhushan, B. Thermodynamics of surface degradation, self-organization and self-healing for biomimetic surfaces. *Philos. Transactions. Ser. A Math. Phys. Eng. Sci.* **2009**, *367*, 1607–1627. [CrossRef]
17. Naderi, M. On the evidence of thermodynamic self-organization during fatigue: A review. *Entropy J.* **2020**, *22*, 372. [CrossRef]
18. Gershman, I.S.; Bushe, N.A. Elements of thermodynamics and self-organization during friction. In *Self-Organization during Friction. Advanced Surface-Engineered Materials and Systems Design*; Fox-Rabinovich, G.S., Totten, G.E., Eds.; CRC Taylor & Francis: Boca Raton, FL, USA, 2006.
19. Fox-Rabinovich, G.; Gershman, I.S.; Yamamoto, K.; Biksa, A.; Veldhui, S.C.; Beak, B.D.; Kovalev, A.I. Self-Organization during Friction in Complex Surface Engineered Tribosystems. *Entropy* **2010**, *12*, 275–288. [CrossRef]
20. Bakaeen, F.G.; Gaudino, M.; Whitman, G.; Doenst, T.; Ruel, M.; Taggart, D.P.; Stulak, J.M.; Benedetto, U.; Anyanwu, A.; Chikwe, J.; et al. 2021: The American Association for Thoracic Surgery Expert Consensus Document: Coronary artery bypass grafting in patients with ischemic cardiomyopathy and heart failure. *J. Thorac. Cardiovasc. Surg.* **2021**, *162*, 829–850.e1. [CrossRef]
21. Lawton, J.S.; Tamis-Holland, J.E.; Bangalore, S.; Bates, E.R.; Beckie, T.M.; Bischoff, J.M.; Bittl, J.A.; Cohen, M.G.; DiMaio, J.M.; Don, C.W.; et al. 2021 ACC/AHA/SCAI Guideline for Coronary Artery Revascularization: Executive Summary: A Report of the American College of Cardiology/American Heart Association Joint Committee on Clinical Practice Guidelines. *J. Am. Coll. Cardiol.* **2022**, *79*, 197–215. [CrossRef]
22. Stehlik, J.; Kobashigawa, J.; Hunt, S.A.; Reichenspurner, H.; Kirklin, J.K. Honoring 50 Years of Clinical Heart Transplantation in Circulation: In-Depth State-of-the-Art Review. *Circulation* **2018**, *137*, 71–87. [CrossRef]
23. Kostetsky, B. Structural -energetic adaptation of materials at friction. *J. Fric Wear* **1985**, *6*, 201.
24. Fox-Rabinovich, G.; Totten, G.E. Self-organization during friction: Advanced surface-engineered materials and systems design. In *Endotext*; CRC Press: BocaRaton, FL, USA, 2006.
25. Mortazavi, V.; Nosonovsky, M. Friction-induced pattern formation and Turing systems. *Langmuir ACS J. Surf. Colloids* **2011**, *27*, 4772–4779. [CrossRef]
26. Glansdorff, P.; Prigogine, I. *Thermodynamics of Structure Stability and Fluctuations*; Wiley: New York, NY, USA, 1971.
27. Nicolis, G.; Prigogine, I. *Self-Organization in Non-Equilibrium Systems: From Dissipative Structures to Order through Fluctuations*; John & Wiley & Sons: New York, NY, USA, 1977.

28. Palatnik, L.S. *Pores in the Films*; Energoizdat: Moscow, Russia, 1982.
29. Davies, P.C.; Rieper, E.; Tuszynski, J.A. Self-organization and entropy reduction in a living cell. *Bio. Syst.* **2013**, *111*, 1–10. [CrossRef] [PubMed]
30. Goel, N.S.; Leith, A.G. Self-sorting of anisotropic cells. *J. Theor. Biol.* **1970**, *28*, 469–482. [CrossRef]
31. Gordon, R.; Goel, N.S.; Steinberg, M.S.; Wiseman, L.L. A rheological mechanism sufficient to explain the kinetics of cell sorting. *J. Theor. Biol.* **1972**, *37*, 43–73. [CrossRef]
32. Schrödinger, E. What is Life? The Physical Aspect of the Living Cell. In *StemBook*; Cambridge University Press: Cambridge, UK, 1967.
33. Lecarpentier, Y.; Chemla, D.; Blanc, F.X.; Pourny, J.C.; Joseph, T.; Riou, B.; Coirault, C. Mechanics, energetics, and crossbridge kinetics of rabbit diaphragm during congestive heart failure. *FASEB J. Off. Publ. Fed. Am. Soc. Exp. Biol.* **1998**, *12*, 981–989. [CrossRef]
34. Brutsaert, D.L.; Sonnenblick, E.H. Force-velocity-length-time relations of the contractile elements in heart muscle of the cat. *Circ. Res.* **1969**, *24*, 137–149. [CrossRef]
35. Hill, A.V. The heat of shortening and the dynamic constants of muscle. *Proc. R. Soc. Lond. Biol. Sci.* **1938**, *126*, 136–195.
36. Woledge, R.C.; Curtin, A.N.; Homsher, E. *Energetic Aspects of Muscle Contraction*; Academic Press: London, UK, 1985; Volume 41, pp. 1–357.
37. Veech, R.L.; Lawson, J.W.; Cornell, N.W.; Krebs, H.A. Cytosolic phosphorylation potential. *J. Biol. Chem.* **1979**, *254*, 6538–6547. [CrossRef]
38. Lecarpentier, Y.; Claes, V.; Krokidis, X.; Vallée, A.A. Comparative Statistical Mechanics of Muscle and Non-Muscle Contractile Systems: Stationary States of Near-Equilibrium Systems in A Linear Regime. *Entropy J.* **2017**, *19*, 558. [CrossRef]
39. Dowson, D. Bio-tribology. *Faraday Discuss.* **2012**, *156*, 9–30. [CrossRef] [PubMed]
40. Jin, Y.; Wei, L.; Cai, W.; Lin, Z.; Wu, Z.; Peng, Y.; Kohmoto, T.; Moss, R.L.; Ge, Y. Complete Characterization of Cardiac Myosin Heavy Chain (223 kDa) Enabled by Size-Exclusion Chromatography and Middle-Down Mass Spectrometry. *Anal. Chem.* **2017**, *89*, 4922–4930. [CrossRef] [PubMed]
41. Jin, Z.; Dowson, D. Bio-friction. *Friction* **2013**, *1*, 100–113. [CrossRef]
42. Suda, H. Molecular friction in an actomyosin molecular machine. *J. Theor. Biol.* **1990**, *146*, 341–346. [CrossRef]

International Journal of
Molecular Sciences

MDPI

Review

Detailing Protein-Bound Uremic Toxin Interaction Mechanisms with Human Serum Albumin in the Pursuit of Designing Competitive Binders

Vida Dehghan Niestanak [1] and Larry D. Unsworth [2,*]

1 Department of Biomedical Engineering, University of Alberta, Edmonton, AB T6G 2G4, Canada; vida1@ualberta.ca
2 Department of Chemical and Materials Engineering, University of Alberta, Edmonton, AB T6G 1H9, Canada
* Correspondence: lunswort@ualberta.ca

Abstract: Chronic kidney disease is the gradual progression of kidney dysfunction and involves numerous co-morbidities, one of the leading causes of mortality. One of the primary complications of kidney dysfunction is the accumulation of toxins in the bloodstream, particularly protein-bound uremic toxins (PBUTs), which have a high affinity for plasma proteins. The buildup of PBUTs in the blood reduces the effectiveness of conventional treatments, such as hemodialysis. Moreover, PBUTs can bind to blood plasma proteins, such as human serum albumin, alter their conformational structure, block binding sites for other valuable endogenous or exogenous substances, and exacerbate the co-existing medical conditions associated with kidney disease. The inadequacy of hemodialysis in clearing PBUTs underscores the significance of researching the binding mechanisms of these toxins with blood proteins, with a critical analysis of the methods used to obtain this information. Here, we gathered the available data on the binding of indoxyl sulfate, p-cresyl sulfate, indole 3-acetic acid, hippuric acid, 3-carboxyl-4-methyl-5-propyl-2-furan propanoic acid, and phenylacetic acid to human serum albumin and reviewed the common techniques used to investigate the thermodynamics and structure of the PBUT–albumin interaction. These findings can be critical in investigating molecules that can displace toxins on HSA and improve their clearance by standard dialysis or designing adsorbents with greater affinity for PBUTs than HSA.

Keywords: chronic kidney disease; protein-bound uremic toxins; human serum albumin; binding affinity

1. Introduction

Kidney dysfunction ultimately leads to the retention of otherwise cleared metabolites (i.e., uremic toxins [1]) in the blood compartment. The increased retention of these uremic toxins further exacerbates kidney health and contributes to the development of chronic kidney disease (CKD), as well as being correlated with other major co-morbidities [2]. CKD is among the top 20 fatal diseases recognized globally, and by 2040 it is estimated to be the fifth largest health issue faced worldwide [3,4]. According to the Canadian Institute for Health Information [5], the number of Canadians who start renal replacement therapy increases by 1.1% annually. Within two decades, twice as many people will be prescribed dialysis due to renal dysfunction in Canada. Worldwide, more than 10% of the population suffers from CKD, accounting for 1.2 million deaths and 28 million years of life lost each year [3].

Hemodialysis is the conventional method for clearing uremic toxins from the blood [1]. However, despite decades of research, the average life expectancy of CKD patients receiving hemodialysis has neither increased beyond three years nor has the incidence or severity of CKD co-morbidities been significantly reduced [6,7]. Protein-bound uremic toxins (PBUTs) are thought to play a crucial role in the health of patients with kidney dysfunction. These small molecular weight toxins (Mw < 500 Da) become bound to human blood proteins (i.e., albumin) and, consequently, cannot be cleared from the blood compartment using membrane-based hemodialysis (see Table 1, Figure 1) [1,8]. Co-morbidities associated with the accumulation of known PBUTs, namely indoxyl sulfate (IS), p-cresyl sulfate (PCS), indole 3-acetic acid (IAA), hippuric acid (HA), 3-carboxyl-4-methyl-5-propyl-2-furan propanoic acid (CMPF), and phenylacetic acid (PAA), are summarized in Table 1. The difference between the blood concentration of these molecules in normal kidneys or those with failing kidneys can be as high as 400 times. That said, the interaction mechanisms between well-known uremic toxins and blood proteins remain largely ill-defined. Elucidating these mechanisms will allow innovation in developing means to remove these toxins, which may include competitive inhibitors and adsorption-based techniques: the former leading to increased clearance efficiency using membrane dialysis, the latter focusing on directly capturing the toxins [1,4,9].

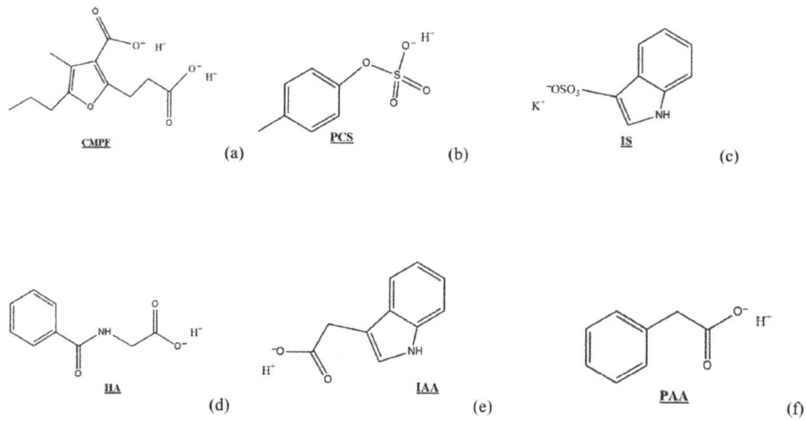

Figure 1. Chemical structure and charge location for (**a**) 3−carboxyl−4−methyl−5−propyl−2−furan propanoic acid (CMPF), (**b**) p−cresyl sulfate (PCS), (**c**) indoxyl sulfate (IS), (**d**) hippuric acid (HA), (**e**) indole−3−acetic acid (IAA), and (**f**) phenylacetic acid (PAA).

Human serum albumin (HSA) is the most abundant blood protein (~60% of total protein, at 0.6 mM) with a net negative charge of −19, at a pH of 7.4, an isoelectric point of 5.4, 585 amino acids, and three structurally similar, homologous α-helical domains (I, II, and III), each with ten helices divided into 6-helix and 4-helix subdomains (A and B) [10–12]. HSA is crucial in binding organic and inorganic molecules, including hydrophobic and anionic substances, either covalently or non-covalently. In patients with CKD, HSA binding capacity is significantly lowered [13,14]. This is thought to occur due to the saturation of HSA binding sites by uremic toxins and HSA conformational changes during post-translational modification such as oxidation, glycosylation, and carbamylation [14,15]. Although ligand–HSA interactions are complicated by both specific and non-specific binding sites that may be involved, depending on the uremic toxin and concentration, it has been determined that PBUT bindings essentially occur at Sudlow's site I (subdomain IIA) and site II (subdomain IIIA), a common location where aromatics and heterocyclic ligands are known to bind [10,11]. Bulky heterocyclic anions charged in the middle generally bond to site I, while aromatic carboxylic acids with charges at one end are more likely to bind to site II [16]. Furthermore, ligand binding with subdomains IB, IIA, and IIIB and these subdomains' interface is vital as these regions hold seven binding sites for fatty acids [11]. Davilas A. et al. [13] highlighted that since there is a common interface between site I and site II, the binding of uremic toxins to each subdomain likely results in conformational and binding behavior changes, and HSA structure changes become more significant while studying several PBUT competitions for the same binding site on HSA.

Table 1. Physical properties of protein-bound uremic toxins, significance in CKD patients, and major co-morbidities caused by the accumulation in the blood compartment.

PBUT	Chemical Group	Molecular Weight (Da)	Average Normal Concentration (mg/L)	Average Uremic Concentration (mg/L)	Average Hemodialysis Reduction (%)	Major Co-Morbidities	Refs.
3-Carboxyl-4-Methyl-5-Propyl-2-Furan Propanoic Acid (CMPF)	Furan Dicarboxylic Acids	240	4.6 (±1.80)	25.95 (±10.23)	20	Altered drug pharmacokinetics, increased intestinal permeability, neurological abnormalities, renal tubule damage, anemia	[4,17–20]
1-8 Indoxyl Sulfate (IS)	Indoles	212	0.53 (±0.29)	3.83 (±2.46)	36	Altered drug pharmacokinetics, bone and cardiovascular diseases, oxidative stress, muscle weakness, endothelial cell, wound repair inhibition	[15,17,18,21,22]
1-8 p-Cresyl sulfate (PCS)	Phenols	187	0.08 (±0.09)	2.60 (±5.10)	31	All-cause mortality, oxidative stress, inflammation, cardiovascular disease	[17,18,22,23]
1-8 Hippuric Acid (HA)	Hippurates	179	3.00 (±2.00)	109.43 (±64.66)	67	Endothelial dysfunction, renal tubule damage	[17,18,24,25]
1-8 Phenylacetic Acid (PAA)	Others	136	1.4	474.64 (±44.88)	35	All-cause mortality, oxidative stress, inflammation, atherosclerosis, cardiovascular disease, increased immunodeficiency	[17,18,26–28]
1-8 Indole-3-acetic acid (IAA)	Indoles	174	0.50 (±0.30)	1.26 (±0.83)	45	Altered drug pharmacokinetics, bone and cardiovascular diseases, oxidative stress, muscle weakness, neuropathy, cytotoxicity, cognitive impairment	[4,17,18,29,30]

Herein, we specifically review the experimental techniques used in the context of understanding the binding of PBUTs with HSA. The accuracy of the standard methods applied to obtain thermodynamic data and investigate protein structure changes during the binding event will be discussed. Additionally, we review the literature regarding PBUT binding to HSA, highlighting relevant physicochemical properties, binding affinities, and binding mechanisms. This information is critical for developing novel pathways to remove these uremic toxins from the blood compartment.

2. Experimental Techniques Commonly Employed

Isothermal Titration Calorimetry (ITC): An ITC instrument consists of a thermally equilibrated syringe, sample, and reference cells [31,32]. The interaction between the sample and sequentially injected titrant causes temperature changes between the reference and experimental cell, where the energy required to balance these changes is integrated to obtain changes in heat as a function of titrant concentration. This information can be used to describe the complete thermodynamic profile of molecular interaction, namely the association binding constant (K_a), enthalpy change (ΔH), stoichiometry (n), Gibb's free energy (ΔG), and entropy change (ΔS) [31,33]. Moreover, the heat of capacity (ΔC_p) can be obtained by changing the temperature to study the binding event [32,34,35].

In order to collect accurate isotherms, protein and ligand concentrations need to be studied within c-values between 1 and 1000 [36], as determined using the expected K_a and the number of ligands (n) in the chamber:

$$c = nK_a \tag{1}$$

Analysis of collected isotherms for determining thermodynamic parameters relies on several underlying assumptions. Firstly, the binding reaction is assumed to be reversible and in equilibrium. Although this assumption is acceptable for most binding reactions, one should note that allosteric changes caused by binding may directly affect the reversibility of the binding event itself [31,32,35–37]. For example, PBUT buildup in the blood has been shown to affect plasma protein conformation, where HSA α-helicity drops from its native 54–56% to 8% when PCS is present at physiological conditions for CKD patients, with the subsequent strongest response by IS, followed by HA, IAA, and CMPF [32,38]. IS and PCS presence yielded an increase in the β-sheet content of HSA by 15.6% and 23%, respectively. This is a clear case where the protein's conformation changes may affect the PBUT binding characteristics. Secondly, all molecules remain soluble, and the attributed ΔH of dissolution is detected and subtracted from the total ΔH measured [36]. Finally, the macromolecule solution is assumed to be ideal, where no macromolecule–macromolecule/co-solute interactions occur. However, this is practically impossible as macromolecules interact with co-solutes and each other by direct binding and indirect modification of vicinal hydration layers [36]. Hence, an additional blank experiment, which can be either a buffer-to-protein solution or toxin-to-buffer titration, is set. This isotherm is subtracted from the ligand–protein data to remove any heat generated/absorbed in secondary interactions, hydration layer modifications, and mixing into the solution [31,32]. With this in mind, it is possible to model the isotherms collected for these molecular interactions as outlined below.

ITC Model 1: The Single Set of Independent Binding Sites Model: This model is used when only one binding site is assumed to be accessible to the PBUT for HSA. The model is derived from the Langmuir adsorption theorem, which is based on the equilibrium between the number of available and occupied binding sites on protein in a solution and can also be correlated to the concentration of free titrant molecules (i.e., uremic toxins) in the media [31,37,39]. As there is often more than one binding region, the fittings provide thermodynamic insights into the overall interaction of the protein with the toxin [37]. That said, even if the model used has a good fit on ITC isotherms, it does not necessarily mean that it represents the actual situation; care must be taken to understand the effect the binding has on the protein as a function of concentration. Moreover, the returned ΔH and further energy calculations are usually considered much more accurate than the returned number of binding sites (n) [31]. Nevertheless, the independent model was found to be most appropriate for IS and PCS since these uremic toxins predominantly interact with a single binding site on HSA (see below) [13,32,40].

ITC Model 2: The Sequential Binding Sites Model: This model is used for more complicated interaction events where more than one site on HSA accepts the toxin. Each binding site's thermodynamic properties are obtained based on the equilibrium stage sequences. It should be noted that this sequence is not necessarily linked with the binding affinity of the toxin to the region on HSA [37]. Zaidi et al. found the sequential binding model to be the best fit for the ITC results of the HA and HSA interaction. However, they observed deviation in ΔC from the literature, which may result from changes in protein structure as a function of solution temperature. They pointed out that the ITC result is limited to whether the interaction or temperature change is causing protein conformational alterations. Hence, it is necessary to combine ITC results with an analysis of HSA conformation under similar solution conditions [35].

In most cases, there is more than one binding site available on HSA for uremic toxins, and in sequential binding, it is assumed that previous binding events affect subsequent interactions, which seems more likely because toxin binding may influence HSA structure [41]. If the binding sites are identical (i.e., similar values for K and ΔH), the isotherm is commonly described using the single independent site model. Indeed, prior investigation on the structure and chemistry of the toxins will be beneficial in preventing misinterpretation of the data [31]. For instance, at low PAA molar ratios, independent and sequential binding site models were fitted ideally on PAA-HSA ITC results. Yu et al. showed that for molar ratios higher than 15 n(PAA)/n(HSA), these two models deviate, and the sequential model seems to be a better candidate, indicating that at a threshold concentration, two binding sites on HSA arise for PAA [37].

ITC Model 3: The Competitive Binding Model: In addition to understanding how toxins interact with HSA, competitive binding mechanisms can be elucidated using ITC isotherms by fitting competitive binding models. An example is when Li et al. investigated the displacement of IS and PCS from HSA by either warfarin or ibuprofen. These drugs were considered ligands for the toxin–HSA complex. The blank isotherms were subtracted from the displacement results for the toxin–HSA complex and buffer, in this case. In the final isotherm profile, both endothermic and exothermic injections are usually seen, which are claimed to result from displacing the toxin by the binding competitor. The prior ligand–protein ITC isotherms and thermodynamic profile provide insights into further interpreting the competitive binding results [32]. Multiple independent binding site models can also be used in similar cases where more than one ligand interacts with the protein. The binding constants and energy involved are independently obtained for each toxin regarding their high-affinity site on HSA [37].

The advantage of ITC over other thermodynamic techniques is the ease of collecting data in the least number of experiments and that no probe or ligand immobilization at a surface is required [31,32]. Due to its extensive measurement range, K_a from 10^3 to 10^9 M^{-1} [31], ITC is utilized in various applications for different molecule sizes down to nanoparticles and small solutes and solvents [33]. While other techniques such as UV spectroscopy, surface plasmon resonance (SPR), or quartz crystal microbalance (QCM) may be more affordable, they do not possess high sensitivity or resolution. Moreover, using these techniques, the samples must undergo adaptations, such as immobilization [31,42].

ITC data can suffer due to its sensitivity to temperature, pH, and other environmental factors [36,43]. In fact, one issue with this technique is that poor or variable sample preparation can lead to drastically different isotherm profiles and inconsistent data analysis [37]. Sample concentrations must satisfy the c-value window specifications as they impact the isotherm shape and the trend reaching saturation. As mentioned above, while using ITC, it must be assured that the thermodynamic assumptions are valid. One of the most common errors in ITC data analysis is the failure to allow signal equilibrium, as evidenced by the return to baseline before the next injection [33,36]. The fact that ITC data analysis cannot overlook the non-ideality of the macromolecules remains unavoidable; thus, precise experimental conditions need to be included in the report [36]. It should also be highlighted that studying weak interactions, $K_a < 10^3$ M^{-1} [31], remains challenging using ITC [33]. Although ITC results show whether an interaction is happening with minor enthalpy change, entropy-dominated interactions cannot easily be studied using this method [31].

Fluorescence Spectroscopy: PBUT accumulation in the blood compartment can cause protein conformational changes that may affect calorimetric results [32]. Fluorescence spectroscopy has been widely used to investigate ligand binding [44,45] and changes in protein tertiary structure during these events [32]. The fluorescence wavelength is sensitive to alterations in the immediate environment of the fluorescent probe. The high signal-to-noise ratio afforded by low amounts of fluorescent probes and rapid data acquisition is why this technique has been used for thermodynamic and kinetic studies of molecular interactions [46]. Moreover, the problematic requirement for a fluorescent probe is overcome by the presence of intrinsic fluorescence originating from the aromatic amino acids already present within the protein. Although there are three intrinsically fluorescent amino acids in HSA [46,47], tryptophan provides enough quantum yield and sensitivity for studying toxin–HSA interactions [46,48]. The fluorescence intensity emitted from tryptophan residue in HSA changes in the presence of toxins [47]. Although fluorescence optical techniques are fast, sensitive, and highly applicable in protein structure studies, their main limitation is their localized signal. The data given are limited to the environment close to the probe and the extent of its mobility. Therefore, the fluorescence spectra data are often contextualized using circular dichroism (CD), FT-IR, or NMR results. Moreover, fluorescence results do not explicitly describe pathways involved in conformational changes [46].

Fluorescence Quenching: The fluorescence quenching technique uses the intrinsic tryptophan fluorescence of proteins and provides data on the fluorophore location within a protein using quenchers that interact with external or internal tryptophan only [32]. The dependence of the emission intensity on quencher concentration (Q) is given by the Stern–Volmer equation (Equation (2)), where F_0 and F represent the fluorescence intensity, τ_0 and τ are fluorescent lifetimes, and the 0 subscripts show the absence of the quencher while the other parameters are the values in the presence of the quencher. K_{sv} is the Stern–Volmer constant [44].

$$\frac{F_0}{F} = \frac{\tau_0}{\tau} = 1 + K_{sv} \cdot Q \tag{2}$$

The association constant (K_a) and the number of binding sites (n) can also be found by analyzing the quenching data [32]:

$$\log \frac{F_0 - F}{F} = n\log\ K_a - n\log \frac{1}{[Q] - \frac{(F_0 - F)[P]}{F}} \tag{3}$$

Dynamic and static quenching modes can be clarified using the shape and position of fluorescence band changes with protein concentration, curvature, or linearity of the Stern–Volmer plots and order of magnitude for the Stern–Volmer constant [44,49]. These data can be used to distinguish between dynamic and static quenching on temperature and viscosity, affecting the fluorescence lifetime measurements [32,44]. Moreover, static quenching is observed when there is a strong coupling between HSA and toxins, where increasing the temperature causes a decrease in K_{sv} [32].

Fluorescence probe displacement: This method is another form of fluorescence spectroscopy that detects changes in the fluorescence characteristics associated with probe binding to macromolecules and assists in identifying the binding location and potential competition in binding events [14,45,50,51]. The probes can possess either single or two binding sites and are expected to bind to specific locations on proteins (e.g., HSA). It is suggested that the concentration of ligands in the sample be as low as possible to avoid non-specific site occupation. This technique highly depends on the availability of binding sites on HSA. When the ligand concentration is high enough for binding sites to be saturated, interpretation of fluorescence changes in terms of individual binding sites is rather difficult. Moreover, several unrelated events, such as competitive displacements, might affect the probe's fluorescence signal. The percentage of the probe's displacement correlates with the fluorescence of the probe and HSA with and without the ligand present (F_2 and F_1, respectively), as shown in Equation (4)) [45]:

$$\frac{F_1 - F_2}{F_1} \times 100 \tag{4}$$

The binding constant of HSA and toxin can be further obtained from the relation between the probe displacement and the fraction of sites occupied [49].

A decrease in probe–HSA complex fluorescence in the presence of the toxin is evidence that the toxin replaces the probe, and as the probes are site-specific, the results can provide an insight into the probable region that the toxin has an affinity on HSA. Warfarin and dansylsarcosine (DNSS) are generally used to locate uremic toxin binding on site I and site II, respectively. For instance, IS and IAA presence in the warfarin–HSA complex had an insignificant impact on the detected fluorescence intensity, as these two toxins have a higher affinity for site II [14,51]. Notably, the association and capacity of toxin to bind to HSA must be higher than that of the probes to result in a detectable change in fluorescence intensity. As HA interaction with HSA is relatively weaker on both sites, insignificant changes were observed using warfarin and DNSS as the probe [14].

Circular Dichroism (CD): If a substance absorbs circularly polarized light, either right-handed or left-handed, it exhibits circular dichroism [52]. CD is often used to determine protein secondary structure [32,38,42,53] and is highly accurate for α-helices; however, β-sheets, β-turns, and random coil contents are also obtained [52]. Changes in the CD spectrum as a function of ligand concentrations can be achieved and analyzed using widely available software that has been extensively used and reviewed [32]. However, one of the main problems with using CD for these PBUT studies is sample preparation, including the effect that salts have on the CD spectra [52,54].

On the CD spectrum of free HSA, the α-helical structure is indicated by two minima at 208 and 222 nm. In case of binding or presence of uremic toxins and HSA conformational changes, the mean residue ellipticity at each minimum may decrease, showing that the α-helicity is reduced. This trend was observed for CMPF, IAA, IS, and HA [38]. The relative reduction in helical content of HSA could be as high as 20.50% in the presence of IS [38]. Furthermore, CD was also utilized to observe HSA conformational changes with temperature and pH [42,55].

Nuclear Magnetic Resonance (NMR): NMR has been used to study ligand binding and macromolecular structure [56]. Compared to ITC, NMR gives a broader view of the dynamic of bindings at equilibrium and whether the interaction occurs on specific sites or solely causes HSA conformational changes without being attached [32,57]. In its infancy for studying PBUT interactions with proteins, 2-D NMR monitors all binding sites on the protein and quantitatively measures site-specific binding constants [58,59]. So far, only Li et al. [32] have utilized this technique to identify potential secondary and tertiary structure changes in HSA while interacting with PBUTs using STD (saturated transfer difference) NMR to detect the intensity of bindings through an on-and-off induced magnetic field without affecting the unbonded toxins. Not only could they recognize the binding percentage of PBUTs to HSA, but they could also observe the positional characteristics during the events. Li et al. utilized STD NMR to selectively transfer the induced magnetic field from HSA protons to those inbound IS and PCS without altering the free toxins in the solution; they could comprehend the strength of binding events from the resonance spectra and integrated area for each proton [32]. Furthermore, NMR is applicable in identifying specific sites on HSA where modifications can be performed to reduce the affinity of a compound [58], information that may be useful in designing effective competitors to prevent PBUT bindings to HSA.

Other Methods: Early research on PBUT binding to HSA utilized less sensitive or advanced mechanisms: equilibrium dialysis [14,19,60], ultrafiltration [14,28], Fourier-transform infrared spectroscopy (FT-IR) [47,61], and high-performance liquid chromatography (HPLC) [19,51]. X-ray crystallography has been used to further understand the binding location but is not widely used due to its availability, requirement to form crystals, and cost [50,62].

Equilibrium dialysis and HPLC measure a ligand's binding to a macromolecule using separation. While both methods measure the clearance of free compounds from a ligand–protein solution, two chambers and separation membranes are often involved. Although equilibrium dialysis has been a preferred method for identifying plasma bindings, it is time-consuming and challenging to automate [56].

The rate-of-dialysis technique proposed by Kragh-Hanson [63] has made this technique's performance more convenient. Still, it requires sample radio-labeling, and the working principle of this technique is to correlate the rate of transfer of the labeled ligand to the unbound fraction of ligands. In contrast, in HPLC, the mass transfer happens in micro-columns as measured using a differential mass detector [56,64,65]. HPLC is usually combined with spectroscopy techniques such as UV–vis [19] and fluorescence [51] in PBUT-HSA studies. Therefore, it has higher precision and could be automated to run relatively fast [65].

Ultrafiltration is a rapid method of measuring the binding extent of ligands to proteins [14]. However, binding constants cannot be quantified [56]. The non-specific binding (NSB) of toxins to HSA can be obtained by measuring the initial concentration of a ligand (C_D) and that of it after ultrafiltration (C_F) [19]:

$$NSB = \frac{C_D - C_F}{C_D} \tag{5}$$

The applications of molecular docking tools in binding studies of HSA and toxins are also worth mentioning [35,42,62,66,67]. Although they involve specific assumptions and protocols depending on the working algorithms, molecular docking and molecular dynamic simulations have not only assessed researchers to obtain insight into the ligand–protein binding and complex structures but also provided tools for understanding the PBUT chemistry, production, and physiology [35,66]. A remarkable dataset on HSA residues participating in binding events for IS, PCS, IAA, and HA is provided using molecular dynamic simulation [62]. In more recent work, the same information for hydrophobic and hydrophilic residue receptors on HSA is given via molecular docking studies [66].

Amongst the methods mentioned above for obtaining the binding thermodynamics, association constants, and stochiometric data, ITC is often used [32,35,37,38,40,55]. Although ultrafiltration is as common, it requires separative membranes, while ITC allows affinity determination in a free medium [40]. Differential scanning calorimetry can be used in combination with ITC to obtain information on the thermostability of the protein [35,38,56]. Investigating the structure of HSA and evaluating the location of PBUT binding can be performed using NMR [32]; nevertheless, CD and fluorescence techniques are commonly employed due to their flexibility and affordability [14,28,32,35,38,42,45,50,51,55,68].

3. Insights into PBUT-HSA Interactions

Thermodynamic and kinetic parameters for PBUT binding with HSA differ based on the type of toxin involved and the interaction site (Table 2). Differences in HSA's free fatty acid content, sample preparation methods, temperature, and salt concentration of the buffer could be some of the reasons for these differences [14]. The salt concentration in the buffer can have a significant impact because changes in ionic strength are observed to substantially affect the binding mechanism and potentially alter the electrostatic forces involved [15,37,62]. 3-carboxyl-4-methyl-5-propyl-2-furan propanoic acid (CMPF), phenylacetic acid (PAA), indoxyl sulfate (IS), indole 3-acetic acid (IAA), p-cresyl sulfate (PCS), and hippuric acid (HA) (Figure 1, [14]) have been shown to have a high affinity for at least either site I or II on HSA, if not for more locations depending on solution concentration [28].

Binding mechanisms are determined from changes in thermodynamic parameters (H, S, and G): negative ΔH indicates an exothermic event arising from electrostatic interactions, and negative ΔS suggests the involvement of hydrogen bonding [31,38].

3-carboxyl-4-methyl-5-propyl-2-furan propanoic acid: CMPF is a strongly lipophilic urofuranic acid that exclusively binds to site I and is known to impact the elimination of bilirubin and thyroxine, affecting the progression of liver damage and cardiovascular disease [18,50,69,70]. In CKD patients, the buildup of CMPF in the blood inhibits renal elimination of other compounds [70], including inhibiting warfarin and dansylsarcosine (DNSS) binding to HSA [14]. CMPF binding to HSA is exothermic, arising from both electrostatic interactions and the hydrophobic effect. The latter is thought to be due to bulky hydrophobic dicarboxylic acid groups on CMPF [38,50]. HSA binding of CMPF is pH dependent; pH increase leads to a change in the charge properties of HSA, altering its confirmation and yielding an electrostatic repulsion that impedes CMPF binding [19,50]. HSA binding of CMPF is also mildly sensitive to increased temperature [19].

Based on the fluorescence displacement data, CMPF and Bilirubin compete for the same site on HSA, with association constants of 130×10^5 and 500×10^5 M^{-1}, respectively [14,50]. Moreover, prostaglandin shares the same binding site as CMPF on HSA as an endogenous substance [14]. However, the association constant obtained by ITC in a much more recent study is 5.76×10^6 M^{-1} on a higher affinity site [38]. Observed thermodynamic parameters differed between the two studies of CMPF-HSA interaction [19,38]. The significant difference observed was that the older work using equilibrium dialysis (37 °C) showed an entropy increase, whereas subsequent work using ITC (25 °C) found that entropy decreased. It has been reported that this temperature difference likely is not the dominant reason for the change in entropy [19].

CMPF is preferentially bound by Sudlow's site I (subdomain IIA), whereas other PBUTs (PCS, IS, IAA, and HA) have a higher affinity for Sudlow's site II (subdomain IIIA). Zaidi et al. [38] showed that among IAA, HA, and IS, the latter has the highest affinity for this site. Due to their carboxylate group, IAA and HA have stronger electrostatic interactions on site II than IS and PCS [62]. However, HA and IAA are replaced by IS and PCS on the high-affinity site in higher concentrations of these four toxins in healthy serum and hemodialysis patients' serum samples [62,71]. Devine et al. suggested that investigating the binding mechanism for purified HSA may overlook the involvement of other competitors, such as fatty acids, as they showed that the IS association constant to human plasma is 7.46×10^4 M^{-1}, much less than the IS affinity for HSA (1.17×10^6 M^{-1}) on only one independent binding site [15]. However, experimenting with native plasma will complicate the sample preparation and locating the binding site on proteins.

Indoxyl Sulfate: IS retention in the blood compartment is correlated to the progression of CKD and many co-morbidities [4]. IS binding to HSA mainly occurs at site II, is exothermic, and involves electrostatic and hydrophobic interactions [18,32,40,62]. The interaction is enthalpy-driven, as the association constant of IS-HSA decreases with increasing temperature; it is also counterbalanced by entropy [15,40]. Entropy changes are considered one of the main factors involved when IS competes with other uremic toxins, especially PCS, for binding site II [40]. Based on ITC results, Bergé-Lefranc et al. claimed that IS interaction with HSA follows "enthalpy-entropy compensation", as they showed a linear relation between ΔH and TΔS at four different temperatures, which is also attributed to the heat capacity change in HSA [40]. Although the HSA conformational changes highly affect the likelihood of bindings, Yu et al. showed that the binding affinity for IS does not significantly change after urea-induced carbamylation of HSA [37]. Observing IS binding to HSA in both sites I and II showed that the binding constants seem relatively insensitive to pH changes [60]. However, this is in contrast to later work that shows IS has a weaker attachment to HSA in acidic and alkaline conditions [55]. Nevertheless, these different association constants for the IS-HSA complex may result from HSA conformation changes in these extreme pH domains, viz., lower than five or higher than ten. That said, using pH changes to alter the binding of toxins may contribute to freeing IS from HSA but likely will have deleterious effects on blood cells and other proteins [55].

Indole 3-acetic Acid: IAA is an essential metabolite for tryptophan and competes with that for binding sites on HSA [72]. IAA binding to HSA is exothermic, with negative and positive ΔS on HSA sites I and II, respectively, and is slightly dependent on pH with the same trend as IS. The primary binding site is considered site II with relatively lower affinity and effect on HSA's secondary structure than IS [38,55]. Although different methods were used to obtain the IAA association constant with HSA, they correlated with sites I and II on HSA [14,38]. Nevertheless, other thermodynamic properties are only reported in Zaidi et al.'s work with ITC and deviate from the molecular docking results, specifically on lower affinity sites [14,38,66].

p-Cresyl sulfate: Up to 90% of PCS is bound by blood proteins and has been correlated to all-cause mortality and the progression of renal damage in patients regardless of dialysis treatment [18,51]. PCS and IS demonstrate competitive interactions with HSA and are often compared thermodynamically [32,40]. IS displacement by PCS is very likely on site II at a PCS concentration of 200–1000 μM, which is relevant to the CKD patients' conditions at the last stages [51]. PCS binding to HSA is exothermic with a decrease in entropy (negative ΔS), indicating that electrostatic forces are involved [32]. Although IS and PCS share the same anionic group, sulfate, PCS forms more hydrogen bonds with HSA on site II [62]. At higher concentrations, PCS may displace IS, accounting for the fluctuation of free PCS concentration in the clinical data of patients under hemodialysis [51]. PCS binding to HSA slightly depends on pH, similar to IS and IAA [55].

Hippuric Acid: HA accumulation causes renal tubule cell damage and advances CKD progression, inhibiting HSA interactions with other compounds [18]. There are few articles on HA interactions with HSA, yet the works reporting HA binding association constants using ITC and equilibrium dialysis are within the same order of magnitude [14,35]. It is worth mentioning that HA has an affinity for both sites on HSA, but its free form is highly accumulated in the blood [14], as it has the highest hemodialysis reduction among the other PBUTs (Table 1). HA binding to HSA is exothermic and involves electrostatic interactions and hydrogen bonds; however, changes in ΔH dominate the binding event occurrence [35,38]. Even though HA binds to site II with higher affinity than site I [35], Davilas A. et al. showed that HA, as well as IS and IAA, reduces the bonded fraction of other drugs, such as diflunisal, to site I on HSA [13,14]. It is worth noting that although HA interaction with HSA is weaker than IS and IAA, it could displace them on HSA site II in high concentrations [14]. Nevertheless, the number of studies reporting the binding parameters is fewer than IS and IAA, which share the same binding site on HSA.

Phenylacetic acid: PAA binding to HSA is exothermic, but the enthalpy change upon binding is relatively less than other PBUTs, suggesting a small hydrophobic effect that may be due to the aromatic structures in the toxin and weak electrostatic forces [37]. Although PAA induces oxidative stress and systematic inflammation that advances CKD progression [18], there are limited studies on thermodynamics and binding mechanisms to HSA. So far, it is known that PAA has relatively less affinity for both sites on HSA, suggesting there might be another site of interaction where it has been claimed that the binding events are sequential on two regions of HSA. Regarding the main binding sites on HSA, as CMPF displaces PAA on site I, binding to site II is tight enough that it is not replaced by either IS or PCS, even though the concentrations of the later toxins are significantly higher than PAA in CKD patients [28,37]. Further data suggest that other plasma proteins may be involved in binding PAA [28].

Table 2. Thermodynamic properties of PBUT-HSA binding for CMPF, IS, IAA, HA, PAA, and PCS with sample properties and conditions.

Toxin	Protein Buffer	pH	T (°C)	Methodology	Results						Discussed Competitors	Ref.
					Binding Site	K_a (M^{-1})	ΔH (kcal mol^{-1})	TΔS (kcal mol^{-1})	ΔG (kcal mol^{-1})	n		
CMPF	**HSA** **Sodium Phosphate**	**7.4**	**25**	**ITC, Far-UV CD, FT-IR, X-ray crystallography, DSC**	**High affinity** **Low affinity**	5.76×10^6 4.57×10^3	−12.89 −14.89	−3.70 −10.21	−9.19 −4.68	-	-	[38]
	HSA (defatted) HSA modifications: HNB-HSA, TNM-HSA, SA-HSA, DEP-HAS Sodium phosphate (dibasic)	6.8–8.2	25	X-ray crystallography, fluorescence quenching, and probe displacement	Site I	130×10^5	-	-	-	1	Bilirubin, warfarin, phenylbutazone	[50]
	HSA BSA Isotonic sodium phosphate-chloride	6.4–8.3	10, 25, 37	Equilibrium dialysis, HPLC	High affinity (at 37 °C, pH 7.4) Low affinity (at 37 °C, pH 7.4)	4.8×10^6 0.6×10^3	(Ref data, 25 °C) -2.15	4.78	−6.93	0.6 10.3	o-methyl red	[19]
	HSA (defatted) Deionized and distilled water, phosphate buffer	7.4	25	Fluorescent probe displacement, ultrafiltration, equilibrium dialysis	High affinity Low affinity	130.5×10^5 33.4×10^4	-	-	-	1 2	Warfarin, DNSS	[14]
	HSA (defatted) Deionized and distilled water, phosphate buffer	7.4	25	Fluorescent probe displacement, ultrafiltration, equilibrium dialysis	High affinity Low affinity	16.1×10^5 8.3×10^3	-	-	-	1 3	Warfarin, DNSS	[14]
	HSA Phosphate buffer	7.2	25, 30, 37	ITC, fluorescence quenching, CD, STD NMR	Overall interaction with HSA (at 25 °C, pH 7.2)	10.06×10^3	−16.8	−11.96	−4.81	1.08	Ibuprofen, warfarin	[32]

Table 2. *Cont.*

Toxin	Protein Buffer	pH	T (°C)	Methodology	Results						Discussed Competitors	Ref.
					Binding Site	K_a (M^{-1})	ΔH (kcal mol^{-1})	TΔS (kcal mol^{-1})	ΔG (kcal mol^{-1})	n		
IS	HSA Phosphate buffer	7.4	25	HPLC, fluorescent probe displacement	High affinity Low affinity	0.98×10^5 8.0×10^3	-	-	-	1 1.6	Warfarin, dansylsarcosine	[51]
	HSA (defatted) Deionized and distilled water Phosphate buffer	6.5, 7.4, 8.5	25	Equilibrium dialysis	High affinity (at pH 7.4) Low affinity (at pH 7.4)	9.1×10^5 0.8×10^3	-	-	-	1 3	Warfarin, DNSA, DNSS	[60]
	HSA (defatted) Modified HSA morpholine-N-oxide	7.4	25, 30, 37	ITC	High affinity (at 25 °C and ionic strength 20—native HSA) Low affinity (at 25 °C and ionic strength 20—native HSA)	-	-	-	−7.55 −4.94	-	-	[37]
	Modeled—HSA			Molecular docking	Overall interaction with HSA	-	−7.5	-	-	-	Laccase 1KYA and 3FU9	[66]
	HSA Phosphate buffer	1.9–12.9	37	CD, ITC	Overall interaction with HSA (at pH 7.1)	1.27×10^5	−20.23	−15.63	−7.25			[55]
	HSA 2-(N-morpholino) ethane sulfonic acid) buffer	7.4	15, 20, 25, 37	ITC	Overall interaction with HSA (at 25 °C)	1750	−8.65	−4.63	−4.35	0.98		[40]
	HSA Sodium phosphate	7.4	25	ITC, Far-UV CD, FT-IR, X-ray crystallography, DSC	High affinity Low affinity	1.04×10^6 6.59×10^5	−20.05 −14.39	−11.80 −6.90	−8.25 −7.49	-	-	[38]

Table 2. *Cont.*

Toxin	Protein Buffer	pH	T (°C)	Methodology	Results						Discussed Competitors	Ref.
					Binding Site	K_a (M^{-1})	ΔH (kcal mol^{-1})	TΔS (kcal mol^{-1})	ΔG (kcal mol^{-1})	*n*		
IAA	HSA (defatted) Deionized and distilled water, phosphate buffer	7.4	25	Fluorescent probe displacement, ultrafiltration, equilibrium dialysis	High affinity Low affinity	2.1×10^5 0.8×10^4	-	-	-	1 5	Warfarin, DNSS	[14]
	Modeled—HSA			Molecular docking	Overall interaction with HSA	-	−6.8	-	-	-	Laccase 1KYA and 3FU9	[66]
	HSA in experiments Model samples—HSA			CD, SPR, molecular docking	Domain 2 Domain 3 Domain 2-3 HSA	5.92×10^3 4.65×10^5 1.25×10^5 2.75×10^4						[42]
	HSA Sodium phosphate	7.4	25	ITC, Far-UV CD, FT-IR, X-ray crystallography, DSC	High affinity Low affinity	1.78×10^5 8.27×10^4	−13.94 −3.8	−6.40 2.8	−7.45 −6.60	-	-	[38]
HA	HSA (defatted) Deionized and distilled water, phosphate buffer	7.4	25	Fluorescent probe displacement, ultrafiltration, equilibrium dialysis	High affinity Low affinity	0.1×10^5 0.3×10^4	-	-	-	1 7	Warfarin, DNSS	[14]
	HSA Sodium phosphate	7.4	25, 30, 37	ITC, DSC, molecular docking, CD, fluorescence quenching	High affinity (at 25 °C) Low affinity (at 25 °C)	2.75×10^4 2.8×10^3	−4.41 −7.58	1.90 2.88	−6.05 −4.69	0.97	-	[35]
	HSA in experiments Model samples—HSA			CD, SPR, molecular docking	Domain 2 Domain 3 Domain 2-3 HSA	9.52×10^3 5.78×10^4 9.26×10^3 2.11×10^4						[42]

Table 2. *Cont.*

Toxin	Protein Buffer	pH	T (°C)	Methodology	Results						Discussed Competitors	Ref.
					Binding Site	K_a (M^{-1})	ΔH (kcal mol^{-1})	TΔS (kcal mol^{-1})	ΔG (kcal mol^{-1})	*n*		
PAA	HSA (defatted) Modified HSA morpholine-N-oxide	7.4	25, 30, 37	ITC	High affinity (at 25 °C and ionic strength 20—native HSA) Low affinity (at 25 °C and ionic strength 20—native HSA)	-	-	-	−5.90 −4.04	-	-	[37]
	BSA HSA (with fatty acids)	7.4	37	Ultrafiltration, fluorescent probe displacement	High affinity Low affinity	3.66×10^5 3.86×10^4				0.07 1.83		[28]
PCS	HSA Phosphate buffer	7.4	25	HPLC, fluorescent probe displacement	High affinity Low affinity	1×10^5 19.6×10^3	-	-	-	1 1.2	Warfarin, dansylsarcosine	[51]
	Modeled—HSA			Molecular docking	Overall interaction with HSA	-	−6.3	-	-	-	Laccase 1KYA and 3FU9	[66]
	HSA Phosphate buffer	7.2	25, 30, 37	ITC, fluorescence quenching, CD, STD NMR	Overall interaction with HSA (at 25 °C, pH 7.2)	1.39×10^3	−27.5	−23.29	−4.23	1.03	Ibuprofen, warfarin	[32]

4. Insights into Indoxyl Sulfate and p-Cresol Sulfate Binding to HSA

Despite more publications for IS and PCS binding to HSA, the resulting high-affinity K_a varies as much as three orders of magnitude. It has been postulated that ultrafiltration results differ due to the chloride ion content of the buffers used, where chloride ions reduce drug binding to HSA, which are effects that may account for lower K_a for PCS and IS using chloride-free buffer [14,51]. Moreover, Bergé-Lefranc et al. suggested that the significant differences in the values obtained for IS are mainly due to the method used, in their case, ITC. They claimed that separative membranes in ultrafiltration obstruct the results as separation methods also lack efficacy in hemodialysis membranes [40].

Studying the STD NMR spectra for IS and PCS interactions with HSA, Li et al. claimed that 32.5% and 23.8% of the molecules participated in the event, respectively. Regarding the IS structure, it was shown that the sulfated group is probably responsible for the high affinity to HSA. Comparing the behavior with the PCS binding event, they proposed that PCS-HSA formation requires a minor "steric hindrance" and results in a more homogenous complex concerning the binding site [32]. Furthermore, STD NMR results showed that ibuprofen or warfarin could displace IS as the IS-HSA signal intensity decreases. These two drugs bind to both Sudlow's sites, suggesting that IS has a relatively strong affinity for both. PCS displacement with ibuprofen was relatively lower, and no significant signal changes were observed by the introduction of warfarin in the NMR signal intensity for this toxin. Although this contrasts with the same work's ITC results, it suggests that there may be another site on HSA, which is a novel finding [32]. Earlier work using warfarin as a displacement probe also claimed that PCS is unlikely to bind to HSA site I, as its presence did not affect the signal intensity [51].

Although on the higher affinity binding site IS and PCS are shown to have comparable K_a values, their binding to HSA is competitive, where introducing either of them to a preformed HSA complex leads to an increase in the free fraction of that previously bound toxin [51]. If PCS does not bind to HSA on site I, the presence of IS is likely compromising PCS binding on an unknown site on HSA.

5. Competitive Binders to Inhibit HSA Uptake of PBUTs

Engineering competitive binders to enhance PBUT clearance from the blood compartment relies on a thorough understanding of how they interact with proteins. HSA is the predominant protein due to its concentration and propensity for binding small molecules. Two main routes can accomplish this outcome: i. exploring molecules that displace toxins from HSA so that they can be cleared using membrane-based dialysis techniques and ii. building surfaces that have higher affinities for these PBUTs than HSA. Herein, we will summarize the significant work on developing molecules that displace toxins from HSA.

Some of the earliest work on competitive binders for CMPF to HSA proposed that drugs with two carboxyl groups and enough hydrophobicity, such as methotrexate, may be an active binder for HSA to displace CMPF [50]. This was based on understanding how bilirubin and CMPF bound to HSA, where the two carboxyl groups and the distance between them were thought to be vital to their binding. In addition to these physicochemical properties of potential competitive binders, the effect of concentration and HSA site saturation must be considered. For instance, o-methyl red and CMPF share the same binding site on HSA, the latter having a higher affinity at low but not higher concentrations; when both are at 200 μM, o-methyl red is nearly 80% bound, compared to only 70% for CMPF [19].

Ibuprofen and warfarin are often considered for displacing PBUT from HSA [14,32,51]. Although Li et al. observed that these two substances could displace IS and PCS on higher affinity sites from ITC, NMR showed that only ibuprofen had this capability [32]. High removal of IS and PCS via ibuprofen infusion to hemodialysis patients also showed the enhanced dialytic removal of these two toxins due to its association constant of $1 \times 10^6\ M^{-1}$ on site II of HSA, which is 10 times higher than those for PCS and IS [51,73]. It is noteworthy that these findings were the first clinical study on ibuprofen. Nevertheless, ibuprofen infusion also displaces albumin-bound tryptophan and lacks the ability to improve the clearance of PBUTs with high affinities to site I [73].

Free fatty acids have also been shown to have the potential to significantly decrease HSA binding of IS, PCS, and IAA and hence increase the efficacy of hemodialysis [74]. However, as CMPF holds the highest binding percentage compared with the named toxins, the increase in the unbound fraction for this toxin in the presence of free fatty acids was insignificant [74,75]. Non-esterified fatty acids, such as octanoate and docosahexaenoic, can lead to total displacement of IS and PCS at a concentration of 0.24 M. These substances strongly interact with HSA and displace uremic toxins bound to site II in in-vitro studies; however, they are toxic and enhance cardiovascular complications and oxidative stress [76].

Thus, the question remains how to design competitive binders that do not have deleterious effects if they are also not cleared. Especially considering that they are bound to HSA so firmly that they cannot be removed from the blood compartment. Additionally, the goal must be to have the least number of molecules introduced to displace the maximum amount of PBUTs from HSA to enhance their clearance and minimize potential adverse health outcomes [74]. Davilas et al. pointed out that while exploring various binding competitors, one must consider that in vivo HSA is surrounded by multiple endogenous and exogenous substances competing for binding, and simulating the same condition to evaluate a specific binder is often problematic [13]. Ideal binding competitors' characteristics include the ability to block both sites on HSA, which are the leading region for PBUT interactions, minimize therapeutic consequences, and utilize clearance pathways available during kidney dysfunction [73]. Perhaps the path forward lies in developing binders that can be consumed by the surrounding cellular milieu to effectively remove them from the blood compartment or that they can be easily displaced from HSA with a minor modification in environmental conditions during dialysis itself.

6. Conclusions

Accumulation of uremic toxins in the blood compartment of CKD patients advances kidney dysfunction and various co-morbidities. The lack of efficacy in hemodialysis for clearing PBUTs highlights the importance of investigating the binding affinity of PBUTs with blood proteins. Human serum albumin has two prominent locations for binding PBUTs. Herein, we summarized the data on binding mechanisms, sites, thermodynamic parameters, and methods used to capture these data to highlight the design criteria for advancing PBUT removal through engineering competitive binders. These competitive binders could act through targeted adsorption of PBUTs from the solution phase or by displacing them from the binding sites found on proteins such as HSA. However, the binding mechanisms of many uremic toxins to plasma proteins are unclear. Common techniques used to understand these interactions are summarized here; advanced calorimetric and spectroscopic techniques are understudied in toxin–protein interactions, e.g., vibrational spectroscopy, and may help achieve valuable insights into the nature of binding. Moreover, the available data on the phenomena in living systems are significantly limited. Amino acids and endogenous molecules surrounding the HSA in the blood compartment largely influence the binding mechanisms. Investigating the sole HSA and toxin interaction omits these side competitions for binding. Molecular docking applications could be one of the theoretical approaches to simulate the in vivo conditions and inform the molecular design for competitive binders. Finally, if multiple competitive binders are added to the blood compartment to dislodge uremic toxins from proteins, such as HSA, there is a risk that they themselves will have deleterious effects on the host. Thus, these binders may need to be designed to remove the maximum number of PBUTs whilst being consumed by localized cellular milieu. Most binders proposed so far are exogenous drugs, and their clearance pathways are still not thoroughly discussed in the literature.

Author Contributions: Conceptualization, V.D.N. and L.D.U.; methodology, L.D.U.; investigation, V.D.N.; writing—original draft preparation, V.D.N.; writing—review and editing, V.D.N. and L.D.U.;

supervision, project administration, funding acquisition, L.D.U. All authors have read and agreed to the published version of the manuscript.

Funding: The authors wish to acknowledge funding from the Alberta Innovates Advance grants, Natural Sciences and Engineering Research Council of Canada (NSERC Alliance).

Institutional Review Board Statement: Not applicable.

Informed Consent Statement: Not applicable.

Data Availability Statement: Not applicable.

Conflicts of Interest: The authors declare no conflict of interest.

References

1. Ma, Y.; Li, S.; Tonelli, M.; Unsworth, L. Adsorption-Based Strategies for Removing Uremic Toxins from Blood. *SSRN Electron. J.* **2020**, *319*, 111035. [CrossRef]
2. William, R.C.; Nader, L.D.; Narsimhan, V.; Ronco, C. Uremic Toxins and Their Relation to Dialysis Efficacy. *Blood Purif.* **2019**, *48*, 299–314. [CrossRef]
3. Kalantar-Zadeh, K.; Jafar, T.H.; Nitsch, D.; Neuen, B.L.; Perkovic, V. Chronic kidney disease. *Lancet* **2021**, *398*, 786–802. [CrossRef] [PubMed]
4. Daneshamouz, S.; Eduok, U.; Abdelrasoul, A.; Shoker, A. Protein-bound uremic toxins (PBUTs) in chronic kidney disease (CKD) patients: Production pathway, challenges and recent advances in renal PBUTs clearance. *NanoImpact* **2021**, *21*, 100299. [CrossRef] [PubMed]
5. Hall, E.D.; Pazara, K.E.; Linseman, K.L. Sex differences in postischemic neuronal necrosis in gerbils. *J. Cereb. Blood Flow Metab.* **1991**, *11*, 292–298. [CrossRef] [PubMed]
6. Agarwal, R.; Burton, J.; Gallieni, M.; Kalantar-Zadeh, K.; Mayer, G.; Pollock, C.; Szepietowski, J.C. Alleviating symptoms in patients undergoing long-term hemodialysis: A focus on chronic kidney disease-associated pruritus. *Clin. Kidney J.* **2022**, *16*, 30–40. [CrossRef]
7. Stokes, J.B. Consequences of frequent hemodialysis: Comparison to conventional hemodialysis and transplantation. *Trans. Am. Clin. Climatol. Assoc.* **2011**, *122*, 124–136.
8. Magnani, S.; Atti, M. Uremic Toxins and Blood Purification: A Review of Current Evidence and Future Perspectives. *Toxins* **2021**, *13*, 246. [CrossRef]
9. Ludwig, J.; Smith, J.; Pfaendtner, J. Analyzing the Long Time-Scale Dynamics of Uremic Toxins Bound to Sudlow Site II in Human Serum Albumin. *J. Phys. Chem. B* **2021**, *125*, 2910–2920. [CrossRef]
10. Varshney, A.; Sen, P.; Ahmad, E.; Rehan, M.; Subbarao, N.; Khan, R.H. Ligand binding strategies of human serum albumin: How can the cargo be utilized? *Chirality* **2010**, *22*, 77–87. [CrossRef]
11. Ahmed-Ouameur, A.; Diamantoglou, S.; Sedaghat-Herati, M.R.; Nafisi, S.; Carpentier, R.; Tajmir-Riahi, H.A. The Effects of Drug Complexation on the Stability and Conformation of Human Serum Albumin: Protein Unfolding. *Cell Biochem. Biophys.* **2006**, *45*, 203–214. [CrossRef] [PubMed]
12. Mapiour, M.; Abdelrasoul, A. Critical Influences of Plasma pH on Human Protein Properties for Modeling Considerations: Size, Charge, Conformation, Hydrophobicity, and Denaturation. *J. Compos. Sci.* **2023**, *7*, 28. [CrossRef]
13. Davilas, A.; Koupparis, M.; Macheras, P.; Valsami, G. In-vitro study on the competitive binding of diflunisal and uraemic toxins to serum albumin and human plasma using a potentiometric ion-probe technique. *J. Pharm. Pharmacol.* **2006**, *58*, 1467–1474. [CrossRef] [PubMed]
14. Sakai, T.; Takadate, A.; Otagiri, M. Characterization of Binding Site of Uremic Toxins on Human Serum Albumin. *Biol. Pharm. Bull.* **1995**, *18*, 1755–1761. [CrossRef] [PubMed]
15. Devine, E.; Krieter, D.; Rüth, M.; Jankovski, J.; Lemke, H.-D. Binding Affinity and Capacity for the Uremic Toxin Indoxyl Sulfate. *Toxins* **2014**, *6*, 416–429. [CrossRef]
16. Zhang, Y.; Wu, S.; Qin, Y.; Liu, J.; Liu, J.; Wang, Q.; Ren, F.; Zhang, H. Interaction of phenolic acids and their derivatives with human serum albumin: Structure–affinity relationships and effects on antioxidant activity. *Food Chem.* **2018**, *240*, 1072–1080. [CrossRef]
17. Faria, M.; de Pinho, M.N. Challenges of reducing protein-bound uremic toxin levels in chronic kidney disease and end stage renal disease. *Transl. Res.* **2021**, *229*, 115–134. [CrossRef]
18. Graboski, A.L.; Redinbo, M.R. Gut-Derived Protein-Bound Uremic Toxins. *Toxins* **2020**, *12*, 590. [CrossRef]
19. Henderson, S.J.; Lindup, W.E. Interaction of 3-Carboxy-4-Methyl-5-Propyl-2-Furanpropanoic ACID, an Inhibitor of Plasma Protein Binding in Uraemia, with Human Albumin. *Biochem. Pharmacol.* **1990**, *40*, 2543–2548. [CrossRef]
20. Sassa, T.; Matsuno, H.; Niwa, M.; Kozawa, O.; Takeda, N.; Niwa, T.; Kumada, T.; Uematsu, T. Measurement of furancarboxylic acid, a candidate for uremic toxin, in human serum, hair, and sweat, and analysis of pharmacological actions in vitro. *Arch. Toxicol.* **2000**, *73*, 649–654. [CrossRef]

21. Ujhelyi, L.; Balla, G.; Jeney, V.; Varga, Z.; Nagy, E.; Vercellotti, G.M.; Agarwal, A.; Eaton, J.W.; Balla, J. Hemodialysis reduces inhibitory effect of plasma ultrafiltrate on LDL oxidation and subsequent endothelial reactions. *Kidney Int.* **2006**, *69*, 144–151. [CrossRef] [PubMed]
22. Mair, R.; Sirich, T.; Meyer, T. Uremic Toxin Clearance and Cardiovascular Toxicities. *Toxins* **2018**, *10*, 226. [CrossRef] [PubMed]
23. Liabeuf, S.; Barreto, D.V.; Barreto, F.C.; Meert, N.; Glorieux, G.; Schepers, E.; Temmar, M.; Choukroun, G.; Vanholder, R.; Massy, Z.A. Free p-cresylsulphate is a predictor of mortality in patients at different stages of chronic kidney disease. *Nephrol. Dial. Transplant.* **2010**, *25*, 1183–1191. [CrossRef]
24. Aronov, P.A.; Luo, F.J.G.; Plummer, N.S.; Quan, Z.; Holmes, S.; Hostetter, T.H.; Meyer, T.W. Colonic Contribution to Uremic Solutes. *J. Am. Soc. Nephrol.* **2011**, *22*, 1769–1776. [CrossRef] [PubMed]
25. Brunet, P.; Dou, L.; Cerini, C.; Berland, Y. Protein-bound uremic retention solutes. *Adv. Ren. Replace* **2003**, *10*, 310–320. [CrossRef]
26. Jankowski, J.; van der Giet, M.; Jankowski, V.; Schmidt, S.; Hemeier, M.; Mahn, B.; Giebing, G.; Tölle, M.; Luftmann, H.; Schlüter, H.; et al. Increased plasma phenylacetic acid in patients with end-stage renal failure inhibits iNOS expression. *J. Clin. Investig.* **2003**, *112*, 256–264. [CrossRef]
27. Schmidt, S.; Westhoff, T.H.; Krauser, P.; Ignatius, R.; Jankowski, J.; Jankowski, V.; Zidek, W.; van der Giet, M. The uraemic toxin phenylacetic acid impairs macrophage function. *Nephrol. Dial. Transplant.* **2008**, *23*, 3485–3493. [CrossRef]
28. Saldanha, J.F.; Yi, D.; Stockler-Pinto, M.B.; Soula, H.A.; Chambert, S.; Fouque, D.; Mafra, D.; Soulage, C.O. Determination of the binding properties of the uremic toxin phenylacetic acid to human serum albumin. *Biochimie* **2016**, *125*, 53–58. [CrossRef]
29. Jourde-Chiche, N.; Dou, L.; Sabatier, F.; Calaf, R.; Cerini, C.; Robert, S.; Camoin-Jau, L.; Charpiot, P.; Argiles, A.; Dignat-George, F.; et al. Levels of circulating endothelial progenitor cells are related to uremic toxins and vascular injury in hemodialysis patients. *J. Thromb. Haemost.* **2009**, *7*, 1576–1584. [CrossRef]
30. Dou, L.; Sallee, M.; Cerini, C.; Poitevin, S.; Gondouin, B.; Jourde-Chiche, N.; Fallague, K.; Brunet, P.; Calaf, R.; Dussol, B.; et al. The cardiovascular effect of the uremic solute indole-3 acetic acid. *J. Am. Soc. Nephrol.* **2015**, *26*, 876–887. [CrossRef]
31. Kabiri, M.; Unsworth, L.D. Application of Isothermal Titration Calorimetry for Characterizing Thermodynamic Parameters of Biomolecular Interactions: Peptide Self-Assembly and Protein Adsorption Case Studies. *Biomacromolecules* **2014**, *15*, 3463–3473. [CrossRef] [PubMed]
32. Li, S.; Tonelli, M.; Unsworth, L.D. Indoxyl and p-cresol sulfate binding with human serum albumin. *Colloids Surf. A Physicochem. Eng. Asp.* **2022**, *635*, 128042. [CrossRef]
33. Falconer, R.J.; Schuur, B.; Mittermaier, A.K. Applications of isothermal titration calorimetry in pure and applied research from 2016 to 2020. *J. Mol. Recognit.* **2021**, *34*, e2901. [CrossRef] [PubMed]
34. Kato, S.; Drout, R.J.; Farha, O.K. Isothermal Titration Calorimetry to Investigate Uremic Toxins Adsorbing onto Metal-Organic Frameworks. *Cell Rep. Phys. Sci.* **2020**, *1*, 100006. [CrossRef]
35. Zaidi, N.; Ajmal, M.R.; Rabbani, G.; Ahmad, E.; Khan, R.H. A Comprehensive Insight into Binding of Hippuric Acid to Human Serum Albumin: A Study to Uncover Its Impaired Elimination through Hemodialysis. *PLoS ONE* **2013**, *8*, e71422. [CrossRef]
36. Falconer, R.J. Applications of isothermal titration calorimetry—The research and technical developments from 2011 to 2015. *J. Mol. Recognit.* **2016**, *29*, 504–515. [CrossRef]
37. Yu, S.; Schuchardt, M.; Tölle, M.; van der Giet, M.; Zidek, W.; Dzubiella, J.; Ballauff, M. Interaction of human serum albumin with uremic toxins: A thermodynamic study. *RSC Adv.* **2017**, *7*, 27913–27922. [CrossRef]
38. Zaidi, N.; Khan, R.H. A biophysical insight into structural and functional state of human serum albumin in uremia mimic milieu. *Int. J. Biol. Macromol.* **2019**, *131*, 697–705. [CrossRef]
39. Langmuir, I. The Adsorption of Gases on Plane Surfaces of Glass, Mica and Platinum. *J. Am. Chem. Soc.* **2002**, *40*, 1361–1403. [CrossRef]
40. Bergé-Lefranc, D.; Chaspoul, F.; Cérini, C.; Brunet, P.; Gallice, P. Thermodynamic study of indoxylsulfate interaction with human serum albumin and competitive binding with p-cresylsulfate. *J. Therm. Anal. Calorim.* **2014**, *115*, 2021–2026. [CrossRef]
41. Monod, J.; Wyman, J.; Changeux, J.P. On the Nature of Allosteric Transitions: A Plausible Model. *J. Mol. Biol.* **1965**, *12*, 88–118. [CrossRef] [PubMed]
42. Nerusu, A.; Vaikuntapu, P.R.; Chinthapalli, D.K.; Podile, A.R.; Subramanyam, R. Truncated domains of human serum albumin improves the binding efficiency of uremic toxins: A surface plasmon resonance and computational approach. *Int. J. Biol. Macromol.* **2020**, *155*, 1216–1225. [CrossRef] [PubMed]
43. Pethica, B.A. Misuse of thermodynamics in the interpretation of isothermal titration calorimetry data for ligand binding to proteins. *Anal. Biochem.* **2015**, *472*, 21–29. [CrossRef]
44. Ghisaidoobe, A.B.T.; Chung, S.J. Intrinsic Tryptophan Fluorescence in the Detection and Analysis of Proteins: A Focus on Förster Resonance Energy Transfer Techniques. *Int. J. Mol. Sci.* **2014**, *15*, 22518–22538. [CrossRef] [PubMed]
45. Sudlow, G.; Birkett, D.J.; Wade, D.N. The Characterization of Two Specific Drug Binding Sites on Human Serum Albumin. *Mol. Pharmacol.* **1975**, *11*, 824–832.
46. Royer, C.A. Probing protein folding and conformational transitions with fluorescence. *Chem. Rev.* **2006**, *106*, 1769–1784. [CrossRef]
47. N'Soukpoé-Kossi, C.N.; Sedaghat-Herati, R.; Ragi, C.; Hotchandani, S.; Tajmir-Riahi, H.A. Retinol and retinoic acid bind human serum albumin: Stability and structural features. *Int. J. Biol. Macromol.* **2007**, *40*, 484–490. [CrossRef]
48. Ross, J.A.; Jameson, D.M. Time-resolved methods in biophysics. 8. Frequency domain fluorometry: Applications to intrinsic protein fluorescence. *Photochem. Photobiol. Sci.* **2008**, *7*, 1301–1312. [CrossRef]

49. Alarcón, E.; Aspée, A.; Abuin, E.B.; Lissi, E.A. Evaluation of solute binding to proteins and intra-protein distances from steady state fluorescence measurements. *J. Photochem. Photobiol. B Biol.* **2012**, *106*, 1–17. [CrossRef]
50. Tsutsumi, Y.; Maruyama, T.; Takadate, A.; Goto, M.; Matsunaga, H.; Otagiri, M. Interaction Between Two Dicarboxylate Endogenous Substances, Bilirubin and an Uremic Toxin, 3-Carboxy-4-Methyl-5-Propyl-2-Furanpropanoic Acid, on Human Serum Albumin. *Pharm. Res.* **1999**, *16*, 916–923. [CrossRef]
51. Watanabe, H.; Noguchi, T.; Miyamoto, Y.; Kadowaki, D.; Kotani, S.; Nakajima, M.; Miyamura, S.; Ishima, Y.; Otagiri, M.; Maruyama, T. Interaction between Two Sulfate-Conjugated Uremic Toxins, p-Cresyl Sulfate and Indoxyl Sulfate, during Binding with Human Serum Albumin. *Drug Metab. Dispos.* **2012**, *40*, 1423–1428. [CrossRef] [PubMed]
52. Ranjbar, B.; Gill, P. Circular Dichroism Techniques: Biomolecular and Nanostructural Analyses—A Review. *Chem. Biol. Drug Des.* **2009**, *74*, 101–120. [CrossRef] [PubMed]
53. Hornok, V.; Juhász, Á.; Paragi, G.; Kovács, A.N.; Csapó, E. Thermodynamic and kinetic insights into the interaction of kynurenic acid with human serum albumin: Spectroscopic and calorimetric approaches. *J. Mol. Liq.* **2020**, *313*, 112869. [CrossRef]
54. Heitz, F.; Van Mau, N. Protein structural changes induced by their uptake at interfaces. *Biochim. Et Biophys. Acta (BBA)—Protein Struct. Mol. Enzymol.* **2002**, *1597*, 1–11. [CrossRef]
55. Yamamoto, S.; Sasahara, K.; Domon, M.; Yamaguchi, K.; Ito, T.; Goto, S.; Goto, Y.; Narita, I. pH-Dependent Protein Binding Properties of Uremic Toxins In Vitro. *Toxins* **2021**, *13*, 116. [CrossRef]
56. Chuang, V.T.G.; Maruyama, T.; Otagiri, M. Updates on Contemporary Protein Binding Techniques. *Drug Metab. Pharmacokinet.* **2009**, *24*, 358–364. [CrossRef] [PubMed]
57. Fielding, L.; Rutherford, S.; Fletcher, D. Determination of protein-ligand binding affinity by NMR: Observations from serum albumin model systems. *Magn. Reson. Chem.* **2005**, *43*, 463–470. [CrossRef]
58. Sun, C.; Huth, J.R.; Hajduk, P.J. NMR in Pharmacokinetic and Pharmacodynamic Profiling. *ChemBioChem* **2005**, *6*, 1592–1600. [CrossRef] [PubMed]
59. Jacobsen, N.E. *NMR Spectroscopy Explained: Simplified Theory, Applications and Examples for Organic Chemistry and Structural Biology*; John Wiley & Sons: Hoboken, NJ, USA, 2007; p. 668.
60. Sakai, T.; Yamasaki, K.; Sako, T.; Kragh-Hansen, U.; Suenaga, A.; Otagiri, M. Interaction Mechanism between Indoxyl Sulfate, a Typical Uremic Toxin Bound to Site II, and Ligands Bound to Site I of Human Serum Albumin. *Pharm. Res.* **2001**, *18*, 520–524. [CrossRef] [PubMed]
61. Li, X.; Wang, S. Binding of glutathione and melatonin to human serum albumin: A comparative study. *Colloids Surf. B Biointerfaces* **2015**, *125*, 96–103. [CrossRef]
62. Smith, J.; Pfaendtner, J. Elucidating the Molecular Interactions between Uremic Toxins and the Sudlow II Binding Site of Human Serum Albumin. *J. Phys. Chem. B* **2020**, *124*, 3922–3930. [CrossRef]
63. Kragh-Hansen, U.; Dørge, E.; Pedersen, A.O. Rate-of-dialysis technique: Theoretical and practical aspects. *Anal. Biochem.* **2005**, *340*, 145–153. [CrossRef]
64. Czaplicki, S. Chromatography in Bioactivity Analysis of Compounds. In *Column Chromatography*; InTech: London, UK, 2013. [CrossRef]
65. Hage, D.S.; Anguizola, J.A.; Jackson, A.J.; Matsuda, R.; Papastavros, E.; Pfaunmiller, E.; Tong, Z.; Vargas-Badilla, J.; Yoo, M.J.; Zheng, X. Chromatographic analysis of drug interactions in the serum proteome. *Anal. Methods* **2011**, *3*, 1449. [CrossRef]
66. Daneshamouz, S.; Saadati, S.; Abdelrasoul, A. Molecular docking study of biocompatible enzyme interactions for removal of indoxyl sulfate (IS), indole-3-acetic acid (IAA), and p-cresyl sulfate (PCS) protein bound uremic toxins. *Struct. Chem.* **2022**, *33*, 1133–1148. [CrossRef]
67. Di Muzio, E.; Polticelli, F.; di Masi, A.; Fanali, G.; Fasano, M.; Ascenzi, P. All-trans-retinoic acid and retinol binding to the FA1 site of human serum albumin competitively inhibits heme-Fe(III) association. *Arch. Biochem. Biophys.* **2016**, *590*, 56–63. [CrossRef] [PubMed]
68. Sudlow, G.; Birkett, D.J.; Wade, D.N. Further Characterization of Specific Drug Binding Sites on Human Serum Albumin. *Mol. Pharmacol.* **1976**, *12*, 1052–1061. [PubMed]
69. Tsutsumi, Y.; Maruyama, T.; Takadate, A.; Shimada, H.; Otagiri, M. Decreased Bilirubin-Binding Capacity in Uremic Serum Caused by an Accumulation of Furan Dicarboxylic Acid. *Nephron* **2000**, *85*, 60–64. [CrossRef] [PubMed]
70. Vanholder, R.; Argilés, A.; Baurmeister, U.; Brunet, P.; Clark, W.; Cohen, G.; Dedeyn, P.P.; Deppisch, R.; Descamps-Latscha, B.; Henle, T.; et al. Uremic Toxicity: Present State of the Art. *Int. J. Artif. Organs* **2001**, *24*, 695–725. [CrossRef]
71. Deltombe, O.; de Loor, H.; Glorieux, G.; Dhondt, A.; Van Biesen, W.; Meijers, B.; Eloot, S. Exploring binding characteristics and the related competition of different protein-bound uremic toxins. *Biochimie* **2017**, *139*, 20–26. [CrossRef]
72. Bertuzzi, A.; Mingrone, G.; Gandolfi, A.; Greco, A.V.; Ringoir, S.; Vanholder, R. Binding of indole-3-acetic acid to human serum albumin and competition with l-tryptophan. *Clin. Chim. Acta* **1997**, *265*, 183–192. [CrossRef]
73. Madero, M.; Cano, K.B.; Campos, I.; Tao, X.; Maheshwari, V.; Brown, J.; Cornejo, B.; Handelman, G.; Thijssen, S.; Kotanko, P. Removal of Protein-Bound Uremic Toxins during Hemodialysis Using a Binding Competitor. *Clin. J. Am. Soc. Nephrol.* **2019**, *14*, 394–402. [CrossRef]
74. Shi, Y.; Zhang, Y.; Tian, H.; Wang, Y.; Shen, Y.; Zhu, Q.; Ding, F. Improved dialytic removal of protein-bound uremic toxins by intravenous lipid emulsion in chronic kidney disease rats. *Nephrol. Dial. Transplant.* **2019**, *34*, 1842–1852. [CrossRef] [PubMed]

75. Nishio, T.; Takamura, N.; Nishii, R.; Tokunaga, J.; Yoshimoto, M.; Kawai, K. Influences of haemodialysis on the binding sites of human serum albumin: Possibility of an efficacious administration plan using binding inhibition. *Nephrol. Dial. Transplant.* **2008**, *23*, 2304–2310. [CrossRef] [PubMed]
76. Florens, N.; Yi, D.; Juillard, L.; Soulage, C.O. Using binding competitors of albumin to promote the removal of protein-bound uremic toxins in hemodialysis: Hope or pipe dream? *Biochimie* **2018**, *144*, 1–8. [CrossRef] [PubMed]

International Journal of
Molecular Sciences

Review

Phase Behavior of Ionic Liquid-Based Aqueous Two-Phase Systems

Lirong Nie [1], Ziwei Zheng [1], Mingxia Lu [1], Shun Yao [2] and Dong Guo [1,*]

[1] School of Health Science and Engineering, University of Shanghai for Science and Technology, Shanghai 200093, China
[2] School of Chemical Engineering, Sichuan University, Chengdu 610065, China
* Correspondence: guodong99@tsinghua.org.cn; Tel./Fax: +86-021-5527-7040

Abstract: As an environmentally friendly separation medium, the ionic liquid (IL)-based aqueous two-phase system (ATPS) is attracting long-term attention from a growing number of scientists and engineers. Phase equilibrium data of IL-based ATPSs are an important basis for the design and optimization of chemical reactions and separation processes involving ILs. This article provides the recent significant progress that has been made in the field and highlights the possible directions of future developments. The effects of each component (such as salting-out agents and ILs) on the phase behavior of IL-based ATPSs are summarized and discussed in detail. We mainly focus on the phase behavior of ATPSs by using ILs, expecting to provide meaningful and valuable information that may promote further research and application.

Keywords: phase equilibrium; ionic liquid; aqueous two-phase system; salting-out agent; mechanism

1. Introduction

As one kind of liquid-liquid extraction system, the aqueous two-phase system (ATPS) was originally discovered by Albertsson [1]. ATPS is composed of two immiscible liquid phases whose solvent in both phases is water. In its early stages, it consisted of two polymers, or polymers and inorganic salts. The two polymers, or a polymer and a salt, were incompatible at certain concentrations due to the repulsive interaction, thus forming two phases. In the ATPS, there is little difference in the properties of the two phases, which is beneficial for maintaining the activities and structures of the extracted substances. Most importantly, water acts as a major component in both phases, which provides a mild environment to effectively avoid the toxicity and environmental pollution caused by volatile organic solvents. As a whole, ATPS is considered as one of the excellent alternatives for traditional liquid-liquid extraction technology. Recently, it has widely been used in the separation and purification of biological macromolecules, metal ions, natural products, carbon nanomaterials, etc. [2–6]. Compared to the traditional liquid-liquid extraction method, ATPS is eliciting increased research interest due to its advantageous characteristics, such as being eco-friendly, having a short extraction time, low energy consumption, good biological compatibility, easy amplification and continuous operation [7–10]. However, the limited polarity interval of the coexistence of two phases has become the bottleneck that has been restricting its application.

Ionic liquids (ILs) are salts that, in contrast to common electrolytes, are liquid at low temperatures. Due to their ionic nature, ILs possess unique properties, such as negligible vapor pressure, non-flammability, thermal and chemical stability, tunable chemical structures and physical properties, and strong solubilization ability [11–13]. As a result of these excellent properties, ILs are often used as entrainers to form the binary or ternary liquid-vapor and liquid-liquid mixtures with various organic compounds, which are applied in the separation processes. Furthermore, the structural characteristics of ILs have an influence on the homogeneous and heterogeneous characteristics of mixtures [14,15]. Recently, with the continuous development of ILs in the field of extraction and separation, ILs are expected to be alternatives of volatile organic solvents to form ATPS. Generally, hydrophobic ILs are often used to form IL/water biphasic systems. However, given the high viscosity of the IL phase, possible denaturation would take place during extraction/separation of biomacromolecules when using simple IL/water biphasic systems. Thus, ILs are more compatible with the advantages of ATPS as an environmentally friendly extraction system. At the same time, ionic liquid-based aqueous two-phase system (IL-based ATPS) can effectively solve the problem of low extraction efficiency resulting from the unregulated polarity of traditional polymer-polymer or polymer-salt ATPS [16]. In 2003, Rogers and co-workers reported, for the first time, a ATPS constructed by hydrophilic IL (1-butyl-3-methylimidazolium chloride, $[C_4mim]Cl$) and inorganic salt (K_3PO_4). Results indicated that the IL-based ATPS could overcome the limitations of the IL/water-extraction systems mentioned above [17]. Since then, significant progress has been made in the field of IL-based ATPSs [18–20]. The current research shows that ILs can form ATPS with salts [21,22], polymers [23–26] or surfactants [27,28]. Furthermore, the properties of ILs can be adjusted to expand the application range of the ATPS extraction system through the structure design in the anions and cations of ILs. The IL-based ATPS possesses shorter phase separation time and clearer two-phase interface than the ATPS composed of polymers. The application process of IL-based ATPS often involves many complex conditions. The stability evaluation of ILs under these conditions is particularly important. Moreover, ILs have high cost and environmental toxicity. The recovery of ILs after use is crucial for the development of green chemistry. However, unfortunately, few studies have reported on the above respects.

Although previous studies have revealed some important aspects of IL-based ATPSs, there are few systematic studies, especially concerning their properties, phase formation ability and phase equilibrium behavior. Moreover, the influence of components in IL-based ATPS on phase behavior are very important for the selection and design of extraction separation process. This review aims to provide a different perspective on IL-based ATPS from those recently published papers, majorly discussing the properties, the phase equilibrium and separation mechanisms. The discovery and development of ATPSs involving ILs are reviewed in detail. To sum up, the review mainly focuses on the following aspects: (1) the effects of the structures of ILs and salting-out agents on the properties of IL-based ATPSs; (2) the phase separation mechanism of IL-based ATPS; (3) the phase equilibrium behavior of various IL-based ATPSs. Furthermore, possible problems or challenges are also discussed, which may provide meaningful and valuable information to the relevant area and thus promote further research and application of IL-based ATPSs.

2. Properties of IL-Based ATPSs

The physical properties of the ATPSs at various concentrations and temperatures are indispensable requirements for the design and scaling up of separation process. The properties of IL-based ATPSs largely depend on the structure and property of ILs. Hydrophilic ILs have the advantages of many kinds of anions, low cost and hypotoxicity in comparison with hydrophobic ILs. Hydrophilic ILs are more widely used in the ATPSs. The chemical structure of ILs commonly used for ATPSs are listed in Figure 1. In the IL-based ATPSs, the chaotropic ILs act as the salting-in species and the anti-chaotropic ILs have the contrary role [29,30]. The use of ILs in ATPS can improve the phase polarities more adequately. These interesting and advantageous properties of the IL-based ATPS have motivated numerous studies of thermodynamic data, modeling, extraction and so on [31–35].

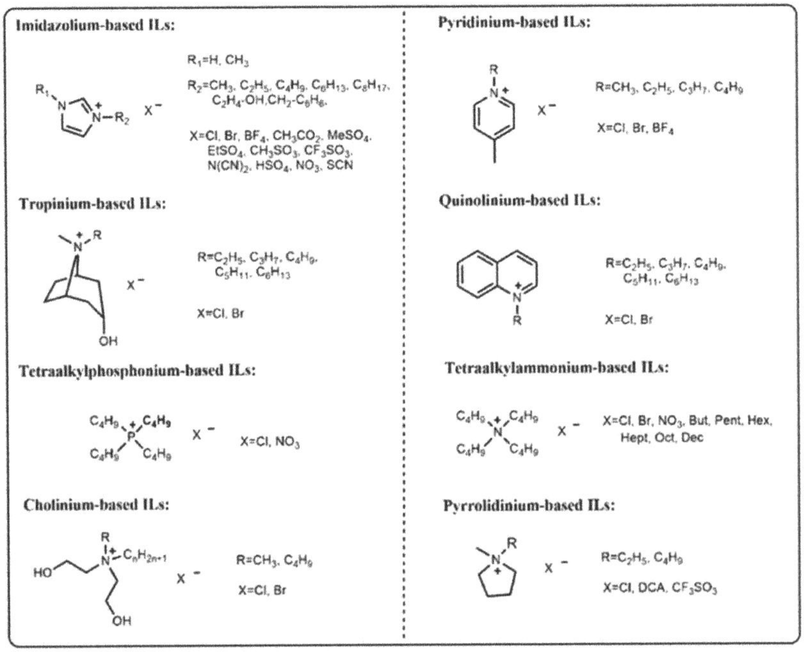

Figure 1. Chemical structure of the ILs commonly used for ATPSs.

2.1. Effects of ILs Cations

The cations of ILs usually consist of parent nucleus and side chains, which are easily modified. Most of the currently studied ILs are designed to have chaotropic cations that are salted-out by kosmotropic salts. For example, Bridges et al. [36] investigated phase behaviors of the IL-based ATPSs formed by [C_4mim]Cl, 1-butyl-2,3-dimethylimidazolium chloride ([C_4mmim]Cl), N-butylpyridinium chloride ([C_4py]Cl), tetrabutylammonium chloride ([$N_{4\,4\,4\,4}$]Cl), and tetrabutylphosphonium chloride ([$P_{4\,4\,4\,4}$]Cl) with different inorganic salts, including K_3PO_4, K_2HPO_4, K_2CO_3, KOH, and $(NH_4)_2SO_4$. The chaotropicity decreased in the order: [$P_{4\,4\,4\,4}$]Cl > [$N_{4\,4\,4\,4}$]Cl >> [C_4py]Cl >> [C_4mmim]Cl $\approx$ C_4mim]Cl. Obviously, this order is due to the increased chaotropic nature of the salts resulting from the chemical differences in the cations. The two quaternary onium salts have highly shielded charge, which is located mostly on the heteroatom surrounded by four butyl chains. Therefore, they are easily salted out by salting agents. Compared to the quaternary onium salts, the charge of the pyridinium cation mostly located on the nitrogen is less shielded. The imidazolium-based ILs have charge diffused cations, leading to depressed melting points. The charge of imidazolium cation is evenly distributed on the two nitrogen atoms and the C2 carbon, which results in multiple interactions between the cation and the molecular water. Therefore, two imidazolium-based ILs are the most difficult ones to be salted out. The systems evaluated are made up of K_3PO_4 and imidazolium-based ILs, including [C_4mim]Cl; [$C_4C_1C_1$im]Cl or [C_6mim]Cl [37]. Results showed that the phase separating ability of [C_4mmim]Cl was between [C_4mim]Cl and [C_6mim]Cl. That is, the hydrogen bonds between the C2 hydrogen atom of imidazolium cation and water have less effect on the phase behavior. By using K_2HPO_4-KH_2PO_4 as the salting agents to control pH, Ventura et al. [38] compared the phase separation ability of four substituted alkyl ILs with the same chain length. The sequence for the phase formation ability for the various families is as follows: [C_4C_1pyr]Cl $\approx$ [C_4mim]Cl < [C_4C_1pip]Cl < [C_4C_1py]Cl. According to the results, the ability of phase formation of above IL-based ATPSs was greatly affected by the IL cation molar volume. ILs cations with six-numbered heterocycles (such pyridiniun and piperidinium) have greater steric hindrance, and the hydrogen bond network of the surrounding water is more easily destroyed. Therefore, they are easier to be salted out than imidazolium- and pyrrolidinium-based ILs, with five-numbered heterocycles. Moreover, the density and viscosity measurements of both phases for ATPS based on phosphonium and imidazolium cations have also been reported. There are no significant differences in density values between IL-based ATPS and typical polymer-based systems. However, the phases containing phosphonium- or imidazolium-based ILs are far less viscous than the typical polymer-rich phases at the condition of close mass fraction compositions [39–41]. Additionally, the hydrophobic of ILs also affects the phase formation of IL-based ATPS. In general, the hydrophobic of ILs increases as the alkyl chain on cations lengthens, meaning stronger phase separation ability. This phenomenon was observed in the ATPSs based on two series of ILs, including n-alkyl-tropinium bromide ([C_nTr]Br, n = 2~5) and n-alkyl-quinolinium bromide ([C_nQ_n]Br, n = 2~6), were formed ATPS with different salts (Figure 2A) [42].

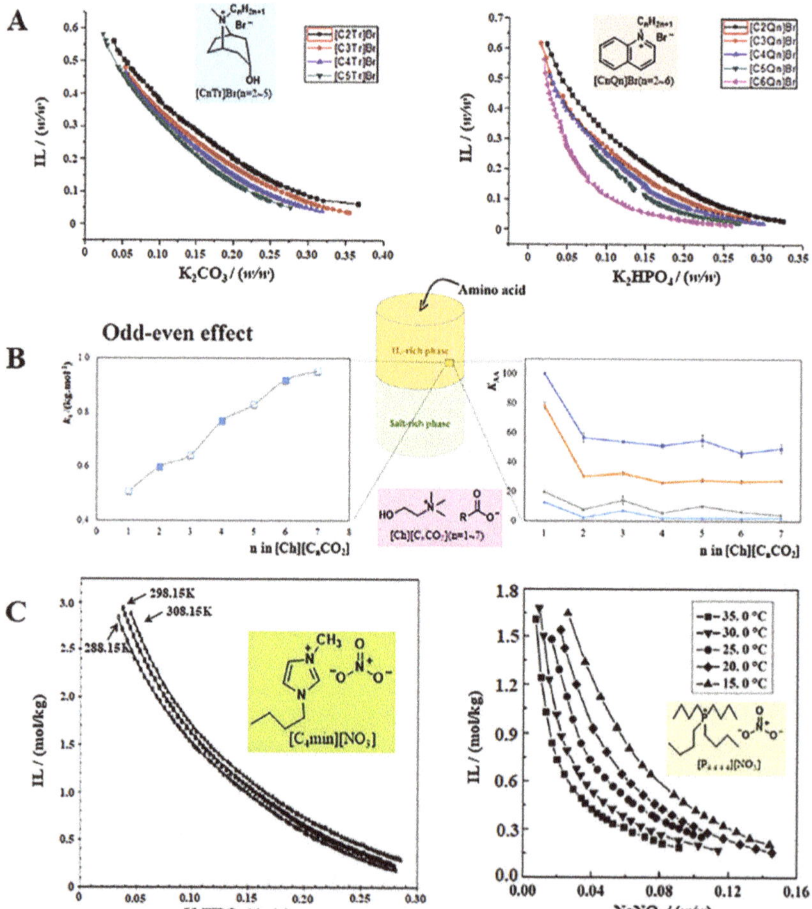

Figure 2. (A) Effect of ILs cations on the binodal curves [42]; (B) Odd-Even effect of the ATPS based on [Ch][CnCO2], comprising anions with odd and even alkyl chain lengths [43]; (C) Effect of temperature on the phase formation ability of different IL-based ATPS [44,45].

Many studies have found that when the carbon number of cationic side chain alkyl group in ILs was less than or equal to 6 ($n \leq 6$), the ability of phase formation of IL-based ATPS was enhanced with the increase in the hydrophobicity of ILs [42,46–48]. However, an over high hydrophilic nature of the IL is not beneficial for the formation of ATPS. The appropriate increase in the length of non-polar alkyl chain can lead to a strengthened hydrophobic nature of ILs, resulting in a weaker affinity with water. Anomalous rules of the phase separation ability of ILs were observed when the carbon number of cationic side chain alkyl group is greater than or equal to 8 ($n \geq 8$) [49,50]. This may be due to the self-aggregation structure formed by ILs in the system. In addition, the substitution of different functional groups on the side chain of ILs also affects the property of the phase forming system. For example, hydroxyl and allyl groups on side chain substituents on imidazolium cation could significantly reduce the phase separation ability of ILs. additionally, the effect of benzyl and n-heptyl groups on phase separation was not significant [47,51,52]. Compared to the benzyl and n-heptyl groups, the hydroxyl and allyl groups on side chain substituents exhibit the higher water affinity. The higher the affinity for water and/or hydrophilic nature of the IL, the less effective is the IL in promoting ATPS.

2.2. *Effects of ILs Anions*

Due to the variety of the structure of ILs anions, there is no unified standard for comparison. Therefore, the effects of ILs anions on the ability of phase formation have rarely been studied. In 2007, Pei and co-workers [49] compared, for the first time, the phase separation ability of the halogenide anions (Br^- and Cl^-). When IL cations were 1-butyl-3-methylimidazolium ($[C_4mim]^+$) and 1-hexyl-3-methylimidazolium ($[C_6mim]^+$), the ability of phase formation of Br^- was obviously better than that of Cl^-. This can be explained by the different Gibbs energies of hydration ($\Delta_{hyd}G$) of Cl^- and Br- anions. The absolute value of $\Delta_{hyd}G$ of Cl^- (−340 kJ/mol) is larger than that of Br^- (−315 kJ/mol). Therefore, the ILs with the Cl^- anion hydrate more water molecules than the ILs with the Br^- anion, resulting in difficulty in the phase formation as salt was added. Subsequently, a wide range of the ILs anions were studied, aiming at obtaining new insights regarding their ability toward the formation of IL-based ATPSs. On the basis of the IL cations, including 1-ethyl-3-methylimidazolium and 1-butyl-3-methylimidazolium, the influence of IL anions on the ATPS formation was assessed through their combination with chloride, bromide, acetate, hydrogensulfate, methanesulfonate, methylsulfate, ethylsulfate, trifluomethanesulfonate, trifluoroacetate, and dicyanamide. The results indicated that the ability of ILs for ATPS formation followed the order: $[C_2mim][CF_3SO_3]$ > $[C_2mim][C_2SO_4]$ > $[C_2mim][MeSO_4]$ > $[C_2mim][Br]$ > $[C_2mim]Cl$ ≈ $[C_2mim][CH_3CO_2]$ > $[C_2mim][CH_3SO_3]$ and $[C_4mim][CF_3SO_3]$ > $[C_4mim][N(CN)_2]$ > $[C_4mim][HSO_4]$ > $[C_4mim][TFA]$ > $[C_4mim]Br$ > $[C_4mim]Cl$ ≈ $[C_4mim][CH_3CO_2]$ ≈ $[C_4mim][CH_3SO_3]$ [53]. With the increasing capacity of anions as hydrogen bond acceptor (i.e., hydrogen-bond alkalinity), the ability of ILs with the same cation for ATPS formation decreased [38]. This may be because the strong hydrogen-bond alkalinity of ILs is favorable to the interaction with water to form hydrate, which will lead to the weak phase separation ability. Moreover, most anions are small in size and do not have long hydrophobic alkyl chains. Therefore, the hydrogen bond interaction is considered to be the main factor affecting the phase separation process.

In some cases, the weak hydrogen-bond alkalinity of ILs can significantly affect the distribution of solute between two phases and reduce the extraction capacity of the ATPSs, which restricts to some extent the application of IL-based ATPS in the separation field [54,55]. Deive et al. [56] constructed the IL-based ATPSs consisting of high charge-density inorganic salts and several ethyl-methylimidazolium alkylsulfate ILs, $[C_2mim][C_nSO_4]$ (n = 2, 4, 6, or 8), at room temperature. The influence of different alkyl chain lengths in the anion on the formation of ATPS was investigated. The results indicated that the phase separation ability of ILs increased with the increase in the length of the alkyl chain in the IL-anion. Similarly, a set of ILs containing tetrabutylammonium cation and carboxylate anions were synthesized by Basaiahgari et al. [57] to form ATPS in presence of strong salting-out agent

K_3PO_4. Furthermore, the influence of alkyl chain length of ILs' anions on the phase splitting ability was evaluated. The resultant trend of phase formation among studied ILs was as follows: [TBA][But] < [TBA][Pent] ≈ [TBA]Br < [TBA][Hex] < [TBA][Hept] ≈ [TBA][Dec] < [TBA][Oct]. It can be assumed that the hydrophobicity of anions plays a dominant role in the phase formation. Jimenez et al. [58] used imidazolium-based ILs with different alkyl side chains and anions (chloride, bromide, acetate or dicyanamide) as phase-forming components of ATPS. It is observed that anions have a dramatic effect on ATPS immiscibility when a random co-polymer of ethylene oxide and propylene oxide monomers (namely, UCON) was used to form ATPS with different ILs. The area of the heterogeneous region follows the series $OAc^- > Cl^- > Br^- > SCN^-$. At the same time, this order is consistent with the absolute value of Gibbs energy of hydration of these anions [59]. Moreover, amino acid ILs (AAILs) have the advantages of the tunable hydrophobicity and acid/base behavior, the low toxicity and biodegradability, which are widely used in ATPS [43,60–63]. Most recently, Korchak et al. [62] compared the phase separation ability of several AAILs with different amino acid anions, such as L-Leucine ([L-Leu]), L-Valine ([L-Val]), L-Lysine ([L-Lys]). Based on the lengths and the slopes of the tie lines, the ability of the studied systems (AAILs + inorganic salts) to phase separation increased in the following sequence: [Lys] > [Val] > [Leu] > Cl > Br. This result indicated that the heterogeneous region for amino acid ILs is wider than for halide ILs. However, an odd-even effect resulting from their structures was observed in the ATPS based on cholinium carboxylate ILs ([Ch][C_nCO_2] with n = 1~7, comprising anions with odd and even alkyl chain lengths) [43]. The Setschenow salting-out coefficients (k_s) was determined to quantitatively describe the two-phase formation ability, which mainly depends on the properties of salts. The value of k_s is proportional to the ion valence. The higher the k_s value, the better salting out effect. As shown in Figure 2B, these ILs comprising even alkyl chains display slightly higher k_s values, meaning that they are more prone to being salted out or more easily phase separated. Moreover, the existence of an odd-even effect is also visible in the extraction performance of ATPS for four amino acids (L-tryptophan, L-phenylalanine, L-tyrosine, and L-3,4-dihydroxyphenylalanine). The result is the opposite. The ATPS formed by ILs with anions comprising odd alkyl chains lead to slightly higher partition coefficients of amino acids (K_{AA}). The differences in partitioning could mainly arise from the dispersive interactions between these hydrophobic amino acids the IL anion aliphatic moieties.

To sum up, the overall phase separation ability and extraction efficiency of IL-based ATPS are governed by multiple factors prevailing at a microscopic level such as hydrophobicity, hydrogen bond accepting ability and dispersive interaction. The above research can provide new strategies for enhancing the phase separation ability of ILs.

2.3. Effects of Temperature

Temperature is a significant factor that affects the formation of the ATPSs. Generally, the heterogeneous region of the IL-based ATPSs varies to a certain degree as the temperature increases or decreases [18]. The influence of temperature on the properties of IL-based ATPS was assessed by several researchers. For instance, a recent review was reported by Chakraborty and co-workers [64]. Furthermore, the Merchuk equation and its fitting parameters were expressed as a function of temperature in the linear form with ($T-T_0$) K as a variable. T_0 was assumed as the reference temperature, 273.15 K [64–67]. The nonlinear expression of Merchuk equation is as follows:

$$w_1 = a \times exp\left(bw_2{}^{0.5} - cw_2{}^3\right) \tag{1}$$

where w_1 and w_2 represent the concentrations (in weight percent) of IL and salt, respectively. a, b, and c represent the fitting parameters.

It must be emphasized that the effect of temperature on ATPS is quite complex, varying differently for different systems. Based on the salting-out coefficient (k_s) obtained from fitting the tie-line data, Zafarani-Moattar et al. explained the effect of temperature on the phase-forming ability of the IL [C_4mim]Br + tri-potassium citrate system [65,66]. The results indicated that the value of k_s increased with the decreasing of temperature. Higher phase-forming ability has a larger value of k_s. Similarly, the ATPS composed of [C_4mim][BF_4] + $(NH_4)_2SO_4$ + H_2O at three different temperatures (298.15, 308.15, 318.15 K) was investigated by Wang et al. [67]. The solubility of IL decreases with the decrease in temperature. Therefore, the binodal curve shifts down, resulting in an expansion in the heterogeneous region. That is, a decrease in temperature leads to an increase in phase-forming ability. Novel ATPS composed of *N*-butylpyridinium tetrafluoroborate ([C_4py][BF_4]) and inorganic salts (Na_2SO_4 and $(NH_4)_2SO_4$) were studied [68]. The two-phase region expanded with the decrease in temperature, which indicated that the low temperature was favorable for the phase separation. Moreover, the reliability of tie-lines was evaluated by Othmer-Tobias and Bancroft equations. The tie lines at low temperatures possessed longer TLL, which mean better phase separation ability at low temperatures. Therefore, the studied ATPS is beneficial for the extraction of targets that are inactivated at higher temperature, such as proteins. This trend has also been observed for the ATPSs constituted by ILs and carbohydrates, such as [C_4mim][BF_4] and glucose [69], fructose [70], sucrose [71,72] or maltose [73]. The strong interaction between the IL and carbohydrates is not conducive to the formation of ATPS. This may be because low temperature can destroy the interaction between the IL and carbohydrates, leading to expansion of the biphasic region. Malekghasemi et al. [44] studied the effect of temperature on the phase formation ability of the IL [C_4mim][NO_3] + K_2HPO_4 + water ATPS. As shown in Figure 2C, the decrease in the temperature slightly caused the expansion of the two-phase area. Although the above phenomenon is different from the traditional polymer + salt system [74–77], it can be found in most ILs + salt ATPSs. However, there are a few ATPSs composed of IL and inorganic salt, which undergo the expansion of immiscibility regime with the increase in temperature [45,78]. This trend mainly occurs in some ILs containing tetrabutylphosphonate cations. For example, a hydrophilic IL tetrabutylphosphonate nitrate ([P_{4444}][NO_3]) was used for the formation of an ATPS with $NaNO_3$. It is clear from Figure 2C that the phase separation ability of IL was significantly enhanced with an increase in temperature, suggesting that the lower concentration of [P_{4444}][NO_3] and $NaNO_3$ was required for the formation of ATPSs as the temperature increased [45].

Additionally, the different trends to induce the liquid-liquid demixing at different temperatures were observed for the IL-based ATPS containing polymers, such as polyethylene glycol (PEG) and polypropylene glycol (PPG) [79–82]. The solubility of PEG and PPG in water mainly depends on the mutual PEG/PPG-water hydrogen bond interactions. In conventional ATPS composed of PEGs and ILs, the temperature affects the liquid-liquid demixing process by disturbing the hydrogen-bonding interactions. Moreover, there are lower critical solution temperatures (LCST) and upper critical solution temperatures (UCST); it refers to two different stimuli-responsive behaviors of the system at different temperatures. For the LCST system, the phase separation occurs when the temperature drops to a certain level. Whereas the UCST system shows the opposite thermal response behavior, that is, phase separation occurs when the temperature rises to a certain degree. Therefore, some ILs-PEG/PPG ATPSs have phase behaviors in response to external thermal stimuli. At a low IL concentration, the system is the type of LCST. Additionally, the system transforms into UCST type at high IL concentration [83]. For example, aqueous solutions of PEG polymers with different molecular weights (600, 1000, 2000, and 3400 g mol^{-1}) and several protic ILs were mixed and their ability to form ATPS at several temperatures was assessed [84]. An increase in the immiscibility region or the phase-forming ability with the increase in temperature for ammonium acetate ([NH_4][OAc]), propylammonium acetate ([C_3NH_3][OAc]) and butylammonium acetate ([C_4NH_3][OAc])-based ATPSs was observed. This belongs to the typical LCST. The high temperature can decrease the mutual IL–PEG/PPG hydrogen-bonding interactions, which facilitates the creation of ATPS. However, an increase in the temperature reduced the biphasic region of the hexylammonium acetate ([C_6NH_3][OAc]) and butylammonium butanoate ([C_4NH_3][But])-based ATPSs. In brief, the temperature influence on the phase separating ability is a quite complex phenomenon. Multiple factors contribute to the result, such as hydrogen-bonding interaction, concentration of ILs and the types of coexisting phase forming components.

3. Mechanism of Phase Separation

The phase separation mechanism of the ATPS composed of ILs and salts has been well established. The salting-out effect is the major force affecting the phase separation and the extraction [85]. In general, ILs are in a stable dispersed state in water. However, the abilities of salts and ILs to adsorb water molecules are different. When an aqueous solution containing salt is added to the IL aqueous solution, the formation of IL-based ATPS is a process of the competition between IL and salt for water molecules. On the one hand, the hydrophilic and hydrophobic properties of ILs affect their phase forming ability. For example, Ren et al. used five ILs to construct ATPS with K_3PO_4, $K_3C_6H_5O_7$, and K_2CO_3 [86]. The results showed that phase forming ability of ILs increased with the increase in their hydrophobicity. That is, the hydrophobicity of ILs is beneficial to the formation of ATPS [3,87,88]. On the other hand, the property of salts has an influence on the phase formation process. When the added salts contain kosmotropic ions (such as CH_3COO^-, SO_4^{2-}, HPO_4^{2-}, Mg^{2+}, Ca^{2+}, Li^+, H^+, OH^-, etc.), water molecules around kosmotropic ions have a more regular arrangement order and lower free energy due to the polarization of kosmotropic ions. The tendency of two-phase separation is consistent with the ability of salt ions to form hydration complexes. Moreover, the stronger the hydration ability of salt ions, the easier it is to repel IL and form the second phase. The salting-out ability can be related to the Gibbs free energy ($\Delta_{hyd}G$) and the entropy ($\Delta_{hyd}S$) of hydration of the salt ions. The salting-out strength of the kosmotropic salts follows the Hofmeister series [89–92]. Compared to the chaotropic ions, the kosmotropic ions have large, negative $\Delta_{hyd}G$ values, due to the formation of the structured water around them. Therefore, it is easier to form the ATPS by adding kosmotropic salts rather than chaotropic salts. As the kosmotropic ions and ILs move closer to each other, H_2O molecules near ILs are gradually taken away by kosmotropic ions, resulting in the decrease in solubility of ILs. Then, the phase separation of IL-based ATPS is completed.

Compared to the IL-based ATPS formed with salts, the mechanism of IL-polymer ATPS is more complex. Although many studies have confirmed that the salting-out effect is the main force for IL-based ATPS with PEG/PPG [93–97], the phase separation principle of some ATPSs constructed by ILs and polymers does not completely follow the salting-out phenomenon. For instance, the phase separation mechanism of ATPS based on imidazolium-based ILs and PEG was investigated by Freire and co-workers [30]. The IL or polymer can form independent solvent effect with H_2O molecules, caused by hydrogen bonds between water and IL or polymer. Additionally, the solvation determines the phase separation mechanism of this system. Similarly, Mourao et al. [98] also constructed the ATPS by mixing cholinium-based ILs and PEG with different molecular weight, respectively. In this system, the interaction between IL and PEG-600 is stronger than that between PEG-600 or IL and water. Thus, the phase separation mechanism is mainly determined by the interaction between PEG-600 and IL. The solvation of ILs in PEG aqueous solutions is very complex and the solvation process depends on the balance of all possible interactions (PEG-water, PEG-ILs, water-ILs) [84,99].

Additionally, the H-bond alkalinity and aggregation structures of ILs regulated by their components and functional groups play an important role for forming ATPS. Here, the aggregation structures often refers to the nature of poly(ionic liquid)s. Poly(ionic liquid)s are a type of ionic polymer obtained by polymerization of IL monomers, or grafting IL units onto the polymer backbone. Generally, the functional groups (such as double bond) on the structure of IL monomers are required to form poly(ionic liquid) by the polymerization. The hydrogen-bond interaction between ILs and water will be enhanced as the increase in hydrogen-bond alkalinity, which is not conducive to phase separation. Furthermore, the aggregation structures of ILs can change the interaction between ILs and water molecules, thus affecting the phase separation rule of ATPS [100]. The physicochemical property of poly(ionic liquid) can be tailored by changing the characteristic functional group of IL monomers. The hydrophobic interaction is the main driving force for micelle formation of ILs in water. When the IL monomer contains hydrophobic groups, the hydrophobicity of poly(ionic liquid) increases, resulting in an increase in its aggregation and phase separation ability.

4. Phase Equilibrium of the IL-Based ATPSs

The IL-based ATPS is formed by two phases, both consisting mainly of water. These phases are composed by water and IL or a salting-out agent. Under specific thermodynamic conditions, the IL-based ATPS typically has a top phase (TP) rich in IL (extract phase) and a bottom phase (BP) rich in salting-out agent. As shown in Figure 3, the formation of the ATPS is based on the composition of the two-phase system through a rectangular phase diagram that expresses the TP and BP concentrations in mass percentage % (w/w) [101]. The binodal curve refers to the line to separate the homogeneous from the heterogeneous regions. The upper side of the binodal curve is a two-phase system, and the lower side is a single-phase system. Under the same conditions, the larger the area of the biphasic region is, the stronger the phase separation ability is. For an ATPS prepared at the composition X, TPC and BPC reflect the top phase composition and the bottom phase composition, respectively. CP indicates the critical point. In addition, the tie lines (TL) are also obtained from the phase diagram. These lines are straight lines that connect global mixing points to their corresponding phase compositions (TPC and BPC). As the length of TL decreases in the diagram, the compositions become closer and closer. This behavior continues until the compositions become equal in the CP and the system is monophasic [102].

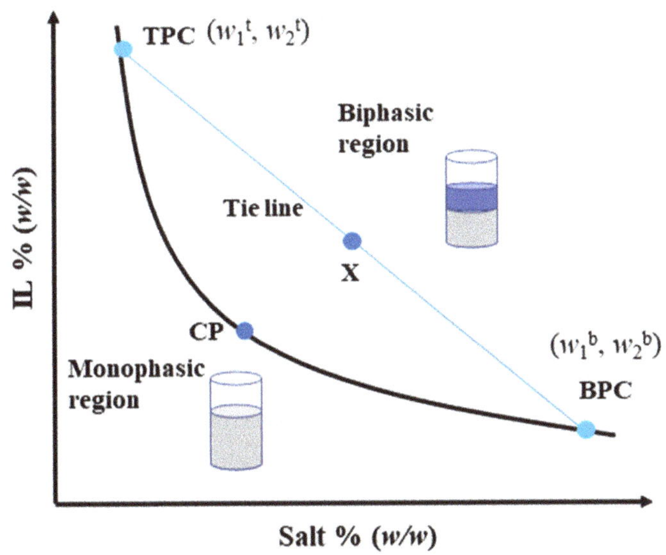

Figure 3. Conceptual diagram of an ATPS constituted of IL and salt, where the ordinate axis represents the IL concentration in % (w/w) and the abscissa axis the salt concentration in % (w/w).

The phase diagram of the ATPS is instructive for its application in the extraction. It is of great significance to understand the genesis and influencing factors of IL-based ATPS for its innovative development. There are many factors affecting the phase diagram of IL-based ATPS, which not only depend on the structure and properties of ILs, but also are restricted by other coexisting phase forming components, such as salts, polymers, surfactants, amino acids and saccharides. The following categories of these influencing additives are discussed.

4.1. Salts

Within the scope of this work, salts are the most commonly used second phases formers besides ILs. As one of the major components of ATPSs, the effect of inorganic salts on the phase equilibrium was investigated by many researchers. It was reported that IL-based ATPSs could be formed by adding appropriate amounts of inorganic and organic salts (such as K_2HPO_4, K_3PO_4, K_2CO_3, KOH, NaOH, Na_2HPO_4, $(NH_4)_2SO_4$, KCl, NaCl, $K_3C_6H_5O_7$ and $Na_3C_6H_5O_7$) into the aqueous solution of IL. In the research of Tanimura et al., two hydrophilic ILs, including 1-allyl-3-methylimidazolium chloride ([Amim]Cl) and [C_4mim]Cl, were used to from the ATPS by adding the inorganic salts (K_2CO_3, K_2HPO_4) (Figure 4A). The phase separation mainly depends on the salting-out effect of inorganic salt in the bottom phase. In the same composition, the [C_4mim]Cl/K_2HPO_4 system is more likely to undergo a two-phase separation at lower temperature conditions, suggesting that the phase-separation ability of the inorganic salt K_2HPO_4 is stronger than that of K_2CO_3 [103]. As a good indicator of the molecular partitioning behaviors in conventional ATPSs, the hydrophobicity index (HF) between the top and bottom phases was characterized by the distribution coefficients (K_{aa}), of amino acids. Results showed that the HF value of this IL-based ATPS was almost consistent with an ATPS composed of PEG and salt. Recently, as a new class of the ILs, choline amino acid ILs ([Ch][AA]) were explored so as to construct the ATPS with different salts including $(NH_4)_2SO_4$, KH_2PO_4, Na_2CO_3, and K_3PO_4 [104]. Compared to traditional ILs, [Ch][AA] are less toxic, more biodegradable, and have good air and water stability. Thus, the ATPS composed of [Ch][AA] and salt are often applied to the extraction and separation of food samples as an environmentally friendly extraction method [105].

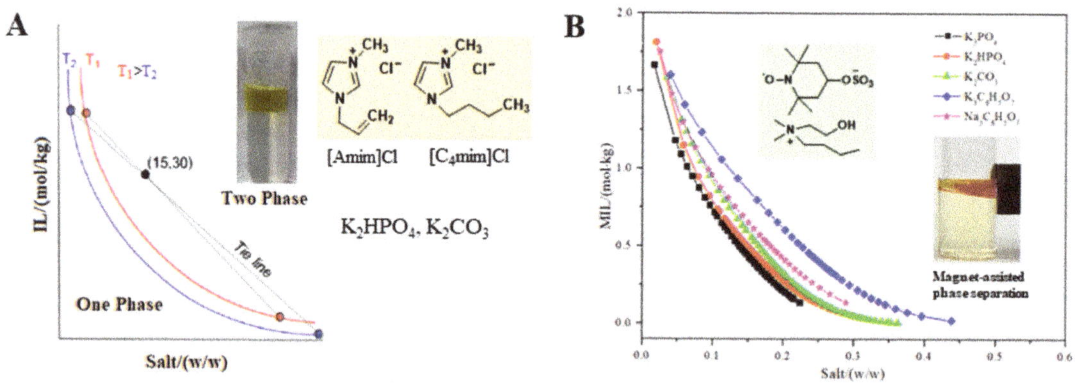

Figure 4. Effect of different salts on phase equilibrium behavior of imidazolium-based IL ATPS (**A**) and cholinium-based MIL ATPS (**B**) [103].

More significantly, as one of functionalized ILs, magnetic ionic liquid (MIL) demonstrates a response to an external magnetic field. Imidazolium-based MIL [C_4mim][$FeCl_4$] was firstly synthesized by Hayashi et al. [106] in 2004. Since then, MILs have become a hotspot of research in field of extraction and separation [107–111]. Recently, remarkable progress has been made in the field of MILs and the related ATPS are stimulating interest [112–115]. Yao et al. prepared three MILs based on the tetramethylguanidinium ([C_nTMG][TEMPO-OSO_3], n = 2,3,4) to form ATPS with different inorganic salts. The authors found that $(NH_4)_2SO_4$, Na_2SO_4, KOH, NaOH, NaCl as well as KCl could not drive any MIL aqueous solution to separate into two phases. Therefore, K_2HPO_4, K_2CO_3 and Na_2CO_3 were used as the salting-out agents for this ATPS. Among them, both K_2HPO_4 and K_2CO_3 have wide range of biphasic regions, indicating that they are better phase forming salts in comparison with Na_2CO_3 [114]. Most recently, a MIL-based ATPS was investigated by our group [115]. In our study, five cholinium-based MILs with the piperidinyloxy radical anion were synthesized for the first time and formed ATPS with inorganic or organic salts. For the five investigated MILs, the abilities of salts for phase-separation are in the following order: $K_3PO_4 > K_2HPO_4 > K_2CO_3 > Na_3C_6H_5O_7 > K_3C_6H_5O_7$. The salting-out ability of the anions follows the order: $PO_4^{3-} > HPO_4^{2-} > CO_3^{2-} > C_6H_5O_7^{3-}$ when they share the same cation. Additionally, the salting-out ability of cations follows the order: $Na^+ > K^+$ for the citrate-based salts. The results are in accordance with the order of the absolute values of $\Delta_{hyd}G$ and $\Delta_{hyd}S$ for these ions [32]. That is, the increasing entropy may be the driving force for the two-phase formation [116]. Moreover, it is clear from Figure 4B that the MIL-rich phase (top phase) can be easily attracted by a magnet. Under an external magnetic field, the phase assembly and separation become more time-saving and easier, which make these systems superior to the common IL-based ATPS.

Although ILs and salts are present in the system as ions, the two phases of the ATPS remain electrically neutral, which is confirmed by Bridges et al. [36]. Similar to the polymer-salt ATPS, the salting-out ability of kosmotropic salt is consistent with the Hofineister series. Thus, it is directly related to their $\Delta_{hyd}G$. Most ILs as chaotropic salts are easily salted out by kosmotropic salts. In the IL-salt ATPSs, the main force of phase formation is salting-out effect, resulting from the formation of water-ion complexes and the increase in cavity surface tension [117,118]. Compared to the water molecules, the kosmotropic ions (e.g., HPO_4^{2-}, SO_4^{2-}, OH-, CO_3^{2-} and PO_4^{3-}) exhibit the stronger interaction with water molecules, which are beneficial to the formation ATPS. Conversely, chaotropic ions (e.g., Cl^-, NH_4^-, K^+, and $H_2PO_4^-$) find difficulty in forming ATPSs with ILs because the interactions between these anions and water molecules are weaker than those between water molecules. Finally, some of the salts commonly used in the IL-based ATPSs are listed in Table 1.

Table 1. Some salts commonly used in the IL-based ATPSs.

Salts	ILs	Temperature/K	Refs.
K_3PO_4, K_2HPO_4, K_2CO_3, KOH, and $(NH_4)_2SO_4$	[C_4mim]Cl, [C_4mmim]Cl, [C_4py]Cl, [N_{4444}]Cl, [P_{4444}]Cl	298.15	[36]
K_2HPO_4, KH_2PO_4	[C_4C_1pyr]Cl, [C_4mim]Cl, [C_4C_1pip]Cl, [C_4C_1py]Cl	298.15	[38]
K_3PO_4, K_2CO_3, K_2HPO_4, $K_3C_6H_5O_7$, $Na_3C_6H_5O_7$, NaH_2PO_4	[C_2Tr]Br, [C_3Tr]Br, [C_4Tr]Br, [C_5Tr]Br, [C_2Qn]Br, [C_2Qn]Br, [C_3Qn]Br, [C4Qn]Br, [C_5Qn]Br, [C_6Qn]Br,	298.15~318.15	[42]
K_3PO_4, K_2CO_3, Na_2CO_3, $(NH_4)_2SO_4$	[C_2mim][C_nSO$_4$] (n = 2, 4, 6, or 8)	298.15	[56]
K_3PO_4, K_2CO_3	[Cnmim]X (n = 4, 8; X = [Lys], [Val], [Leu], Cl, Br)	298.15	[62]
K_2CO_3, Na_2HPO_4, Na_2SO_4, $Na_3C_6H_5O_7$, Na_2CO_3, K_2HPO_4, KH_2PO_4, $(NH_4)_2SO_4$, $(NH_4)_3PO_4$, K_3PO_4, $NaNO_3$	[Ch][L-Pro], [Ch][L-Cys], [Ch][L-His], [Ch][L-Val], [Ch][L-Ser], [Ch][L-Met], [Ch][L-Ala]	298.15	[63]
K_2HPO_4	[Ch][C_nCO$_2$] (n = 1~7), [Ch]Cl	298.15	[43,119,120]
K_2CO_3	[C_4mim][NO$_3$], [C_6mim][NO$_3$], [C_2mim]OAc	288.15~308.15	[76,121]
$NaNO_3$, NH_3NO_3	[P_{4444}][NO$_3$], [N_{4444}][NO$_3$]	288.15~308.15	[45]
K_3PO_4, $K_3C_6H_5O_7$, K_2CO_3	[C_4MDEA]Br, [C_6MDEA]Br, [C_8MDEA]Br, [C_{10}MDEA]Br, [C_4BDEA]Br	298.15	[86]
K_2CO_3, K_2HPO_4	[Amim]Cl, [C_4mmim]Cl	278.15~318.15	[103]
$(NH_4)_2SO_4$, KH_2PO_4, Na_2CO_3, K_3PO_4	[Ch][Ala], [Ch][Gly]; [Ch][Lys], [Ch][Arg]	298.15~338.15	[104]
K_2HPO_4, K_2CO_3, Na_2CO_3	[C_2TMG][TEMPO-OSO$_3$], [C_3TMG][TEMPO-OSO$_3$], [C_4TMG][TEMPO-OSO$_3$]	298.15	[114]
K_3PO_4, K_2HPO_4, K_2CO_3, $K_3C_6H_5O_7$, $Na_3C_6H_5O_7$	[N_{1122OH}][TEMPO-OSO$_3$], [N_{1132OH}][TEMPO-OSO$_3$], [N_{1142OH}][TEMPO-OSO$_3$], [N_{1152OH}][TEMPO-OSO$_3$]	298.15~318.15	[115]

Table 1. *Cont.*

Salts	ILs	Temperature/K	Refs.
Na_3PO_4, $(NH_4)_3PO_4$	$[C_4mim]BF_4$	288.15~318.15	[122]
$NaNO_3$	$[N_{4441}][NO_3]$	298.15	[123]
$(NH_4)_2SO_4$, NaH_2PO_4, Na_2SO_4, Na_2HPO_4, K_2HPO_4, NaCl, $Na_3C_6H_5O_7$	$[C_4mim][CF_3SO_3]$, $[C_4mim]Cl$, $[C_4mim][BF_4]$, $[C_2mim]Br$, $[C_4mim]Br$, $[C_6mim]Br$, $[C_8mim]Br$	278.15~318.15	[124]
Na_3PO_4, Na_2CO_3, Na_2SO_4, $Na_3C_6H_5O_7$, $K_3C_6H_5O_7$, NaH_2PO_4, NaCl, $MgCl_2$, $CaCl_2$	$[C_4mim][CF_3SO_3]$	278.15~318.15	[125]
$K_3C_6H_5O_7$, $(NH_4)_3C_6H_5O_7$, $K_2C_4H_4O_6$	[EOMiM]Br	288.15~308.15	[126]
$(NH_4)_3C_6H_5O_7$, $Na_2C_4H_4O_4$	$[C_4py]OTF$	298.15~328.15	[127]
$Na_3C_6H_5O_7$	$[C_2mim]DCA$, $[C_3mim]DCA$, $[C_4mim]DCA$, $[C_6mim]DCA$, $[C_4C_1pyr]DCA$	298.15	[128]
K_3PO_4	$[C_4mim]Cl$, $[C_4mim]Cl$; $[C_4C_1C_1im]Cl$, $[C_6mim]Cl$, $[C_2mim][CF_3SO_3]$, $[C_2mim][C_2SO_4]$, $[C_2mim][MeSO_4]$, $[C_2mim][Br]$, $[C_2mim]Cl$, $[C_2mim][CH_3CO_2]$, $[C_2mim][CH_3SO_3]$, $[C_4mim][CF_3SO_3]$, $[C_4mim][N(CN)_2]$, $[C_4mim][HSO_4]$, $[C_4mim][TFA]$, $[C_4mim]Br$, $[C_4mim]Cl$, $[C_4mim][CH_3CO_2]$, $[C_4mim][CH_3SO_3]$, [TBA][But], [TBA][Pent], [TBA]Br, [TBA][Hex], [TBA][Hept], [TBA][Dec], [TBA][Oct], [Ch][Leu], [Ch][Ala], [Ch][Gly], [Ch][Lys]	298.15	[17,37,53,57,129]
K_3PO_4, K_2HPO_4	[Ch][Pro], [Ch][Cys], [Ch][Ala], [Ch][His], [Ch][Met]	298.15	[130]

4.2. Polymers

Compared to common IL-based ATPS, the polymer-IL-based ATPS is more special. In the IL-based ATPS containing polymer, the neutral molecules (such as PEG and PPG) are the most commonly used polymers. Moreover, ILs are electrically charged salts. Therefore, ILs are the true sense of the salting-out agents in this ATPS, which is similar to the conventional polymer-salt ATPS. In 2007, Visak et al. reported the phase behavior of imidazolium-based ILs ($[C_nmim]Cl$, n = 2~10) and PEG-3500 (M_w = 3500 g·mol^{-1}) in water for the first time [131]. The ILs with long side chains in cation ($[C_nmim]Cl$, n = 6, 8, 10) can promote the hydrogen bond interaction between PEG and water, resulting in the solubilization effect for PEG. At the same time, the presence of long alkyl chains makes the IL form micelles and other self-aggregation structures, which also acts as a cosolvent effect. ILs with short side chains ($[C_nmim]Cl$, n = 2, 4) also could solubilize PEG at low concentrations, but could precipitate PEG at high concentrations. This may be due to the fact that ILs are both hydrophobic and hydrophilic, which are caused by the non-polar alkyl chain and the polar part of the anion or cation, respectively. Subsequently, the phase diagrams of various ILs and PEGs with different molecular weights in water were determined by Freire et al. [30]. The ability of phase formation was closely related to the hydrophobicity of PEG. The phase separation ability was enhanced with the increase in the molecular weight of PEG as a consequence of the increased hydrophobic properties of PEG. In the study of Tomé et al. [132], the experimental results combined with molecular dynamics (MD) simulations and density functional theory (DFT) calculations were provided to understand the molecular-level mechanisms behind the formation of ATPS composed of ILs and polymers. As a proof of principle, the experimental ternary phase diagrams composed of an IL ($[C_4mim]Cl$), PEG-1500 (M_w = 1500 g·mol^{-1}) and water were determined at two distinct temperatures. The results showed that this ternary IL-PEG-water system was of Type "0". That is, the binary pairs (such as IL-PEG, IL-water and PEG-water) are completely miscible yet are able to form two-phase systems at given compositions of the ternary mixtures. Distinct from what happens in IL-salt-based ATPS, the formation of IL-PEG-based ATPS was controlled by the IL anion solvation by water, which resulted in the destruction of the hydrogen bond interaction between the IL anion and the hydroxyl groups of the polymer. When water is introduced into the PEG-IL binary system, the hydrogen bonds formed by IL and PEG are replaced by the stronger water-IL anion hydrogen bonds. Therefore, this mechanism of the ATPS formation is here labelled as a "washing-out" phenomenon, given the analogy with the washing process. In brief, it has herein been shown for the first time that IL-polymer-based ATPS was a result of a "washing-out" phenomenon, and not of a salting-out effect of the IL over the polymer as assumed in the past few years.

A recent extensive study on the formation of aqueous biphasic systems (ABS) using aqueous solutions of protic ILs and PEG was performed by Cláudio and co-workers [84]. A series of different molecular weight PEGs (600, 1000, 2000, 3400 g·mol^{-1}) were selected to study the effect of the molecular weight of PEG for the formation of IL-based ATPS. Among them, the PEG-600 was unable to induce ATPS formation when mixed with the IL octylammonium acetate ([C_8NH_3][OAc]). For other polymers, the ability of the polymer to induce liquid-liquid demixing follows the order: PEG-3400 > PEG-2000 > PEG-1000 (Figure 5A). The results obtained indicated that polymers with the higher molecular weight were more able to promote phase separation. Ola et al. employed an ATPS formed by the IL 1-hexyl-3-methyl imidazolium dodecyl sulfonate ([C_6mim][$C_{12}SO_3$]) and PEGs with a higher molecular weight [133]. The same trend was presented for the above ATPS. It is well-known that the closer a curve is to the origin of the coordinates, the lower the polymer concentration required for the formation of two phases and the stronger is the phase-forming ability of the polymer. As shown in Figure 5B, the phase formation ability was in the following order: PEG-8000 > PEG-6000 > PEG-4000. That is, the larger the molecular weight of PEG, the smaller the concentration of both PEG and IL necessary to form a two-phase system. The tendencies observed might be related to two factors: (i) the increase in the molecular weight increased the hydrophobicity of the phases, leading to consequent reduction in its water solubility/affinity; (ii) the PEG-IL miscibility increased by increasing the number of terminal –OH groups per PEG molecule.

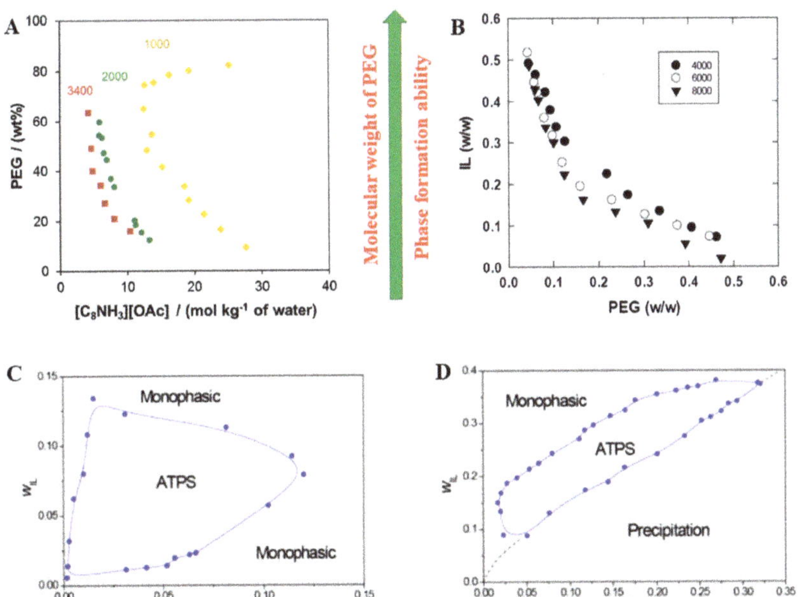

Figure 5. Effect of PEG molecular weight on phase formation ability of IL-based ATPS composed of [C_8NH_3][OAc] + water + PEG (**A**) and [C_6mim][$C_{12}SO_3$] + water+ PEG (**B**) [84,133]; Phase diagrams (25 °C) of ternary systems of [C_2pi][BF_4] + SDS + H_2O (**C**) and [Phpi][BF_4] + SDBS + H_2O (**D**) [134].

Alongside PEG, PPG is also one of the commonly used polymers. Different from PEG-IL-based ATPS, PPG is more hydrophobic. So, the increase in hydrophobicity of cationic parent core and side chain of IL can enhance the interaction between IL and PPG, which is not conducive to phase separation [135,136]. Liu et al. employed two PPGs with distinct molar mass and poly(ethylene glycol)-block-poly(propylene glycol)-poly(ethylene glycol) ($EO_{10}PO_{90}$) to form environmentally friendly ATPS with four cholinium-based ILs, such as cholinium glycollate, cholinium propionate, cholinium lactate and choline chloride [81]. The results showed that the phase-forming ability of the polymers in cholinium IL-based ATPSs follows the order: $EO_{10}PO_{90}$ > PPG-1000 > PPG-400. Obviously, the order of their hydrophobic ability and molar mass is in agreement with the order of their phase-forming ability. Such a phenomenon is also observed for the PEG + IL ATPS. In the research of Neves et al. [79], a vast number of imidazolium-based ILs and PPG-400 were formed ATPS. Then, ^{1}H NMR (Nuclear Magnetic Resonance) spectroscopy and COSMO-RS (Conductor-like Screening Model for Real Solvents) were used to obtain the molecular-level mechanisms which rule the phase splitting. For some systems, the IL-PPG-400 pairs were completely miscible, revealed to be of type "0". All evidence suggested that the formation of PPG-IL-based ATPS was controlled by the interactions established between the IL and PPG. For a given IL concentration, we can move from the monophasic-biphasic-monophasic regimes only by increasing the amount of PPG in the system, or vice-versa. Therefore, these phase diagrams can be recognized as interesting separation approaches. In addition, the rise in temperature was beneficial to the formation of ATPS, which was consistent with the LCST phase behavior. Of course, the cholinium-based ILs with low toxicity and high biodegradability can also form ATPS with suitable polymers, such as polyethylene glycol dimethyl ether 250 (PEGDME-250) or PPG-400 [24,137]. It can be summarized as follows: among all PPG polymers, the PPG-400 are the most commonly used in IL-based ATPS. Since it is thermoresponsive, PPG-400 can be recovered from aqueous solution simply by heating the solution above the LCST.

Among all the IL-based ATPS studied hitherto, the more recent polymer-IL-based ATPS boost their applicability since they increase the hydrophilic and hydrophobic range of the coexisting phases, thus allowing for more selective separations to be achieved. Compared to other ATPS, the polymer-IL-based ATPS showed a different partitioning behavior. Moreover, both species (polymers and ILs) are able to act as salting-out agents. This property opens possibilities for the design of an appropriate and selective ATPS [138,139]. Some polymers used in the IL-based ATPSs are summarized in Table 2.

Table 2. Some polymers commonly used in the IL-based ATPSs.

Polymers	ILs	Temperature/K	Ref.
PEGDME-250, PPG-400	Ch[L-Ala]	298.15~318.15	[24]
UCON	$[C_2mim]Cl$, $[C_2mim]Br$, $[C_2mim][SCN]$, $[C_2mim][OAc]$	288.15~308.15	[58]
PPG-400	$[N_{4444}]Cl$, $[P_{4\,4\,4\,4}]Cl$, $[C_4mim][C_2H_5SO_4]$, $[C_4mim][CF_3SO_3]$, $[C_4mim][N(CN)_2]$, $[C_4mim][SCN]$, $[C_5mim]Cl$, $[C_6mim]Cl$, $[C_7mim]Cl$	298.15~318.15	[79]
PPG-400, PPG-700, PPG-1000	DIMCARB, DPCARB, DACARB, DBCARB	288.15~308.15	[80]
$EO_{10}PO_{90}$, PPG-1000, PPG-400	Cholinium glycollate, cholinium propionate, cholinium lactate, choline chloride	288.15	[81]
PEG-600, PEG-1000, PEG-2000, PEG-3400	$[C_8NH_3][OAc]$	323.15	[84]
PEG-3500	$[C_nmim]Cl$ (n = 2~10)	298.15	[131]
PEG-1500	[C4mim]Cl	323.15, 333.15	[132]
PEG-4000, PEG-6000, PEG-8000,	$[C_6mim][C_{12}SO_3]$	298.15	[133]
PEG-400, PPG-400	[Ch]Cl, [Ch][DHcit], [Ch][Bit], [Ch][Bic], [Ch][DHp], [Ch][Ac]	298.15	[137]
PPG-400	[Ch][BES]	298.15	[140]
PEG-6000	$[C_2mim][BF_4]$, $[C_4mim][BF_4]$, $[C_2mim]Br$, $[C_4mim]Br$	298.15	[141]
PEG-2000, PEG-4000, PEG-6000	$[C_6mim][C_{12}SO_3]$	298.15	[142]

4.3. Surfactants

For the ATPS composed of surfactants and ILs, there was very limited research until 2011. In 2011, Wei et al. reported for the first time that the IL $[C_4mim][BF_4]$ and a surfactant sodium dodecyl benzene sulfonate ($C_{18}H_{29}SO_3Na$, SDBS) were mixed in water at a certain concentration to form the ATPS [143]. The borderlines of the different specific regions in the phase diagrams were determined using the turbidity titration method. In their work, the formation of SDBS-IL-based ATPS may be due to the formation of very large size micelle aggregates. The authors confirmed the result by Transmission Electron Microscopy (TEM) and Steady-state Fluorescence Quenching Measurements (SFQM) methods. The average size of the micelle aggregates in the upper phase and bottom phase was much larger than that in water. The larger micelles possessing smaller density aggregate in the upper phase, while the relatively small micelles remain in the bottom phase. Therefore, the bottom phase was slightly smaller than the upper phase in the micelle size. Moreover, this phase separation phenomenon was found to be likely due to the existence of micelle aggregates with quite a large size. The phenyl group attached to the -SO_3^- in SDBS may conduct cation-π interaction with the cation ($[C_4mim]^+$) of $[C_4mim][BF_4]$. This interaction makes the butyl side chain align along the alkyl chain of the surfactant in the palisade layer of the micelle surface close to the phenyl moieties, which increases the size of micelle aggregates. In order to verify this mechanism, the author also studied the phase forming ability of different surfactants and found that only the surfactant containing a benzene ring can form a two-phase system with IL. Subsequently, the above ATPS was successfully used to extract the Sudan dyes from food samples [144]. Furthermore, inorganic salts played pivotal role on forming the ATPS because of intermicellar and intramicellar interaction. Thus, the inorganic salt $(NH_4)_2SO_4$ was selected to maintain the stability of ATPS. This was mainly because the interaction between water molecule and SO_4^{2-} was stronger than that between water molecules. Similarly, the ATPS formed by the IL $[C_6mim]Cl$ and SDS was reported for the determination of sulfonamides in blood [145]. In their work, four kinds of salts including K_2CO_3, $(NH_4)_2SO_4$, K_2HPO_4 and NaCl were chosen to investigate the effect of ionic strength on the ATPS. Among them, Cl^- as chaotropic ions, the interaction with water molecules was relatively weak and solubilizing effect may be produced. With HPO_4^{2-} as the kosmotropic ion, the interaction with water molecules was stronger than the interaction between water molecules, thereby facilitating the formation of ATPS [17,146]. Therefore, K_2HPO_4 was added to this system.

More recently, Lu et al. [134] constructed two kinds of ATPSs formed by 1-ethylpiperazinium tertrafluoroborate ($[C_2pi][BF_4]$)/sodium dodecyl sulfonate (SDS) and 1-phenylpiperazinium tertrafluoroborate ($[Phpi][BF_4]$)/SDBS. At room temperature, their phase diagrams of two systems were shown in Figure 5C,D. Moreover, the water solubility of two anionic surfactants with the solution maintaining clear and transparent was determined by mass fraction. Although the water solubility of SDBS is lower than that of SDS, the addition of IL effectively increases the solubility of SDBS compared with that of SDS, while the experimental temperature is below the Krafft point of neat SDBS (38 °C) but higher than that of neat SDS (9 °C). Additionally, the critical micelle concentration (CMC) value of surfactants was also measured, and the CMC value for neat SDBS (1.2×10^{-3} mol/L) is slightly lower than that of SDS (8.2×10^{-3} mol/L). The lower the CMC is, the lower the concentration required for micelle formation is. The addition of $[C_2pi][BF_4]$ into aqueous solution of SDS can significantly reduce the CMC of SDS. The cation ($[C_2pi]^+$) with positive charge is able to influence the electric double layer of the micelles formed by anionic surfactant, reducing the minimum concentration for micelle formation. During the micelle formation, the $[C_2pi]^+$ may participate in the self-assembly of micelles as cosurfactant, which is distributed in the interfacial layer mainly through electrostatic interactions with the negatively charged headgroup of surfactant. In a word, the packing effect induced by the re-assembly of micelles formed by IL cation and surfactant anions is a critical factor. That is, the presence of $[C_2pi][BF_4]$ can improve the surfactant efficiency by reaching the minimum surface tension at a lower SDS concentration. However, as shown in Figure 5D, an insufficient amount addition of [Phpi][BF_4] into the aqueous solution of SDBS leads to precipitation.

Furthermore, nonionic surfactants, such as octylphenol polyethoxylene (Triton X and Tween family, were also considered to be able to form ATPS with ILs. In H_2O-IL-Triton X ternary system, H_2O-IL, H_2O-Triton X, and IL-Triton X can be completely miscible under room temperature conditions. However, an immiscibility window in the ternary region is observed, whose area is enlarged with the increase in temperature (known as island-type ternary system) [147,148]. For example, the detailed phase diagram of this island-type ternary system ([C_2mim][C_2SO_4]+Triton X+water) has been published by Alvarez et al. [147]. The immiscibility window occurred only in the ternary region, while binary mixtures involved in the system were completely miscible. Moreover, the area of the immiscibility window increased obviously with the rising of temperature. This group further studied the phase separation of [Ch]Cl-Triton X-water ternary system. Compared with the IL [C_2mim][C_2SO_4], the hydrophilic IL [Ch]Cl and nonionic surfactant leaded to a relatively larger two-phase region [148]. It is called IL-nonionic surfactant ATPS, in which the two-phase system consists of an IL-rich phase and a nonionic surfactant-rich phase [149]. More importantly, anionic species exhibit a high partitioning coefficient between the IL-rich phase and the surfactant-rich phase, which may be useful for stripping the anionic species off nonionic surfactant aqueous solution. As well as this, the novel IL-nonionic surfactant ATPS was formed in the [Ch]Cl-Tween 80-water ternary system [150]. In IL-surfactant-water ATPS, the IL was often used as salting out agent. The ability of the IL to salt out aqueous solutions of polymers such as PEG-600, organic compounds such as tetrahydrofuran or nonionic surfactant such as Tween 20 and Tween 80, has been researched and published [82,151,152]. Recently, the ATPS for the ternary mixtures (Tween 20 or Tween 80 + [$N_{1\,1\,1\,2OH}$][$C_4H_5O_6$] + H_2O) were constructed to evaluate the segregation capacity of IL [152]. In terms of the IL chemical properties, the lower hydrophilicity prefers to a reduced capacity to establish hydrogen bonds with water molecules existing in the nonionic surfactant aqueous solutions, resulting in a smaller biphasic region, as can be checked from the experimental data previously reported [150,153]. Moreover, the role of the surfactant mainly depends on its hydrophobicity. It is expected that a surfactant with lower hydrophilicity is more favorable to the formation of two phases. This analysis can also be quantitatively performed by checking the HLB, which is a dimensionless parameter going from 0 to 20 (from high to low hydrophobicity). Hence, the area of the biphasic regions follows the order: Tween 80 (HLB = 15.0) > Tween 20 (HLB = 16.7). Apart from that, the interest of the present IL can be revealed by the comparison of the size of the immiscibility region with those provided by conventional salts such as K_3PO_4 or $K_3C_6H_5O_7$ [154,155]. Finally, some surfactants recently used in the IL-based ATPSs were summarized in Table 3.

Table 3. Some surfactants/saccharides commonly used in the IL-based ATPSs.

Surfactants/Saccharides	ILs	Temperature/K	Refs.
Glucose	[C_3mim][BF_4], [C4mim][BF_4]	242.15~308.15	[69]
Fructose	[C_4mim][BF_4]	298.15~318.15	[70]
Glucose, sucrose, maltose, and xylose	[$C_n$mim][BF_4] (n = 3~10)	242.15~373.15	[73]
SDBS	[C_4mim][BF_4]	283.15~303.15	[143]
SDS	[C_4min]Cl, [C_6mim]Cl, [C_8mim]Cl	303.15	[145]
SDS, SDBS	[C_2pi][BF_4], [Phpi][BF_4]	298.15~323.15	[134]
Triton X-100, Triton X-102	[C_2mim][C_2SO_4], [Ch]Cl	298.15~333.15	[147,148]
Tween 20, Tween 80	[Ch]Cl, [N_{1112OH}][$C_4H_5O_6$]	293.2~333.15	[149,152]
Triton X-100	[C_2mim]Cl	288.15~308.15	[156]
D-(+)-Glucose, D-(+)-galactose, D-(+)-fructose, D-(+)-mannose, D-(-)-arabinose, L-(+)-arabinose, D-(+)-xylose, D-(+)-maltose, D-sorbitol, maltitol, xylitol, sucrose	[C_2C_1py][$C_4F_9SO_3$], [N_{1112OH}][$C_4F_9SO_3$], [C_2mim][$C_4F_9SO_3$], [C_2mim][CF_3SO_3], [C_4mim][CF_3SO_3], [C_2C_1pyr][CF_3SO_3]	298.15	[157]

4.4. Amino Acids

In addition to inorganic salts with a high charge density, some small molecular organic compounds with a low charge density can also be used as the salting agents to build the ATPS with ILs, such as amino acid organic compounds. In 2007, Zhang et al. [158] reported for the first time that a hydrophilic IL ([C_4mim][BF_4]) could form ATPS with three amino acids (including glycine, L-serine, and L-proline), respectively. Different from the ATPS composed of K_3PO_4 and [C_4mim]Cl [17], the upper phase is the glycine-rich phase, and the lower phase is the [C_4mim][BF_4]-rich phase. By forming ATPS with the IL at different temperatures (298.15K, 308.15K and 318.15K), it was found that the ATPS was more easily formed at lower temperatures. This may be due to the decreased miscibility between the [C_4mim][BF_4]-rich phase and the amino acid-rich phase with the increasing of temperature. Moreover, the ATPSs of three amino acids at 298.15 K were compared in their study. As a result, their tendency to form an ATPS with the IL [C_4mim][BF_4] is in the order of glycine > L-serine > L-proline. What's more, the authors investigated another two hydrophilic ILs ([C_2mim][BF_4] and [C_4mim]Cl) to form ATPS with the above three amino acids. Unfortunately, however, they failed to form an ATPS with these amino acids no matter how the ratio between the IL and amino acids was adjusted. Due to the stronger hydration ability of the above two hydrophilic ILs, amino acids seemed to salt out from the transparent phase when both ILs were added. Therefore, not all hydrophilic ILs can form ATPS with amino acids. On the basis of the above research, Domínguez-Pérez et al. [159] selected the ILs with the same cation 1-butyl-3-methylimidazolium ([C_4mim]$^+$) to form ATPS with different amino acids, such as L-lysine, D, L-lysine HCl and L-proline. The IL anion influence on ATPS formation was assessed through its combination with [BF_4]$^-$, [CF_3SO_3]$^-$, and [N(CN_2)]$^-$ anions. Results indicated that the phase formation ability of three ILs was in the order: [C_4mim][BF_4] > [C_4mim][CF_3SO_3] > [C_4mim][N(CN)$_2$]. The above sequence was also observed for IL-based ATPS formed by the addition of K_3PO_4 [53]. Thus, the amino acids under study behave similarly to salting-out inducing salts. The phase separation mechanism of IL-based ATPS formed by the addition of amino acids can be explained by the competition between the amino acids and the IL ions for the creation of water-ion hydration complexes. This is closely related to the anion's hydrogen-bond basicity of the ILs. The hydrogen-bond basicity (β) data of three [C_4mim]-based ILs were reported by Lungwitz et al. [160]. The smaller hydrogen-bond basicity value of IL is, the stronger its phase forming ability is. As the value of β increases, the hydrogen-bond accepting strength of the IL anion increases, which enhances its ability to be preferentially hydrated, resulting in a lower ATPS formation capacity of the IL. Additionally, the phase formation ability of amino acids to form ATPS follows the order: L-lysine $\approx$ D, L-lysine HCl > L-proline. It is not difficult to find that the more water-soluble amino acid is, the easier it is to form hydrate with IL, and thus the stronger the ability to separate out IL.

Compared with inorganic salts, amino acids have weak salting-out ability as salting-out agents. They promote phase separation by forming a water-amino acid complex to build an ATPS. Amino acids reduce the ionic strength of the solution, which can weaken or prevent the ion exchange between the two phases. This not only contributes to the practical application of IL-based ATPS, but also enables more efficient and convenient recycling of IL. Further insight into the salting-out inducing mechanism of amino acids can be obtained in the previous research [159,161]. According to the effects of a series of amino acids on the mutual solubilities of water and imidazolium-based IL, the salting-in and salting-out phenomena are the result of a delicate balance among water-amino acid side chain, IL-amino acid side chain and water-IL interactions, which are determined by the relative affinities of the biomolecule side chains to water and to IL. That is to say, the strength of the phase separation ability of amino acids greatly depends on their hydrophilicity. The stronger the hydrophilicity of an amino acid is, the higher its solubility in water is, and the stronger its ability to form hydrate is, which can then precipitate IL from aqueous solution. Moreover, the IL-amino acid ATPS has the same temperature sensitivity as IL-inorganic salt ATPS. Generally, the immiscibility region formed by IL and amino acid decreases with the increase in temperature, meaning that other phase forming component need to be added to facilitate phase separation. Therefore, the lower temperature is favorable for the formation of ATPS containing IL and amino acid. At present, only imidazolium-based ILs have been found to form ATPS with amino acids, which need further study.

4.5. Saccharides

Low molecular weight saccharides with specific structures also can be mixed with IL in water at a certain concentration and then phase separation can occur to form an ATPS. Similarly, the hydrophilic IL ([C_4mim][BF_4]) was firstly used to form ATPS with the fructose by Zhang and co-workers [70]. It was found that the ATPS can be formed over a wide component range and the effect of the temperature on the phase equilibria is obvious within the fructose concentration changing from 3 to 40%. Unfortunately, what they reported is just focused on the [C_4mim][BF_4]. The experiment was relatively simple and unrepresentative. Later, Chen et al. [69,73] systemically investigated the phase behaviors of 1-alkyl-3-methylimidazolium derivatives [$C_n$mim]X (n = 2 to 10, X = Cl^-, Br^-, BF_4^-)-carbohydrate-H_2O system. The results found that [$C_n$mim]Cl (n = 2 to 10) and [$C_n$mim]Br (n = 2 to 10) aqueous solutions never formed ATPSs with carbohydrates (such as glucose, sucrose, maltose, and xylose) in a wide temperature range (242.15K to 373.15K). However, carbohydrates can induce phase separation of not only the reported hydrophilic IL ([C_4mim][BF_4]), but also another homologous hydrophilic IL ([C_3mim][BF_4]) aqueous solution in the investigated conditions. The effect of carbon number of alkyl chain on imidazolium ring on phase separation ability was also studied. The two-phase area of [C_4mim][BF_4] system was larger than that of [C_3mim][BF_4] system. For hydrophobic ILs ([$C_n$mim][BF_4], n = 5 to 10), the addition of carbohydrate can reduce their mutual solubility with water, and therefore promote the formation of two-phase system. Moreover, the more glucose is added, the lower mutual solubility of IL and water is. By comparing the slope of tie-line (STL), the phase separation ability of the investigated carbohydrates followed the order: glucose > maltose > sucrose > xylose. More importantly, it is also found that ILs-rich phase and glucose-rich phase can be reversed by adjusting the amounts of glucose.

Obviously, all of the above-mentioned ILs are fluorinated-based ILs, which have emerged recently due to their remarkable performances, namely in the recovery of contaminants either in gaseous mixtures or in liquid effluents [162]. In addition to the commonly used imidazolium-based ILs, other fluorinated-based ILs are also used to form the ATPSs with different carbohydrates. In the research of Ferreira et al. [157], the novel ATPSs were developed by mixing a series of perfluoromethanesulfonate- and perfluorobutanesulfonate-based ILs and a large number of carbohydrates (monosaccharides, disaccharides and polyols) aiming at establishing more benign alternatives to the salts commonly used. In order to gain a better insight into ATPS formation involving ILs and carbohydrates, different ILs and carbohydrates were used to evaluate the influence of the structures of both components on the phase formation. The influence of the IL anion, in particular the size of the fluorinated alkyl chain length, was investigated by using the ILs comprising the same cation. Compared to anion [CF_3SO_3], the IL anion [$C_4F_9SO_3$] behaved as the stronger two-phase promoter, indicating that the size of the fluoroalkyl chain of the anion has an effect on the formation of two-phase systems with carbohydrates. The longer fluorinated alkyl chain of the anion renders it more hydrophobic, and diminishes its affinity towards water, thus making it easier to separate out of the aqueous media. Then, three kinds of IL cations, including $[N_{1\,1\,1\,2OH}]^+$, $[C_2C_1py]^+$ and $[C_2mim]^+$, were selected to understand the role of the cation core in the process of phase splitting. Due to the higher hydrophilicity of cholinium-based ILs [163], it is not easy to induce phase separation in presence of a carbohydrate aqueous solution. In previous studies, cholinium-based ILs usually only form ATPS with strong salting-out salts, such as K_3PO_4 [163] or polymers [82]. The ability to form two phase systems is higher for $[C_2C_1py]^+$ than for $[C_2mim]^+$. IL cations containing 6-carbon rings (such as pyridinium and piperidinium) formed ATPS more easily than those cations containing 5-carbon rings (such as imidazolium and pyrrolidinium). The above results suggested that the steric hindrance of the cation is also suggested to play important roles in water/IL interactions in addition to the hydrophobicity of ILs.

Nevertheless, the ability of carbohydrates to induce liquid-liquid demixing is mainly related to their characteristics and particularly to its hydration extension, which is determined by the interactions between carbohydrate and water, especially hydrogen-bonding. All the structures of carbohydrates have diverse –OH groups with dual donor-acceptor character and can establish hydrogen bonds with water and act as salting-out/sugaring-out species. In general, the more hydroxyl groups there are in a carbohydrate structure, the easier it is to form the hydration complexes by hydrogen bonding with water molecules, the stronger its phase separation ability is [164]. In this line, arabinose bearing 4 hydroxyl groups is expected to be a weaker two-phase promoter than all the other monosaccharides investigated with 5 hydroxyl groups. Likewise, noticeably different phase formation ability of polyols was observed following the order: maltitol > D-sorbitol > xylitol. The above trend is exactly consistent with the order of the number of hydroxyl groups they have. Maltitol is the largest polyol with more –OH groups, followed by D-sorbitol that has one more –OH group than xylitol. These experimental results also confirmed this expectation. Specially, for the disaccharides, including maltose and sucrose, despite they have the same number of –OH groups, maltose exhibits a higher ability to salt-out the IL when compared to sucrose. It could be because the six-membered pyranose rings in maltose makes it more likely to interact with water than the five-membered furanose ring in sucrose. Finally, some saccharides commonly used in the IL-based ATPSs were listed in Table 3.

Based on some relevant research, some major insights into the phase formation mechanism can be derived. The ability of ILs and saccharides to promote phase separation in aqueous media mainly depends on their hydration ability. Essentially, the phase separation is the process in which ILs and saccharides compete for the water molecules. When the saccharide aqueous solution is added to homogeneous IL solution, the two solutes compete for the water molecules. There is a "migration" of solvent molecules away from the ions of the IL towards the carbohydrate, which result in the decreasing of the solubility of the ionic solute in water. The water molecules may be structured around the saccharide and the hydrogen-bond interaction between saccharide and water is reinforced. That is, the sugar molecules are likely to win the competition since they have the innate higher affinity for water molecules than the IL and, hence, can establish stronger interactions with the solvent. On basis of their structures, the interaction of first-shell water molecules with solute primarily is with the hydroxyl groups of saccharides. When the saccharide concentration reaches a certain level, saccharide and water dissociate because of the interactions between saccharide and saccharide [165]. On the other hand, H_2O-mediated clusters is formed by the attraction between the ions of IL as the concentration of IL increases [166]. The phase separation occurs when the cluster size is sufficiently large. In brief, two differently structured microphases of water are formed with the addition of saccharide to homogeneous IL solution. As the amount of IL or saccharide or in the solution increases, the stability of this microemulsion is disrupted, leading to the coagulation of the droplets (turbidity) and occurrence of phase separation. The competition of IL and saccharide for water molecules can cause the dehydration of IL ions. In this case, the saccharides are usually employed as salting-out agents [167].

5. Conclusions

ATPS formulation combines two chemicals whose aqueous solutions are immiscible under certain conditions (composition and temperature). The traditional ATPS is a liquid-liquid two-phase system which is formed by two polymers or polymer and salt in water. Recently, the rise of ILs provides a new direction for the exploration of new ATPS forming agents. As a structurally designable solvent, ILs can be adapted to different ATPSs by tailoring their structures. Taking into account the promising results reported in the literature reviewed, IL-based ATPS has many advantages shared by ILs (e.g., non-volatility, non-flammability, high thermal stability, structural designability and no need to use volatile organic solvent) and ATPSs (e.g., simple, high-efficiency, quick phase separation and gentle biocompatible environment). It is an excellent alternative to the traditional liquid-liquid two-phase system with competitive advantages at present when environmental problems are paid more and more attention.

Although ILs have good effects and industrial prospects in the ATPS extraction techniques, there are still some challenges associated with the IL-based ATPS. For example, the high viscosity and high costs as well as the recovery of the ILs after extraction process could limit their industrial application. Generally, high temperature can reduce the viscosity of ILs but may affect the property of the ATPS. It is also unsuitable for the extraction of heat-sensitive substances. Furthermore, more attention should be paid to the economics of recycling and reusing ILs. On the one hand, it is important to control the fabricating cost and energy consumption throughout the process, such as reducing the raw material cost and optimizing the process route. On the other hand, recycling ILs after use is another cost-saving approach. What's more, the following knowledge is urgently needed before practical application of the IL-based ATPSs: (1) the basic physicochemical data, microstructure and properties for the IL-based ATPSs; (2) the phase separation mechanism; (3) the rules for the ILs selection for specific analytes in IL-based ATPSs.

The research results so far may expand the application of ILs, and more importantly, the work may provide new separation systems which may be potentially applied in biological, pharmaceuticals, and environmental engineering. As a kind of acknowledged extraction separation medium, ILs are actually not totally environmentally friendly and pollution-free. Moreover, ILs are difficult to be biodegraded due to their stable chemical properties. Current trends in IL-based ATPS gear toward employing more environmentally friendly approaches to comply with green analytical chemistry requirements. Therefore, there is a lot of hard work to be carried out on the development of IL-based ATPSs. It is expected that an environmentally benign and potentially ATPS will grow out of this review in the near future.

Author Contributions: Conceptualization, Writing—Original Draft Preparation (Sections 1, 2, 4 and 5), L.N.; Writing—Original Draft Preparation (Section 3), Z.Z.; Table editing, drawing and literature retrieval, M.L.; Writing—Review & Revision, S.Y.; Writing—Editing & Supervision, D.G. All authors have read and agreed to the published version of the manuscript.

Funding: This work was supported by the National Natural Science Foundation of China (Grant No. 11574346).

Institutional Review Board Statement: Not applicable.

Informed Consent Statement: Not applicable.

Data Availability Statement: Not applicable.

Acknowledgments: Thanks to Runpeng Guo for her great assistance in this article.

Conflicts of Interest: The authors declare no conflict of interest.

Abbreviation

[C_2mim]Cl	1-Ethyl-3-methylimidazolium chloride
[C_4mim]Cl	1-Butyl-3-methylimidazolium chloride
[C_5mim]Cl	1-Pentyl-3-imidazolium chloride
[C_6mim]Cl	1-Hexyl-3-methylimidazolium chloride
[C_7mim]Cl	1-Heptyl-3-methylimidazolium chloride
[C_4mmim]Cl	1-Butyl-2,3-dimethylimidazolium chloride
[C_4py]Cl	*N*-Butylpyridinium chloride
[N_{4444}]Cl	Tetrabutylammonium chloride
[P_{4444}]Cl	Tetrabutylphosphonium chloride
[C_4C_1pyr]Cl	1-Butyl-1-methylpyrrolidinium chloride
[C_4C_1pip]Cl	1-Butyl-1-methylpiperidinium chloride
[C_4C_1py]Cl	1-Butyl-4-methylpyridinium chloride
[C_2mim]Br	1-Ethyl-3-methylimidazolium bromide
[C_4mim]Br	1-Butyl-3-methylimidazolium bromide

$[C_6mim]Br$	1-Hexyl-3-methylimidazolium bromide
$[C_8mim]Br$	1-Octyl-3-methylimidazolium bromide
$[C_4C_1pyr]Cl$	1-Butyl-1-methylpyrrolidinium chloride
$[C_4C_1pip]Cl$	1-Butyl-1-methylpiperidinium chloride
$[C_4C_1py]Cl$	1-Butyl-4-methylpyridinium chloride
$[C_2mim]Br$	1-Ethyl-3-methylimidazolium bromide
$[C_4mim]Br$	1-Butyl-3-methylimidazolium bromide
$[C_6mim]Br$	1-Hexyl-3-methylimidazolium bromide
$[C_8mim]Br$	1-Octyl-3-methylimidazolium bromide
$[C_2mim][CH_3CO_2]$	1-Ethyl-3-methylimidazolium acetate
$[C_2mim][MeSO_4]$	1-Ethyl-3-methylimidazolium methylsulfate
$[C_2mim][C_2SO_4]$	1-Ethyl-3-methylimidazolium ethylsulfate
$[C_2mim][CH_3SO_3]$	1-Ethyl-3-methylimidazolium methanesulfonate
$[C_2mim][CF_3SO_3]$	1-Ethyl-3-methylimidazolium trifluoromethanesulfonate
$[C_4mim][CH_3CO_2]$	1-Butyl-3-methylimidazolium acetate
$[C_4mim][N(CN)_2]$	1-Butyl-3-methylimidazolium dicyanamide
$[C_4mim][CH_3SO_3]$	1-Butyl-3-methylimidazolium methanesulfonate
$[C_4mim][HSO_4]$	1-Butyl-3-methylimidazolium hydrogensulfate
$[C_4mim][CF_3SO_3]$	1-Butyl-3-methyl-imidazolium trifluoromethanesulfonate
$[C_4mim][TFA]$	1-Butyl-3-methylimidazolium trifluoroacetate
[TBA][But]	Tetrabutylammonium butanoate
[TBA][Pent]	Tetrabutylammonium pentanoate
[TBA][Hex]	Tetrabutylammonium hexanoate
[TBA][Hept]	Tetrabutylammonium heptanoate
[TBA][Oct]	Tetrabutylammonium octanoate
[TBA][Dec]	Tetrabutylammonium decanoate
$[C_4mim][BF_4]$	1-Butyl-3-methylimidazolium tetrafluoroborate
$[NH_4][OAc]$	Ammonium acetate
$[C_3NH_3][OAc]$	Propylammonium acetate
$[C_4NH_3][OAc]$	Butylammonium acetate
$[C_6NH_3][OAc]$	Hexylammonium acetate
$[C_8NH_3][OAc]$	Octylammonium acetate
$[C_4NH_3][But]$	Butylammonium butanoate
[Amim]Cl	1-Allyl-3-methylimidazolium chloride
[Ch]Cl	Cholinium chloride
$[N_{4\,4\,4\,1}][NO_3]$	Tributylmethylammonium nitrate
[EOMiM]Br	1-(2-Methoxyethyl)-3-methylimidazolium bromide
$[C_4py]OTF$	*N*-Butylpyridinium trifluoromethanesulfonate
$[C_4py][BF_4]$	*N*-Butylpyridinium tetrafluoroborate
$[C_4mim][NO_3]$	1-Butyl-3-methylimidazolium nitrate
$[C_6mim][NO_3]$	1-Hexyl-3-methylimidazolium nitrate
$[C_2mim]DCA$	1-Ethyl-3-methylimidazolium dicyanamide
$[C_3mim]DCA$	1-Propyl-3-methylimidazolium dicyanamide
$[C_4mim]DCA$	1-Butyl-3-methylimidazolium dicyanamide
$[C_6mim]DCA$	1-Hexyl-3-methylimidazolium dicyanamide
$[C_4C_1pyr]DCA$	1-Butyl-1-methylpyrrolidinium dicyanamide
[Ch][L-Pro]	Choline L-proline
[Ch][L-Cys]	Choline L-cysteine
[Ch][L-His]	Choline L-histidine
[Ch][L-Val]	Choline L-valine
[Ch][L-Ser]	Choline L-serine
[Ch][L-Met]	Choline L-methionine
[Ch][L-Ala]	Choline L-alanine
$[C_2Tr]Br$	*N*-Ethyl-tropinium bromide
$[C_3Tr]Br$	*N*-Propyl-tropinium bromide
$[C_4Tr]Br$	*N*-Butyl-tropinium bromide
$[C_5Tr]Br$	*N*-Pentyl-tropinium bromide
$[C_2Qn]Br$	*N*-Ethyl-quinolinium bromide
$[C_3Qn]Br$	*N*-Propyl-quinolinium bromide
$[C_4Qn]Br$	*N*-Butyl-quinolinium bromide
$[C_5Qn]Br$	*N*-Pentyl-quinolinium bromide
$[C_6Qn]Br$	*N*-Hexyl-quinolinium bromide
$[C_2mim][OAc]$	1-Ethyl-3-methylimidzaolium acetate
$[C_4MDEA]Br$	*N*-methyl-*N*-butyl-*N*,*N*-di(hydroxyethyl)ammonium bromide
$[C_6MDEA]Br$	*N*-methyl-*N*-hexyl-*N*,*N*-di(hydroxyethyl)ammonium bromide
$[C_8MDEA]Br$	*N*-methyl-*N*-octyl-*N*,*N*-di(hydroxyethyl)ammonium bromide
$[C_{10}MDEA]Br$	*N*-methyl-*N*-decyl-*N*,*N*-di(hydroxyethyl)ammonium bromide
$[C_4BDEA]Br$	*N*,*N*-di(butyl)-*N*,*N*-di(hydroxyethyl)ammonium bromide

$[P_{4444}][NO_3]$	Tetrabutylphosphonate nitrate
$[N_{4444}][NO_3]$	Tetrabutylammonium nitrate
[Ch][Leu]	Choline leucine
[Ch][Gly]	Choline glycine
[Ch][Lys]	Choline lysine
$[C_2TMG][TEMPO\text{-}OSO_3]$	Ethyl-1,1,3,3-tetramethylguanidinium 4-sulfonatooxy-2,2,6,6-tetramethyl piperidine-1-yloxyl
$[C_3TMG][TEMPO\text{-}OSO_3]$	Propyl-1,1,3,3-tetramethylguanidinium 4-sulfonatooxy-2,2,6,6-tetramethyl piperidine-1-yloxyl
$[C_4TMG][TEMPO\text{-}OSO_3]$	Butyl-1,1,3,3-tetramethylguanidinium 4-sulfonatooxy-2,2,6,6-tetramethyl piperidine-1-yloxyl
$[N_{1\,1\,2\,2OH}][TEMPO\text{-}OSO_3]$	Ethyl-(2-hydroxyethyl)-dimethylammonium 4-sulfonatooxy-2,2,6,6-tetramethyl piperidine-1-yloxyl
$[N_{1\,1\,3\,2OH}][TEMPO\text{-}OSO_3]$	Propyl-(2-hydroxyethyl)-dimethylammonium 4-sulfonatooxy-2,2,6,6-tetramethyl piperidine-1-yloxyl
$[N_{1\,1\,4\,2OH}][TEMPO\text{-}OSO_3]$	Butyl-(2-hydroxyethyl)-dimethylammonium 4-sulfonatooxy-2,2,6,6-tetramethyl piperidine-1-yloxyl
$[N_{1\,1\,5\,2OH}][TEMPO\text{-}OSO_3]$	Pentyl-(2-hydroxyethyl)-dimethylammonium 4-sulfonatooxy-2,2,6,6-tetramethyl piperidine-1-yloxyl
$[C_6mim][C_{12}SO_3]$	1-Hexyl-3-methyl imidazolium dodecyl sulfonate
$[C_2mim][SCN]$	1-Ethyl-3-methylimidazolium thiocyanate
$[C_4mim][SCN]$	1-Butyl-3-methylimidazolium thiocyanate
[Ch][L-Ala]	Cholinium L-alaninate
[Ch][DHcit]	Cholinium dihydrogen citrate
[Ch][Bit]	Cholinium bitartrate
[Ch][Bic]	Cholinium bicarbonate
[Ch][DHp]	Choline dihydrogen phosphate
[Ch][Ac]	Cholinium acetate
[Ch][BES]	Cholinium 2-[bis(2-hydroxyethyl)amino]ethanesulfonate
UCON	Poly(ethylene glycol-ran-propylene glycol) monobutyl ether
SDBS	Sodium dodecyl benzene sulfonate
SDS	Sodium dodecyl sulfonate
$[C_2pi][BF_4]$	1-Ethylpiperazinium tertrafluoroborate
$[Phpi][BF_4]$	1-Phenylpiperazinium tertrafluoroborate
$[N_{1\,1\,1\,2OH}][C_4H_5O_6]$	Cholinium bitartrate
$[C_2mim][BF_4]$	1-Ethyl-3-methylimidazolium tetrafluoroborate
$[C_2C_1py][C_4F_9SO_3]$	1-Ethyl-3-methylpyridinium perfluorobutanesulfonate
$[C_2mim][C_4F_9SO_3]$	1-Ethyl-3-methylimidazolium perfluorobutanesulfonate
$[N_{1\,1\,1\,2OH}][C_4F_9SO_3]$	Cholinium perfluorobutanesulfonate
$[C_2C_1pyr][CF_3SO_3]$	1-Ethyl-1-methylpyrrolidinium triflate
DIMCARB	*N,N*-dimethylammonium *N′,N′*-dimethylcarbamate
DPCARB	*N,N*-dipropylammonium *N′N′*-dipropylcarbamate
DACARB	*N,N*-diallylammonium *N′N′*-diallylcarbamate
DBCARB	Bis(2-ethylhexyl)ammonium bis(2-ethylhexyl)carbamate

References

1. Albertsson, P.A. Partition of cell particles and macromolecules in polymer two-phase systems. *Adv. Protein Chem.* **1970**, *24*, 309–341. [PubMed]
2. Rosa, P.A.J.; Azevedo, A.M.; Sommerfeld, S.; Bäcker, W.; Aires-Barros, M.R. Aqueous two-phase extraction as a platform in the biomanufacturing industry: Economical and environmental sustainability. *Biotechnol. Adv.* **2011**, *29*, 559–567. [CrossRef] [PubMed]
3. Liang, Q.; Zhang, J.; Su, X.; Meng, Q.; Dou, J. Extraction and separation of eight ginsenosides from flower buds of panax ginseng using aqueous ionic liquid-based ultrasonic-assisted extraction coupled with an aqueous biphasic system. *Molecules* **2019**, *24*, 778. [CrossRef] [PubMed]
4. Molino, J.V.D.; Marques, D.V.; Júnior, A.P.; Mazzola, P.G.; Gatti, M. Different types of aqueous two-phase systems for biomolecule and bioparticle extraction and purification. *Biotechnol. Progr.* **2013**, *29*, 1343–1353. [CrossRef] [PubMed]
5. Tang, M.; Ng, E.P.; Juan, J.C.; Wei, O.C.; Ling, T.C.; Woon, K.L.; Show, P.L. Metallic and semiconducting carbon nanotubes separation using an aqueous two-phase separation technique: A review. *Nanotechnology* **2016**, *27*, 332002. [CrossRef]
6. Grilo, A.L.; Aires-Barros, M.R.; Azevedo, A.M. Partitioning in aqueous two-phase systems: Fundamentals, applications and trends. *Sep. Purif. Method.* **2016**, *45*, 68–80. [CrossRef]
7. He, J.L.; Zhang, S.S.; Zheng, Q.; Lei, Q.F.; Fang, W.J. Phase property, composition and temperature-induced phase inversion of ATPS-C formed by aqueous cationic-anionic surfactant mixtures. *Colloid. Surface A* **2013**, *436*, 193–200. [CrossRef]
8. Teng, H.N.; Wang, X.Y.; Hou, Y.X.; Chen, Y.; Yang, C.Y.; Shen, T. Properties and extraction for $[Ni(NH_3)_6]^{2+}$ of ATPS-a formed by aqueous cationic-anionic surfactant mixtures. *J. Disper. Sci. Technol.* **2016**, *37*, 830–835. [CrossRef]
9. Quintao, J.C.; Patricio, P.R.; Veloso, A.C.G.; Carvalho, R.M.M.; Silva, L.H.M.; Hespanhol, M.C. Liquid-liquid equilibrium of the ternary ammonium salt plus poly(propylene glycol) plus water system. *Fluid Phase Equilibr.* **2017**, *442*, 96–103. [CrossRef]

10. Jamshidi, S.; Pazuki, G. Effect of hybrane hyperbranched polymer additive on partitioning of cephalexin antibiotic in aqueous biphasic systems. *J. Mol. Liq.* **2018**, *259*, 48–54. [CrossRef]
11. Vioux, A.; Taubert, A. Ionic liquids 2014 and selected papers from ILMAT 2013: Highlighting the ever-growing potential of ionic liquids. *Int. J. Mol. Sci.* **2014**, *15*, 22815–22818. [CrossRef] [PubMed]
12. Wasserscheid, P.; Keim, W. Ionic liquids-new "Solutions" for transition metal catalysis. *Angew. Chem. Int. Edit.* **2000**, *39*, 3772–3789. [CrossRef]
13. Dupont, J.; Souza, R.F.S.; Suarez, P.A.Z. Ionic liquid (molten salt) phase organometallic catalysis. *Chem. Rev.* **2002**, *102*, 3667–3692. [CrossRef] [PubMed]
14. Reshetov, S.A.; Frolkova, A.K. Ionic liquids as entrainers. *Fine Chem. Tech.* **2009**, *4*, 27–44.
15. Reshetov, S.A.; Frolkova, A.K. Influence of some structural characteristics of ionic liquids on homogeneous and heterogeneous characteristics of their mixtures with organic compounds. *Fine Chem. Tech.* **2010**, *5*, 73–80.
16. Lee, S.Y.; Khoiroh, I.; Ooi, C.W.; Ling, T.C.; Show, P.L. Recent advances in protein extraction using ionic liquid-based aqueous two-phase systems. *Sep. Purif. Rev.* **2017**, *46*, 291–304. [CrossRef]
17. Gutowski, K.E.; Broker, G.A.; Willauer, H.D.; Huddleston, J.G.; Swatloski, R.P.; Holbrey, J.D.; Rogers, R.D. Controlling the aqueous miscibility of ionic liquids: Aqueous biphasic systems of water-miscible ionic liquids and water-structuring salts for recycle, metathesis, and separations. *J. Am. Chem. Soc.* **2003**, *125*, 6632–6633. [CrossRef]
18. Du, Z.; Yu, Y.L.; Wang, J.H. Extraction of proteins from biological fluids by use of an ionic liquid/aqueous two-phase system. *Chem. Eur. J.* **2007**, *13*, 2130–2137. [CrossRef]
19. Wu, B.; Zhang, H.; Wang, J. Phase behavior for ternary systems composed of ionic liquid + saccharides + water. *J. Phys. Chem. B* **2008**, *112*, 6426–6429. [CrossRef]
20. Zeng, Q.; Wang, Y.; Li, N.; Huang, X.; Ding, X.; Lin, X.; Huang, S.; Liu, X. Extraction of proteins with ionic liquid aqueous two-phase system based on guanidine ionic liquid. *Talanta* **2013**, *116*, 409–416. [CrossRef]
21. Pratiwi, A.I.; Yokouchi, T.; Matsumoto, M.; Kondo, K. Extraction of succinic acid by aqueous two-phase system using alcohols/salts and ionic liquids/salts. *Sep. Purif. Technol.* **2015**, *155*, 127–132. [CrossRef]
22. Tian, H.Z.; Bai, X.S.; Xu, J. Extraction of bisphenol A in honey samples using aqueous biphasic systems coupled with high-performance liquid chromatography. *Sep. Sci. Plus* **2018**, *1*, 374–381. [CrossRef]
23. Santos, N.V.D.; Martins, M.; Ventura, S.P.M.; Coutinho, J.A.P.; Valentini, S.R.; Pereira, J.F.B. Aqueous biphasic systems composed of cholinium chloride and polymers as effective platforms for the purification of recombinant green fluorescent protein. *ACS Sustain. Chem. Eng.* **2018**, *6*, 9383–9393. [CrossRef]
24. Zafarani-Moattar, M.T.; Shekaari, H.; Jafari, P. Design of novel biocompatible and green aqueous two-phase systems containing cholinium L-alaninate ionic liquid and polyethylene glycol di-methyl ether 250 or polypropylene glycol 400 for separation of bovine serum albumin (BSA). *J. Mol. Liq.* **2018**, *254*, 322–332. [CrossRef]
25. Tian, H.Z.; Fu, H.L.; Xu, C.Q.; Xu, C.H. Simultaneous determination of three herbicides in honey sSamples using an aqueous biphasic system coupled with HPLC-MS/MS. *Chromatographia* **2019**, *82*, 1571–1577. [CrossRef]
26. Tian, H.; Berton, P.; Rogers, R.D. Aqueous biphasic systems composed of random ethylene/propylene oxide copolymers, choline acetate, and water for triazine-based herbicide partitioning study. *Solvent Extr. Ion Exch.* **2018**, *36*, 602–616. [CrossRef]
27. Tian, H.Z.; Xu, C.Q.; Cai, J.N.; Xu, J. The aqueous biphasic system based on cholinium ionic liquids and nonionic surfactant and its application for triazine-based herbicides extraction. *J. Chem. Thermodyn.* **2018**, *125*, 41–49. [CrossRef]
28. Escudero, N.; Morandeira, L.; Sanromán, M.; Deive, F.J.; Rodríguez, A. Salting out potential of cholinium dihydrogen citrate in aqueous solution of Triton surfactants. *J. Chem. Thermodyn.* **2018**, *118*, 235–243. [CrossRef]
29. Caban, M.; Stepnowski, P. The antagonistic role of chaotropic hexafluorophosphate anions and imidazolium cations composing ionic liquids applied as phase additives in the separation of tri-cyclic antidepressants. *Anal. Chim. Acta* **2017**, *967*, 102–110. [CrossRef]
30. Freire, M.G.; Pereira, J.F.; Francisco, M.; Rodríguez, H.; Rebelo, L.P.N.; Rogers, R.D.; Coutinho, J.A. Insight into the interactions that control the phase behaviour of new aqueous biphasic systems composed of polyethylene glycol polymers and ionic liquids. *Chem. Eur. J.* **2012**, *18*, 1831–1839. [CrossRef]
31. Han, J.; Yu, C.; Wang, Y.; Xie, X.; Yan, Y.; Yin, G.; Guan, W. Liquid-liquid equilibria of ionic liquid 1-butyl-3-methylimidazolium tetrafluoroborate and sodium citrate/tartrate/acetate aqueous two-phase systems at 298.15 K: Experiment and correlation. *Fluid Phase Equilibr.* **2010**, *295*, 98–103. [CrossRef]
32. Shahriari, S.; Neves, C.M.; Freire, M.G.; Coutinho, J.O.A. Role of the Hofmeister series in the formation of ionic-liquid-based aqueous biphasic systems. *J. Phys. Chem. B* **2012**, *116*, 7252–7258. [CrossRef] [PubMed]
33. Tang, X.; Han, J.; Hu, Y.; Wang, Y.; Lu, Y.; Chen, T.; Ni, L. The study of phase behavior of aqueous two-phase system containing [$C_n$mim]BF4 (n = 2,3,4) + $(NH_4)_2SO_4$ + H_2O at different temperatures. *Fluid Phase Equilibr.* **2014**, *383*, 100–107. [CrossRef]
34. Havlová, M.; Dohnal, V. Phase equilibria and thermophysical properties of aqueous solutions of two bis(fluorosulfonyl)imide-based ionic liquids. *Fluid Phase Equilibr.* **2021**, *547*, 113137. [CrossRef]
35. Lin, X.; Wang, Y.; Zeng, Q.; Ding, X.; Chen, J. Extraction and separation of proteins by ionic liquid aqueous two-phase system. *Analyst* **2013**, *138*, 6445–6453. [CrossRef]
36. Bridges, N.J.; Gutowski, K.E.; Rogers, R.D. Investigation of aqueous biphasic systems formed from solutions of chaotropic salts with kosmotropic salts (salt-salt ABS). *Green Chem.* **2007**, *9*, 177–183. [CrossRef]

37. Freire, M.G.; Cláudio, A.F.M.; Araújo, J.M.M.; Coutinho, J.A.P.; Rebelo, L. Aqueous biphasic systems: A boost brought about by using ionic liquids. *Chem. Soc. Rev.* **2012**, *41*, 4966–4995. [CrossRef]
38. Ventura, S.; Sousa, S.G.; Serafim, L.S.; Lima, Á.S.; Freire, M.G.; Coutinho, J. Ionic liquid based aqueous biphasic systems with controlled PH: The ionic liquid cation effect. *J. Chem. Eng. Data* **2012**, *56*, 4253–4260. [CrossRef]
39. Freire, M.G.; Louros, C.L.S.; Rebelo, L.P.N.; Coutinho, J.A.P. Aqueous biphasic systems composed of a water-stable ionic liquid + carbohydrates and their applications. *Green Chem.* **2011**, *13*, 1536–1545. [CrossRef]
40. Neves, C.S.; Granjo, J.O.; Freire, M.; Robertson, A.; Oliveira, N.C.; Coutinho, J.P. Separation of ethanol-water mixtures by liquid-liquid extraction using phosphonium-based ionic liquids. *Green Chem.* **2011**, *13*, 1517–1526. [CrossRef]
41. Claudio, A.F.M.; Freire, M.G.; Freire, C.S.R.; Silvestre, A.J.D.; Coutinho, J.A.P. Extraction of vanillin using ionic-liquid-based aqueous two-phase systems. *Sep. Purif. Technol.* **2010**, *75*, 39–47. [CrossRef]
42. He, A.; Dong, B.; Feng, X.; Yao, S. Extraction of bioactive ginseng saponins using aqueous two-phase systems of ionic liquids and salts. *Sep. Purif. Technol.* **2018**, *196*, 270–280. [CrossRef]
43. Belchior, D.C.; Almeida, M.R.; Sintra, T.E.; Ventura, S.; Duarte, I.F.; Freire, M.G. Odd-even effect in the formation and extraction performance of ionic-liquid-based aqueous biphasic systems. *Ind. Eng. Chem. Res.* **2019**, *58*, 8323–8331. [CrossRef]
44. Malekghasemi, S.; Mokhtarani, B.; Hamzehzadeh, S.; Sharifi, A.; Mirzaei, M. Liquid-liquid equilibria of aqueous biphasic systems of ionic liquids and dipotassium hydrogen phosphate at different temperatures: Experimental study and thermodynamic modeling. *J. Mol. Liq.* **2016**, *219*, 95–103. [CrossRef]
45. Chen, Y.; Wang, H.; Pei, Y.; Wang, J. A green separation strategy for neodymium (III) from cobalt (II) and nickel (II) using an ionic liquid-based aqueous two-phase system. *Talanta* **2018**, *182*, 450–455. [CrossRef]
46. Shill, K.; Padmanabhan, S.; Xin, Q.; Prausnitz, J.M.; Clark, D.S.; Blanch, H.W. Ionic liquid pretreatment of cellulosic biomass: Enzymatic hydrolysis and ionic liquid recycle. *Biotechnol. Bioeng.* **2011**, *108*, 511–520. [CrossRef]
47. Deive, F.J.; Rivas, M.A.; Rodríguez, A. Sodium carbonate as phase promoter in aqueous solutions of imidazolium and pyridinium ionic liquids. *J. Chem. Thermodyn.* **2011**, *43*, 1153–1158. [CrossRef]
48. Lu, Y.; Lu, W.; Wang, W.; Guo, Q.; Yang, Y. Thermodynamic studies of partitioning behavior of cytochrome c in ionic liquid-based aqueous two-phase system. *Talanta* **2011**, *85*, 1621–1626. [CrossRef]
49. Pei, Y.; Wang, J.; Li, L.; Wu, K.; Yang, Z. Liquid-liquid equilibria of aqueous biphasic systems containing selected imidazolium ionic liquids and salts. *J. Chem. Eng. Data* **2007**, *52*, 2026–2031. [CrossRef]
50. Freire, M.G.; Neves, C.; Lopes, J.C.; Marrucho, I.M.; Coutinho, J.; Rebelo, L. Impact of self-aggregation on the formation of ionic-liquid-based aqueous biphasic systems. *J. Phys. Chem. B* **2012**, *116*, 7660–7668. [CrossRef]
51. Neves, C.; Ventura, S.; Freire, M.G.; Marrucho, I.M.; Coutinho, J. Evaluation of cation influence on the formation and extraction capability of ionic-liquid-based aqueous biphasic systems. *J. Phys. Chem. B* **2009**, *113*, 5194–5199. [CrossRef] [PubMed]
52. Claudio, A.; Ferreira, A.M.; Shahriari, S.; Freire, M.G.; Coutinho, J. Critical assessment of the formation of ionic-liquid-based aqueous two-phase systems in acidic media. *J. Phys. Chem. B.* **2011**, *115*, 11145–11153. [CrossRef] [PubMed]
53. Ventura, S.P.M.; Neves, C.M.S.S.; Freire, M.G.; Marrucho, I.M.; Oliveira, J.; Coutinho, J.A.P. Evaluation of anion influence on the formation and extraction capacity of ionic-liquid-based aqueous biphasic systems. *J. Phys. Chem. B* **2009**, *113*, 9304–9310. [CrossRef]
54. Xu, D.; Yang, Q.; Su, B.; Bao, Z.; Ren, Q.; Xing, H. Enhancing the basicity of ionic liquids by tuning the cation-anion interaction strength and via the anion-tethered strategy. *J. Phys. Chem. B* **2014**, *118*, 1071–1079. [CrossRef] [PubMed]
55. Yang, Q.; Xu, D.; Zhang, J.; Zhu, Y.; Xing, H. Long-chain fatty acid-based phosphonium ionic liquids with strong hydrogen-bond basicity and good lipophilicity: Synthesis, characterization, and application in extraction. *ACS Sustain. Chem. Eng.* **2015**, *3*, 309–316. [CrossRef]
56. Deive, F.J.; Rodríguez, A.; Marrucho, I.M.; Rebelo, L. Aqueous biphasic systems involving alkylsulfate-based ionic liquids. *J. Chem. Thermodyn.* **2011**, *43*, 1565–1572. [CrossRef]
57. Basaiahgari, A.; Gardas, R.L. Evaluation of anion chain length impact on aqueous two phase systems formed by carboxylate anion functionalized ionic liquids. *J. Chem. Thermodyn.* **2018**, *120*, 88–96. [CrossRef]
58. Jimenez, Y.P.; Freijeiro, C.R.; Soto, A.; Rodríguez, O. Phase equilibrium for polymer/ionic liquid aqueous two-phase systems. *Fluid Phase Equilibr.* **2020**, *506*, 112387. [CrossRef]
59. Marcus, Y. Thermodynamics of solvation of ions. Part 5.-gibbs free energy of hydration at 298.15 K. *J. Chem. Soc. Faraday Trans.* **1991**, *87*, 2995–2999. [CrossRef]
60. Song, C.P.; Ramanan, R.N.; Vijayaraghavan, R.; MacFarlane, D.R.; Chan, E.S.; Ooi, C.W. Green, aqueous two-phase systems based on cholinium aminoate ionic liquids with tunable hydrophobicity and charge density. *ACS Sustain. Chem. Eng.* **2015**, *3*, 3291–3298. [CrossRef]
61. Wang, R.; Chang, Y.; Tan, Z.; Li, F. Phase behavior of aqueous biphasic systems composed of novel choline amino acid ionic liquids and salts. *J. Mol. Liq.* **2016**, *222*, 836–844. [CrossRef]
62. Korchak, P.A.; Alopina, E.V.; Pukinsky, I.B.; Safonova, E.A. Liquid-liquid equilibria of aqueous biphasic systems containing 1-alkyl-3-methylimidazolium amino acid ionic liquids with different anions (L-Leucine, L-Valine, L-Lysine) and inorganic salt (tripotassium phosphate, potassium carbonate). *Fluid Phase Equilibr.* **2020**, *525*, 112789. [CrossRef]
63. Sun, D.; Wang, R.; Li, F.; Liu, L.; Tan, Z. Enantioselective extraction of phenylalanine enantiomers using environmentally friendly aqueous two-phase systems. *Processes* **2018**, *6*, 212. [CrossRef]

64. Chakraborty, A.; Sen, K. Impact of pH and temperature on phase diagrams of different aqueous biphasic systems. *J. Chromatogr. A* **2016**, *1433*, 41–55. [CrossRef] [PubMed]
65. Zafarani-Moattar, M.T.; Hamzehzadeh, S. Phase diagrams for the aqueous two-phase ternary system containing the ionic liquid 1-butyl-3-methylimidazolium bromide and tri-potassium citrate at T = (278.15, 298.15, and 318.15). *J. Chem. Eng. Data* **2009**, *54*, 833–841. [CrossRef]
66. Zafarani-Moattar, M.T.; Hamzehzadeh, S. Salting-out effect preferential exclusion, and phase separation in aqueous solutions of chaotropic water-miscible ionic liquids and kosmotropic salts: Effects of temperature, anions, and cations. *J. Chem. Eng. Data* **2010**, *55*, 1598–1610. [CrossRef]
67. Wang, Y.; Xu, X.H.; Yan, Y.S.; Han, J.; Zhang, Z.L. Phase behavior for the [Bmim]BF4 aqueous two-phase systems containing ammonium sulfate/sodium carbonate salts at different temperatures: Experimental and correlation. *Thermochim. Acta* **2010**, *501*, 112–118. [CrossRef]
68. Li, Y.; Huang, R.; He, Z.; Li, N.; Lu, X. Phase behavior of an aqueous two-phase ionic liquid containing (n-butylpyridiniumtetrafluoroborate + sulfate salts + water) at different temperatures. *J. Mol. Liq.* **2016**, *216*, 174–184. [CrossRef]
69. Chen, Y.H.; Zhang, S.J. Phase, behavior of (1-alkyl-3-methyl imidazolium tetrafluoroborate+6-(hydroxymethyl)oxane-2,3,4,5-tetrol+water). *J. Chem. Eng. Data* **2010**, *55*, 278–282. [CrossRef]
70. Zhang, Y.Q.; Zhang, S.J.; Chen, Y.H.; Zhang, J.M. Aqueous biphasic systems composed of ionic liquid and fructose. *Fluid Phase Equilibr.* **2007**, *257*, 173–176. [CrossRef]
71. Chen, Y.H.; Meng, Y.S.; Zhang, S.M.; Zhang, Y.; Liu, X.W.; Yang, J. Liquid-liquid equilibria of aqueous biphasic systems composed of 1-butyl-3-methyl imidazolium tetrafluoroborate + sucrose/maltose + water. *J. Chem. Eng. Data* **2010**, *55*, 3612–3616. [CrossRef]
72. Wu, B.; Zhang, Y.M.; Wang, H.P.; Yang, L.L. Temperature dependence of phase behavior for ternary systems composed of ionic liquid + sucrose + water. *J. Phys. Chem. B* **2008**, *112*, 13163–13165. [CrossRef] [PubMed]
73. Chen, Y.H.; Wang, Y.G.; Cheng, Q.Y.; Liu, X.L.; Zhang, S.J. Carbohydrates-tailored phase tunable systems composed of ionic liquids and water. *J. Chem. Thermodyn.* **2009**, *41*, 1056–1059. [CrossRef]
74. Pirdashti, M.; Bozorgzadeh, A.; Ketabi, M.; Khoiroh, I. Phase equilibria of aqueous mixtures of PEG with formate salt: Effects of pH, type of cation, polymer molecular weight and temperature. *Fluid Phase Equilibr.* **2019**, *485*, 158–167. [CrossRef]
75. Li, Y.; Zhang, N.; Xu, S.; Zhu, Q.; Hu, J. The influence of temperature on the phase behavior of ionic liquid aqueous two-phase systems. *J. Disper. Sci. Technol.* **2019**, *40*, 874–883. [CrossRef]
76. Bonifacio, P.L.; Aguiar, C.; Alvarenga, B.G.; Lemes, N.; Virtuoso, L.S. Phase behavior in aqueous two-phase systems based-ionic liquid composed of 1-butyl-3-methylimidazolium tetrafluoroborate and copper sulfate in different temperatures. *J. Chem. Eng. Data* **2019**, *64*, 2143–2152. [CrossRef]
77. Xu, S.; Zhu, Q.; Luo, Q.; Li, Y.L. Influence of ions and temperature on aqueous biphasic systems containing ionic liquid and ammonium sulfate. *J. Chem. Eng. Data* **2019**, *64*, 3139–3147. [CrossRef]
78. Gao, J.; Guo, J.; Nie, F.; Ji, H.; Liu, S. LCST-type phase behavior of aqueous biphasic systems composed of phosphonium-based ionic liquids and potassium phosphate. *J. Chem. Eng. Data* **2017**, *62*, 1335–1340. [CrossRef]
79. Neves, C.M.S.S.; Shahriari, S.; Lemus, J.; Pereira, J.; Freire, M.G.; Coutinho, J. Aqueous biphasic systems composed of ionic liquids and polypropylene glycol: Insights into their liquid–liquid demixing mechanisms. *Phys. Chem. Chem. Phys.* **2016**, *18*, 20571–20582. [CrossRef]
80. Song, C.P.; Yi, Y.Q.; Chong, M.; Ramanan, R.N.; Vijayaraghavan, R.; Macfarlane, D.R.; Chan, E.S.; Ooi, C.W. Environmentally benign and recyclable aqueous two-phase system composed of distillable CO_2-based alkyl carbamate ionic liquids. *ACS Sustain. Chem. Eng.* **2018**, *6*, 10344–10354. [CrossRef]
81. Liu, X.; Li, Z.; Pei, Y.; Wang, H.; Wang, J.J. (Liquid + liquid) equilibria for (cholinium-based ionic liquids + polymers) aqueous two-phase systems. *J. Chem. Thermodyn.* **2013**, *60*, 1–8. [CrossRef]
82. Pereira, J.F.B.; Kurnia, K.A.; Cojocaru, O.A.; Gurau, G.; Rebelo, L.P.N.; Rogers, R.D.; Freire, M.G.; Coutinho, J.A.P. Molecular interactions in aqueous biphasic systems composed of polyethylene glycol and crystalline vs. liquid cholinium-based salts. *Phys. Chem. Chem. Phys.* **2014**, *16*, 5723–5731. [CrossRef] [PubMed]
83. Fischer, V.; Borchard, W.; Karas, M. Thermodynamic properties of poly(ethylene glycol)/water systems. 1. a polymer sample with a narrow molar mass distribution. *J. Phys. Chem.* **1996**, *100*, 15992–15999. [CrossRef]
84. Cláudio, A.F.; Pereira, J.F.B.; Mccrary, P.D.; Freire, M.G.; Coutinho, J.A.P.; Rogers, R.D. A critical assessment of the mechanisms governing the formation of aqueous biphasic systems composed of protic ionic liquids and polyethylene glycol. *Phys. Chem. Chem. Phys.* **2016**, *18*, 30009–30019. [CrossRef] [PubMed]
85. Li, Z.; Pei, Y.; Wang, H.; Fan, J.; Wang, J. Ionic liquid-based aqueous two-phase systems and their applications in green separation processes. *Trac-Trend. Anal. Chem.* **2010**, *29*, 1336–1346. [CrossRef]
86. Ren, J.; Li, Z.Y.; Liu, J.; Pei, Y.C.; Wang, H.Y.; Wang, J.J. Choline derivative ionic liquids-based aqueous two-phase systems: Phase diagrams and partition of purine alkaloids. *J. Chem. Thermodyn.* **2018**, *118*, 51–57. [CrossRef]
87. Shu, Y.; Gao, M.C.; Wang, X.Y.; Song, R.S.; Lu, J.; Chen, X.W. Separation of curcuminoids using ionic liquid based aqueous two-phase system coupled with in situ dispersive liquid-liquid microextraction. *Talanta* **2016**, *14*, 6–12. [CrossRef]
88. Tan, Z.J.; Wang, C.Y.; Yang, Z.Z.; Yi, Y.J.; Wang, H.Y.; Zhou, W.L.; Li, F.F. Ionic liquid based ultrasonic-assisted extraction of secoisolariciresinol diglucoside from flaxseed (*Linum usitatissimum* L.) with further purification by an aqueous two-phase system. *Molecules* **2015**, *20*, 17929–17943. [CrossRef]

89. Tome, L.I.N.; Varanda, F.R.; Freire, M.G.; Marrucho, I.M.; Coutinho, J.A.P. Towards an understanding of the mutual solubilities of water and hydrophobic ionic liquids in the presence of salts: The anion effect. *J. Phys. Chem. B* **2009**, *113*, 2815–2825. [CrossRef]
90. Lv, H.; Guo, D.; Jiang, Z.; Li, Y.; Ren, B. Phase behavior of aqueous two-phase systems composed of 1-ethyl-3-methylimidazolium tetrafluoroborate and phosphate-based salts at different temperatures. *Fluid Phase Equilibr.* **2013**, *341*, 23–29. [CrossRef]
91. Li, C.; Han, J.; Wang, Y.; Yan, Y.; Pan, J.; Xu, X.; Zhang, Z. Phase behavior for the aqueous two-phase systems containing the ionic liquid 1-butyl-3-methylimidazolium tetrafluoroborate and kosmotropic salts. *J. Chem. Eng. Data* **2010**, *55*, 1087–1092. [CrossRef]
92. Holz, M.; Grunder, R.; Sacco, A.; Meleleo, A. Nuclear magnetic resonance study of self-association of small hydrophobic solutes in water: Salt effects and the lyotropic series. *J. Chem. Soc. Faraday Trans.* **1993**, *89*, 1215–1222. [CrossRef]
93. Li, Z.Y.; Liu, X.X.; Pei, Y.C.; Wang, J.J.; He, M.Y. Design of environmentally friendly ionic liquid aqueous two-phase systems for the efficient and high activity extraction of proteins. *Green Chem.* **2012**, *14*, 2941–2950. [CrossRef]
94. Ruiz-Angel, M.J.; Pino, V.; Carda-Broch, S.; Berthod, A. Solvent systems for countercurrent chromatography: An aqueous two phase liquid system based on a room temperature ionic liquid. *J. Chromatogr. A* **2007**, *1151*, 65–73. [CrossRef] [PubMed]
95. Mondal, D.; Sharma, M.; Quental, M.V.; Tavares, A.P.; Prasad, K.; Freire, M.G. Suitability of bio-based ionic liquids for the extraction and purification of IgG antibodies. *Green Chem.* **2016**, *18*, 6071–6081. [CrossRef]
96. Quental, M.V.; Caban, M.; Pereira, M.M.; Stepnowski, P.; Coutinho, J.A.; Freire, M.G. Enhanced extraction of proteins using cholinium-based ionic liquids as phase-forming components of aqueous biphasic systems. *Biotechnol. J.* **2015**, *10*, 1457–1466. [CrossRef]
97. Vahidnia, M.; Pazuki, G.; Abdolrahimi, S. Impact of polyethylene glycol as additive on the formation and extraction behavior of ionic-liquid based aqueous two-phase system. *Aiche J.* **2016**, *62*, 264–274. [CrossRef]
98. Mourao, T.; Tomé, L.; Florindo, C.; Rebelo, L.P.N.; Marrucho, I.M. Understanding the role of cholinium carboxylate ionic liquids in PEG-based aqueous biphasic systems. *ACS Sustain. Chem. Eng.* **2014**, *2*, 2426–2434. [CrossRef]
99. Pereira, J.F.; Vicente, F.; Santos-Ebinuma, V.C.; Araújo, J.M.; Pessoa, A.; Freire, M.G.; Coutinho, J.A. Extraction of tetracycline from fermentation broth using aqueous two-phase systems composed of polyethylene glycol and cholinium-based salts. *Process Biochem.* **2013**, *48*, 716–722. [CrossRef]
100. Wang, Y.; Wang, S.; Liu, L. Recovery of natural active molecules using aqueous two-phase systems comprising of ionic liquids/deep eutectic solvents. *Green Chem. Eng.* **2022**, *3*, 5–14. [CrossRef]
101. Iqbal, M.; Tao, Y.; Xie, S.; Zhu, Y.; Chen, D.; Wang, X.; Huang, L.; Peng, D.; Sattar, A.; Shabbir, M.A.B.; et al. Aqueous two-phase system (ATPS): An overview and advances in its applications. *Biol. Proced. Online* **2016**, *18*, 18. [CrossRef] [PubMed]
102. Assis, R.C.; Mageste, A.B.; Lemos, L.; Orlando, R.M.; Rodrigues, G.D. Application of aqueous two-phase system for selective extraction and clean-up of emerging contaminants from aqueous matrices. *Talanta* **2021**, *223*, 121697. [CrossRef] [PubMed]
103. Tanimura, K.; Amau, M.; Kume, R.; Suga, K.; Umakoshi, H. Characterization of ionic liquid aqueous two-phase systems: Phase separation behaviors and hydrophobicity index between the two phases. *J. Phys. Chem. B* **2019**, *123*, 5866–5874. [CrossRef]
104. Li, Q.; Liu, W.; Zhu, X. Green choline amino acid ionic liquid aqueous two-phase extraction coupled with HPLC for analysis sunset yellow in beverages. *Food Anal. Method.* **2019**, *12*, 2527–2534. [CrossRef]
105. Zafarani-Moattar, M.T.; Shekaari, H.; Jafari, P. The role of water soluble polymers in the phase separation of aqueous cholinium phenylalaninate solution as a green and biocompatible ionic liquid. *Fluid Phase Equilibr.* **2019**, *485*, 199–210. [CrossRef]
106. Hayashi, S.; Hamaguchi, H. Structure of an ionic liquid, 1-n-butyl-3-methylimidazolium iodide, studied by wide-angle X-ray scattering and Raman spectroscopy. *Chem. Phys. Lett.* **2004**, *33*, 1590–1591.
107. Jiang, Y.; Guo, C.; Liu, H. Magnetically rotational reactor for absorbing benzene emissions by ionic liquids. *Particuology* **2007**, *5*, 130–133. [CrossRef]
108. Deng, N.; Li, M.; Zhao, L.; Lu, C.; de Rooy, S.L.; Warner, I.M. Highly efficient extraction of phenolic compounds by use of magnetic room temperature ionic liquids for environmental remediation. *J. Hazard. Mater.* **2011**, *192*, 1350–1357. [CrossRef]
109. Wang, Y.; Sun, Y.; Xu, B.; Li, X.; Jin, R.; Zhang, H.; Song, D. Magnetic ionic liquid-based dispersive liquid-liquid microextraction for the determination of triazine herbicides in vegetable oils by liquid chromatography. *J. Chromatogr. A* **2014**, *1373*, 9–16. [CrossRef]
110. Wang, J.; Yao, H.; Nie, Y.; Bai, L.; Zhang, X.; Li, J. Application of iron-containing magnetic ionic liquids in extraction process of coal direct liquefaction residues. *Ind. Eng. Chem. Res.* **2012**, *51*, 3776–3782. [CrossRef]
111. Clark, K.D.; Nacham, O.; Yu, H.; Li, T.; Yamsek, M.M.; Ronning, D.R.; Anderson, J.L. Extraction of DNA by magnetic ionic liquids: Tunable solvents for rapid and selective DNA analysis. *Anal. Chem.* **2015**, *87*, 1552–1559. [CrossRef] [PubMed]
112. Santos, E.; Albo, J.; Irabien, A. Magnetic ionic liquids: Synthesis, properties and applications. *RSC Adv.* **2014**, *4*, 40008–40018. [CrossRef]
113. Yao, T.; Zang, H.; Yao, S.; Dai, X.; Song, H. Measurement and correlation of phase equilibria in aqueous two-phase systems containing functionalized magnetic ionic liquids and potassium phosphate at different temperatures. *J. Mol. Liq.* **2018**, *263*, 72–80. [CrossRef]
114. Yao, T.; Huang, X.; Zang, H.; Hang, S.; Yao, S. Measurement and correlation of phase equilibria in aqueous two-phase systems containing functionalized magnetic ionic liquids and $K_2HPO_4/K_2CO_3/NaCO_3$ at 298.15K. *J. Mol. Liq.* **2017**, *231*, 411–418. [CrossRef]
115. Nie, L.R.; Song, H.; Alula, Y.; Liang, S.; Yao, S. Extraction in cholinium-based magnetic ionic liquid aqueous two-phase system for the determination of berberine hydrochloride in rhizoma coptidis. *RSC Adv.* **2018**, *8*, 25201–25209. [CrossRef]
116. Zafarani-Moattar, M.T.; Hosseinpour-Hashemi, V. Effect of temperature on the aqueous two-phase system containing poly(ethylene glycol) dimethyl ether 2000 and dipotassium oxalate. *J. Chem. Eng. Data* **2012**, *57*, 532–540. [CrossRef]
117. Freire, M.G.; Carvalho, P.J.; Silva, A.M.; Santos, L.M.; Rebelo, L.P.; Marrucho, I.M.; Coutinho, J.A. Ion specific effects on the mutual solubilities of water and hydrophobic ionic liquids. *J. Phys. Chem. B* **2008**, *113*, 202–211. [CrossRef]

118. Freire, M.G.; Neves, C.M.S.S.; Silva, A.M.S.; Santos, L.M.N.B.F. ^{1}H NMR and molecular dynamics evidence for an unexpected interaction on the origin of salting- in/salting-out phenomena. *J. Phys. Chem. B* **2010**, *114*, 2004–2014. [CrossRef]
119. Wang, Q.; Wei, N.; Wang, Y.; Hou, Y.; Wei, Q. Single-step purification of c-phycocyanin from arthrospira platensis using aqueous two-phase system based on natural deep eutectic solvents. *J. Appl. Phycol.* **2020**, *32*, 1–11. [CrossRef]
120. Osloob, M.; Roosta, A. Experimental study of choline chloride and K_2HPO_4 aqueous two-phase system, and its application in the partitioning of penicillin G. *J. Mol. Liq.* **2019**, *279*, 171–176. [CrossRef]
121. Lima, Á.S.; Soares, C.; Paltram, R.; Halbwirth, H.; Bica, K. Extraction and consecutive purification of anthocyanins from grape pomace using ionic liquid solutions. *Fluid Phase Equilibr.* **2017**, *451*, 68–78. [CrossRef]
122. Bao, X.; Chen, Z.; Han, J.; Wang, Y.; Ni, L. Liquid-liquid equilibrium of imidazolium ionic liquids + phosphate + water aqueous two-phase systems and correlation. *J. Solution Chem.* **2019**, *48*, 1167–1187. [CrossRef]
123. Sun, P.; Huang, K.; Song, W.; Gao, Z.; Liu, H. Separation of rare earths from the transition metals using a novel ionic-liquid-based aqueous two-phase system: Toward green and efficient recycling of rare earths from the NdFeB magnets. *Ind. Eng. Chem. Res.* **2018**, *57*, 16934–16943. [CrossRef]
124. Xu, H.; Li, X.; Hao, Y.; Zhao, X.; Zhang, J. Highly selective separation of acteoside from cistanche tubulosa using an ionic liquid based aqueous two-phase system. *J. Mol. Liq.* **2021**, *333*, 115982. [CrossRef]
125. Li, M.; Yu, X.J.; Zhou, C.S.; Yagoub, A.E.A.; Ji, Q.H.; Chen, L. Construction of an integrated platform for 5-HMF production and separation based on ionic liquid aqueous two-phase system. *J. Mol. Liq.* **2020**, *313*, 113529. [CrossRef]
126. Wang, D.D.; Lu, Y.; Sun, Z.; Liang, W.; Sun, D.S.; Qi, C.L.; Sheng, C.Z.; Yu, X.P. Measurement and correlation of phase equilibria in aqueous two-phase systems containing ionic liquid ([EOMiM]Br) and potassium citrate/ammonium citrate/potassium tartrate at different temperatures. *Korean J. Chem. Eng.* **2020**, *37*, 332–340. [CrossRef]
127. Li, Y.L.; Lu, X.J.; He, W.T.; Huang, R.; Zhao, Y.J.; Wang, Z.H. Influence of the salting-out ability and temperature on the liquid-liquid equilibria of aqueous two-phase systems based on ionic liquid-organic salts-water. *J. Chem. Eng. Data* **2016**, *61*, 475–486. [CrossRef]
128. Gómez, E.; Macedo, E.A. Partitioning of DNP-amino acids in ionic liquid/citrate salt based aqueous two-phase system. *Fluid Phase Equilibr.* **2019**, *484*, 82–87. [CrossRef]
129. Li, Q.; Liu, W.; Zhu, X. Green choline amino acid ionic liquids aqueous two-phase extraction coupled with synchronous fluorescence spectroscopy for analysis naphthalene and pyrene in water samples. *Talanta* **2020**, *219*, 121305. [CrossRef]
130. Liu, L.L.; Sun, D.Y.; Li, F.F.; Ma, S.P.; Tan, Z.J. Enantioselective liquid-liquid extraction of valine enantiomers in the aqueous two-phase system formed by the cholinium amino acid ionic liquid copper complexes and salt. *J. Mol. Liq.* **2019**, *294*, 111599. [CrossRef]
131. Visak, Z.P.; Lopes, J.; Rebelo, L. Ionic liquids in polyethylene glycol aqueous solutions: Salting-in and salting-out effects. *Monatsh. Chem.* **2007**, *138*, 1153–1157. [CrossRef]
132. Tomé, L.I.N.; Pereira, J.F.B.; Rogers, R.D.; Freire, M.G.; Gomes, J.R.B.; Coutinho, J.A.P. "Washing-out" ionic liquids from polyethylene glycol to form aqueous biphasic systems. *Phys. Chem. Chem. Phys.* **2014**, *16*, 2271–2274. [CrossRef] [PubMed]
133. Ola, P.D.; Matsumoto, M. Extraction mechanism of ferric and manganese ions with aqueous two-phase system formed by ionic liquid and polyethylene glycol. *Chem. Biochem. Eng. Q.* **2019**, *33*, 229–234. [CrossRef]
134. Lu, X.X.; Lu, Z.Z.; Zhang, R.; Zhao, L.J.; Xie, H.J. Distribution of pigments in the aqueous two-phase system formed with piperazinium-based ionic liquid and anionic surfactant. *J. Mol. Liq.* **2021**, *330*, 115677. [CrossRef]
135. Wu, C.; Wang, J.; Pei, Y.; Wang, H.; Li, Z. Salting-out effect of ionic liquids on poly (propyleneglycol) (PPG): Formation of PPG + ionic liquid aqueous two-phase systems. *J. Chem. Eng. Data* **2010**, *55*, 5004–5008. [CrossRef]
136. Zafarani-Moattar, M.T.; Hamzehzadeh, S.; Nasiri, S. A new aqueous biphasic system containing polypropylene glycol and a water-miscible ionic liquid. *Biotechnol. Progr.* **2012**, *28*, 146–156. [CrossRef]
137. Ruiz, C.A.S.; Kwaijtaal, J.; Peinado, O.C.; van den Berg, C.; Wijffels, R.H.; Eppink, M.H.M. Multistep fractionation of microalgal biomolecules using selective aqueous two-phase systems. *ACS Sustain. Chem. Eng.* **2020**, *8*, 2441–2452. [CrossRef]
138. Sadeghi, R. *Green Chemistry and Sustainable Technology: Chapter 4-ABS Composed of Ionic Liquids and Polymers;* Springer: Berlin/Heidelberg, Germany, 2016; pp. 61–88.
139. Bubalo, M.C.; Radosevic, K.; Redovnikovic, I.R.; Halambek, J.; Srcek, V.G. A brief overview of the potential hazards of ionic liquids. *Ecotoxicol. Environ. Saf.* **2014**, *99*, 1–12. [CrossRef]
140. Lee, S.Y.; Khoiroh, I.; Ling, T.C.; Show, P.L. Enhanced recovery of lipase derived from burkholderia cepacia from fermentation broth using recyclable ionic liquid/polymer-based aqueous two-phase systems. *Sep. Purif. Technol.* **2017**, *179*, 152–160. [CrossRef]
141. Aziz, N.; Abbasiliasi, S.; Hui, S.N.; Phapugrangkul, P.; Tan, J.S. Purification of β-mannanase derived from bacillus subtilis ATCC 11774 using ionic liquid as adjuvant in aqueous two-phase system. *J. Chromatogr. B* **2017**, *1055*, 104–112. [CrossRef]
142. Zheng, Y.; Tong, Y.; Wang, S.B.; Zhang, H.; Yang, Y.Z. Mechanism of gold (III) extraction using a novel ionic liquid-based aqueous two phase system without additional extractants. *Sep. Purif. Technol.* **2015**, *154*, 123–127. [CrossRef]
143. Wei, X.L.; Wei, Z.B.; Wang, X.H.; Wang, Z.N.; Sun, D.Z.; Liu, J.; Zhao, H.H. Phase behavior of new aqueous two-phase systems: 1-butyl-3-methylimidazolium tetrafluoroborate + anionic surfactants + water. *Soft Matter* **2011**, *7*, 5200–5207. [CrossRef]
144. Yu, W.; Liu, Z.; Li, Q.; Zhang, H.; Yu, Y. Determination of Sudan I-IV in candy using ionic liquid/anionic surfactant aqueous two-phase extraction coupled with high-performance liquid chromatography. *Food Chem.* **2015**, *173*, 815–820. [CrossRef] [PubMed]
145. Yu, W.; Li, K.; Liu, Z.L.; Zhang, H.Q.; Jin, X.Q. Novelty aqueous two-phase extraction system based on ionic liquid for determination of sulfonamides in blood coupled with high-performance liquid chromatography. *Microchem. J.* **2018**, *136*, 263–269. [CrossRef]

146. Li, S.; He, C.; Liu, H.; Kean, L.; Feng, L. Ionic liquid-based aqueous two-phase system, a sample pretreatment procedure prior to high-performance liquid chromatography of opium alkaloids. *J. Chromatogr. B* **2005**, *826*, 58–62. [CrossRef]
147. Alvarez, M.S.; Rivas, M.; Deive, F.J.; Sanroman, M.A.; Rodriguez, A. Ionic liquids and non-ionic surfactants: A new marriage for aqueous segregation. *RSC Adv.* **2014**, *4*, 32698–32700. [CrossRef]
148. Alvarez, M.S.; Patino, F.M.; Deive, F.J.; Sanroman, M.A.; Rodriguez, A. Aqueous immiscibility of cholinium chloride ionic liquid and Triton surfactants. *J. Chem. Thermodyn.* **2015**, *91*, 86–93. [CrossRef]
149. Zhao, L.H.; Zhang, X.H.; Wang, Z.L. Extraction of anionic dyes with ionic liquid-nonionic surfactant aqueous two-phase system. *Sep. Sci. Technol.* **2017**, *52*, 804–811. [CrossRef]
150. Alvarez, M.S.; Esperança, J.M.S.S.; Deive, F.J.; Sanroman, M.A.; Rodriguez, A. A biocompatible stepping stone for the removal of emerging contaminants. *Sep. Purif. Technol.* **2015**, *153*, 91–98. [CrossRef]
151. Souza, R.L.; Lima, R.A.; Coutinho, J.A.P.; Soares, C.M.F.; Lima, A.S. Aqueous two-phase systems based on cholinium salts and tetrahydrofuran and their use for lipase purification. *Sep. Purif. Technol.* **2015**, *155*, 118–126. [CrossRef]
152. Escudero, N.; Deive, F.J.; Sanromán, M.Á.; Álvarez, M.S.; Rodríguez, A. Design of eco-friendly aqueous two-phase systems for the efficient extraction of industrial finishing dyes. *J. Mol. Liq.* **2019**, *284*, 625–632. [CrossRef]
153. Álvarez, M.S.; Gómez, L.; Ulloa, R.G.; Deive, F.J.; Sanromán, M.A.; Rodríguez, A. Antibiotics in swine husbandry effluents: Laying the foundations for their efficient removal with a biocompatible ionic liquid. *Chem. Eng. J.* **2016**, *298*, 10–16. [CrossRef]
154. Palǒić, A.; Puškarić, A.; Mazaj, M.; Žunkovǒ, E.; Logar, N.Z.; Bronić, J. Structural and degradation studies of a new biocompatible chiral Zn-L-tartrate metal-organic framework. *J. Solid State Chem.* **2015**, *225*, 59–64. [CrossRef]
155. Salabat, A.; Moghadam, S.T.; Far, R.M. Liquid-liquid equilibria of aqueous two-phase systems composed of TritonX-100 and sodium citrate or magnesium sulfate salts. *Calphad-Comput. Coupling Phase Diagr. Thermochem.* **2010**, *34*, 81–83. [CrossRef]
156. Li, F.; Liu, Y.; Lin, W. Phase equilibrium and protein partitioning in aqueous two-phase systems containing imidazolium ionic liquids and surfactant at low voltage levels. *J. Mol. Liq.* **2018**, *256*, 372–379. [CrossRef]
157. Ferreira, A.; Esteves, P.D.O.; Boal-Palheiros, I.; Pereiro, A.B.; Freire, M.G. Enhanced tunability afforded by aqueous biphasic systems formed by fluorinated ionic liquids and carbohydrates. *Green Chem.* **2016**, *18*, 1070–1079. [CrossRef]
158. Zhang, J.M.; Zhang, Y.Q.; Chen, Y.H.; Zhang, S.J. Mutual coexistence curve measurement of aqueous biphasic systems composed of [bmim][BF_4] and Glycine, L-Serine, and L-Proline, respectively. *J. Chem. Eng. Data* **2007**, *52*, 2488–2490. [CrossRef]
159. Domínguez-Pérez, M.; Tomé, L.I.; Coutinho, A.P.; Freire, M.G. Extraction of biomolecules using aqueous biphasic systems formed by ionic liquids and amino acids. *Sep. Purif. Technol.* **2010**, *72*, 85–91. [CrossRef]
160. Lungwitz, R.; Friedrich, M.; Linert, W.; Spange, S. New aspects on the hydrogen bond donor (HBD) strength of 1-butyl-3-methylimidazolium room temperature ionic liquids. *New J. Chem.* **2008**, *32*, 1493–1499. [CrossRef]
161. Tomé, L.I.N.; Domínguez-Pérez, M.; Freire, M.G.; Marrucho, I.M.; Cabeza, O.; Coutinho, J.A.P. On the interactions between amino acids and ionic liquids in aqueous media. *J. Phys. Chem. B* **2009**, *113*, 13971–13979. [CrossRef]
162. Pereiro, A.B.; Araújo, J.M.M.; Martinho, S.; Alves, F.; Nunes, S.; Matias, A.; Duarte, C.M.M.; Rebelo, L.P.N.; Marrucho, I.M. Fluorinated ionic liquids: Properties and applications. *ACS Sustain. Chem. Eng.* **2013**, *1*, 427–439. [CrossRef]
163. Shahriari, S.; Tome, L.C.; Araújo, J.M.M.; Rebelo, L.P.N.; Coutinho, J.A.P.; Marrucho, I.M.; Freire, M.G. Aqueous biphasic systems: A benign route using cholinium-based ionic liquids. *RSC Adv.* **2013**, *3*, 1835–1843. [CrossRef]
164. Galema, S.A.; Engberts, J.; Hoiland, H.; Forland, G.M. Informative thermodynamic properties of the effect of stereochemistry on carbohydrate hydration. *J. Phys. Chem.* **1993**, *97*, 6885–6889. [CrossRef]
165. Mason, P.E.; Neilson, G.W.; Enderby, J.E.; Saboungi, M.L.; Brady, J.W. Structure of aqueous glucose solutions as determined by neutron diffraction with isotopic substitution experiments and molecular dynamics calculations. *J. Phys. Chem. B* **2005**, *109*, 13104–13111. [CrossRef]
166. Katayanagi, H.; Nishikawa, K.; Shimozaki, H.; Mike, K.; Weath, P.; Koga, Y. Mixing schemes in ionic liquid-H_2O systems: A thermodynamic study. *J. Phys. Chem. B* **2004**, *108*, 19451–19457. [CrossRef]
167. Shekaari, H.; Kazempour, A.; Ghasedi-Khajeh, Z. Structure-making tendency of ionic liquids in the aqueous D-glucose solutions. *Fluid Phase Equilibr.* **2012**, *316*, 102–108. [CrossRef]

MDPI AG
Grosspeteranlage 5
4052 Basel
Switzerland
Tel.: +41 61 683 77 34

International Journal of Molecular Sciences Editorial Office
E-mail: ijms@mdpi.com
www.mdpi.com/journal/ijms

www.ingramcontent.com/pod-product-compliance
Lightning Source LLC
Chambersburg PA
CBHW061548210926
51850CB00009B/2039